Advances in Mobile Mapping Technologies

Advances in Mobile Mapping Technologies

Editors

Ville Lehtola
Andreas Nüchter
François Goulette

MDPI • Basel • Beijing • Wuhan • Barcelona • Belgrade • Manchester • Tokyo • Cluj • Tianjin

Editors
Ville Lehtola
Department of Earth
Observation Science
University of Twente
Enschede
Netherlands

Andreas Nüchter
Department of Robotics and
Telematics
University of Würzburg
Würzburg
Germany

François Goulette
Department of Robotics
Mines Paris- PSL University
Paris
France

Editorial Office
MDPI
St. Alban-Anlage 66
4052 Basel, Switzerland

This is a reprint of articles from the Special Issue published online in the open access journal *Remote Sensing* (ISSN 2072-4292) (available at: www.mdpi.com/journal/remotesensing/special_issues/RS_mobile_mapping).

For citation purposes, cite each article independently as indicated on the article page online and as indicated below:

LastName, A.A.; LastName, B.B.; LastName, C.C. Article Title. *Journal Name* **Year**, *Volume Number*, Page Range.

ISBN 978-3-0365-3490-9 (Hbk)
ISBN 978-3-0365-3489-3 (PDF)

Contents

About the Editors

Ville Lehtola

Ville LEHTOLA is an assistant professor in mobile mapping at the ITC faculty of University of Twente, Netherlands, formerly a senior researcher at the Finnish Geospatial Research Institute (FGI), and Academy Post-doc 2012-2015 by Academy of Finland. His research interests are in robot perception and close-range remote sensing, with a focus on (1) automation and artificial intelligence, (2) measurements taken from motion, and (3) multi-sensor fusion, especially lidar, inertial, GNSS, and cameras. The relevant applications include digital twins of cities, situational awareness for autonomous ships, forest robotics, and asset management in built environment. Holding a doctorate in computational statistical physics from Aalto University, Finland, he has a background in algorithms, high-performance computing, and superclusters. He is a member of International Society for Photogrammetry and Remote Sensing (ISPRS).

Andreas Nüchter

Andreas NÜCHTER is a tenured professor of telematics at University of Würzburg. Before summer 2013, he headed as an assistant professor the Automation group at Jacobs University Bremen. Before this, he was at University of Osnabrück, the Fraunhofer Institute for Autonomous Intelligent Systems (AIS, Sankt Augustin), the University of Bonn, MSc computer science in 2002 (best paper award by the German society of informatics (GI) for his thesis) and the Washington State University. He holds a doctorate degree (Dr. rer. nat) from University of Bonn. His thesis was shortlisted for the EURON PhD award. Andreas works on robotics and automation, cognitive systems and artificial intelligence. His main research interests include reliable robot control, 3D environment mapping, 3D vision, and laser scanning technologies, resulting in fast 3D scan matching algorithms that enable robots to perceive and map their environments in 3D, representing the pose with 6 degrees of freedom. The capabilities of these robotic SLAM approaches were demonstrated at RoboCup Rescue competitions, ELROB and several other events. He is a member of the GI and the IEEE.

François Goulette

François GOULETTE, a full professor and researcher at the Robotics Lab of Mines Paris / PSL University, graduated as an engineer in 1992 and obtained a PhD in 1997, both from Mines Paris, a French leading institution. He worked in industry (Electricité de France) as a research engineer for a few years and was a Visiting Scientist at MIT over 1993-1995. He joined Mines Paris again as an assistant Professor in 1996 and developed a research theme on laser scanning data acquisition and processing. In 2009, he obtained his habilitation to conduct research (HDR) from University Pierre and Marie Curie. As a researcher, he worked on data acquisition (developing prototypes of Laser Mobile Mapping Systems) and several processing aspects (registration, modelling, and applications), within the scopes of industrial or collaborative partnerships. His interests include mobile mapping systems; 3D point cloud acquisition; 3D modeling; and Lidar.

Preface to "Advances in Mobile Mapping Technologies"

Mobile mapping is applied widely in society, for example, in asset management, fleet management, construction planning, road safety, and maintenance optimization. Yet, further advances in these technologies are called for. Advances can be radical, such as changes to the prevailing paradigms in mobile mapping, or incremental, such as the state-of-the-art mobile mapping methods. With current multi-sensor systems in mobile mapping, laser-scanned data are often registered in point clouds with the aid of global navigation satellite system (GNSS) positioning or simultaneous localization and mapping (SLAM) techniques and then labeled and colored with the aid of machine learning methods and digital camera data. These multi-sensor platforms are beginning to undergo further advancements via the addition of multi-spectral and other sensors and via the development of machine learning techniques used in processing this multi-modal data. Embedded systems and minimalistic system designs are also attracting attention, from both academic and commercial perspectives.

In order to address these topics, we edited the Special Issue Advances in Mobile Mapping Technologies for the *Remote Sensing* journal. This book contains the published articles of this Special Issue and is aimed at those in academia and industry alike. Specifically, it consists of works introducing a new mobile mapping dataset ('Paris CARLA 3D'), system calibration studies, SLAM topics, and multiple deep learning works for asset detection. We wish to thank all the authors who contributed to this collection.

Ville Lehtola, Andreas Nüchter, François Goulette
Editors

Article

Paris-CARLA-3D: A Real and Synthetic Outdoor Point Cloud Dataset for Challenging Tasks in 3D Mapping

Jean-Emmanuel Deschaud [1,*], David Duque [2], Jean Pierre Richa [1], Santiago Velasco-Forero [2], Beatriz Marcotegui [2] and François Goulette [1]

1   MINES ParisTech, PSL University, Centre for Robotics, 75006 Paris, France; jean-pierre.richa@mines-paristech.fr (J.P.R.); francois.goulette@mines-paristech.fr (F.G.)
2   MINES ParisTech, PSL University, Centre for Mathematical Morphology, 77300 Fontainebleau, France; david.duque@mines-paristech.fr (D.D.); santiago.velasco@mines-paristech.fr (S.V.-F.); beatriz.marcotegui@mines-paristech.fr (B.M.)
*   Correspondence: jean-emmanuel.deschaud@mines-paristech.fr

Abstract: Paris-CARLA-3D is a dataset of several dense colored point clouds of outdoor environments built by a mobile LiDAR and camera system. The data are composed of two sets with synthetic data from the open source CARLA simulator (700 million points) and real data acquired in the city of Paris (60 million points), hence the name Paris-CARLA-3D. One of the advantages of this dataset is to have simulated the same LiDAR and camera platform in the open source CARLA simulator as the one used to produce the real data. In addition, manual annotation of the classes using the semantic tags of CARLA was performed on the real data, allowing the testing of transfer methods from the synthetic to the real data. The objective of this dataset is to provide a challenging dataset to evaluate and improve methods on difficult vision tasks for the 3D mapping of outdoor environments: semantic segmentation, instance segmentation, and scene completion. For each task, we describe the evaluation protocol as well as the experiments carried out to establish a baseline.

Keywords: dataset; LiDAR; mobile mapping; laser scanning; 3D mapping; synthetic; point cloud; outdoor; semantic; scene completion

check for
updates

Citation: Deschaud, J.-E.; Duque, D.; Richa, J.P.; Velasco-Forero, S.; Marcotegui, B.; Goulette, F. Paris-CARLA-3D: A Real and Synthetic Outdoor Point Cloud Dataset for Challenging Tasks in 3D Mapping. *Remote Sens.* **2021**, *13*, 4713. https://doi.org/10.3390/rs13224713

Academic Editor: Ayman F. Habib

Received: 15 October 2021
Accepted: 17 November 2021
Published: 21 November 2021

Publisher's Note: MDPI stays neutral with regard to jurisdictional claims in published maps and institutional affiliations.

1. Introduction

Data in the form of a 3D point cloud are becoming increasingly popular. There are mainly three families of 3D data acquisition: photogrammetry (Structure from Motion and Multi-View Stereo from photos), RGB-D or structured light scanners (for small objects or indoor scenes), and static or mobile LiDARs (for outdoor scenes). The advantage of this last family (mobile LiDARs) is their ability to acquire large volumes of data. This results in many potential applications: city mapping, road infrastructure management, construction of HD maps for autonomous vehicles, etc.

There are already many datasets published on the first two families, but few are available on outdoor mapping. However, there are still many challenges in the ability to analyze outdoor environments from mobile LiDARs. Indeed, the data contain a lot of noise (due to the sensor but also to the mobile system) and have significant local anisotropy and also missing parts (due to occlusion of objects).

The main contributions of this article are as follows:

- the publication of a new dataset, called Paris-CARLA-3D (PC3D in short)—synthetic and real point clouds of outdoor environments; the dataset is available at the following URL: https://npm3d.fr/paris-carla-3d, accessed on 15 October 2021;
- the protocol and experiments with baselines on three tasks (semantic segmentation, instance segmentation, and scene completion) based on this dataset.

2. Related Datasets

With the democratization of 3D sensors, there are more and more point cloud datasets available. We can see in Table 1 a list of datasets based on 3D point clouds. We have listed only datasets available in the form of a point cloud. We have, therefore, not listed the datasets such as NYUv2 [1], which do not contain the poses (trajectory of the RGB-D sensor), and thus do not allow for producing a dense point cloud of the environment. We are also only interested in terrestrial datasets, which is why we have not listed aerial datasets such as DALES [2], Campus3D [3] or SensatUrban [4].

First, in Table 1, we performed a separation according to the environment: the indoor datasets (mainly from RGB-D sensors) and the outdoor datasets (mainly from LiDAR sensors). For outdoor datasets, we also made the distinction between perception datasets (to improve perception tasks for the autonomous vehicle) and mapping datasets (to improve the mapping of the environment). For example, the well-known SemanticKITTI [5] consists of a set of LiDAR scans from which it is possible to produce a dense point cloud of the environment with the poses provided by SLAM or GPS/IMU, but the associated tasks (such as semantic segmentation or scene completion) are only centered on a LiDAR scan for the perception of the vehicle. This is very different from the dense point clouds of mapping systems such as Toronto-3D [6] or our Paris-CARLA-3D dataset. For the semantic segmentation and scene completion tasks, SemanticKITTI [7] uses only one single LiDAR scan as input (one rotation of the LiDAR). In our dataset, we wish to find the semantic and seek to complete the "holes" on the dense point cloud after the accumulation of all LiDAR scans.

Table 1 thus shows that Paris-CARLA-3D is the only dataset to offer annotations and protocols that allow for working on semantic, instance, and scene completion tasks on dense point clouds for outdoor mapping.

Table 1. Point cloud datasets for semantic segmentation (SS), instance segmentation (IS), and scene completion (SC) tasks. RGB means color available on all points of the point clouds. In parentheses for SS, we show only the number of classes evaluated (the annotation can have more classes).

Scene	Type	Dataset (Year)	World	# Points	RGB	Tasks SS	IS	SC
Indoor	Mapping	SUN3D [8] (2013)	Real	8 M	Yes	✓(11)	✓	
		SceneNet [9] (2015)	Synthetic	-	Yes	✓(11)	✓	✓
		S3DIS [10] (2016)	Real	696 M	Yes	✓(13)	✓	✓
		ScanNet [11] (2017)	Real	5581 M	Yes	✓(11)	✓	✓
		Matterport3D [12] (2017)	Real	24 M	Yes	✓(11)	✓	✓
Outdoor	Perception	PreSIL [13] (2019)	Synthetic	3135 M	Yes	✓(12)	✓	
		SemanticKITTI [5] (2019)	Real	4549 M	No	✓(25)	✓	✓
		nuScenes-Lidarseg [14] (2019)	Real	1400 M	Yes	✓(32)	✓	
		A2D2 [15] (2020)	Real	1238 M	Yes	✓(38)	✓	
		SemanticPOSS [16] (2020)	Real	216 M	No	✓(14)	✓	
		SynLiDAR [17] (2021)	Synthetic	19,482 M	No	✓(32)		
		KITTI-CARLA [18] (2021)	Synthetic	4500 M	Yes	✓(23)	✓	
	Mapping	Oakland [19] (2009)	Real	2 M	No	✓(5)		
		Paris-rue-Madame [20] (2014)	Real	20 M	No	✓(17)	✓	
		iQmulus [21] (2015)	Real	12 M	No	✓(8)	✓	
		Semantic3D [7] (2017)	Real	4009 M	Yes	✓(8)		
		Paris-Lille-3D [22] (2018)	Real	143 M	No	✓(9)	✓	
		SynthCity [23] (2019)	Synthetic	368 M	Yes	✓(9)		
		Toronto-3D [6] (2020)	Real	78 M	Yes	✓(8)		
		TUM-MLS-2016 [24] (2020)	Real	41 M	No	✓(8)		
		Paris-CARLA-3D (2021)	**Synthetic+Real**	**700 + 60 M**	**Yes**	**✓(23)**	**✓**	**✓**

3. Dataset Construction

This dataset is divided into two parts: a first set of real point clouds (60 M points) produced by a LiDAR and camera mobile system, and a second synthetic set produced by the open source CARLA simulator. Images of the different point clouds and annotations are available in Appendix B.

3.1. Paris (Real Data)

To create the Paris-CARLA-3D (PC3D) dataset, we developed a prototype mobile mapping system equipped with a LiDAR (Velodyne HDL32) tilted at 45° to the horizon and a 360° poly-dioptric camera Ladybug5 (composed of 6 cameras). Figure 1 shows the rear of the vehicle with the platform containing the various sensors.

Figure 1. Prototype acquisition system used to create the PC3D dataset in the city of Paris. Sensors: Velodyne HDL32 LiDAR, Ladybug5 360° camera, Photonfocus MV1 16-band VIR and 25-band NIR hyperspectral cameras (hyperspectral data are not available in this dataset; they cannot be used in mobile mapping due to the limited exposure time).

The acquisition was made on a part of Saint-Michel Avenue and Soufflot Street in Paris (a very dense urban area with many static and dynamic objects, presenting challenges for 3D scene understanding).

Unlike autonomous vehicle platforms such as KITTI [25] or nuScenes [14], the LiDAR is positioned at the rear and is tilted to allow scanning of the entire environment, thus allowing the buildings and the roads to be fully mapped.

To create the dense point clouds, we aggregated the LiDAR scans using a precise SLAM LiDAR based on IMLS-SLAM [26]. IMLS-SLAM only uses LiDAR data for the construction of the dataset. However, our platform is equipped with a high-precision IMU (LANDINS iXblue) and a GPS RTK. However, in a very dense environment (with tall buildings), an IMU + GPS-based localization (even with post-processing) achieves lower performance than a good LiDAR odometry (thanks to the buildings). The important hyperparameters of IMLS-SLAM used for Paris-CARLA are: $n = 30$ scans, $s = 600$ keypoints/scan, $r = 0.50$ m for neighbor search (explanations of these parameters are given in [26]). The drift of the IMLS-SLAM odometry is less than 0.40 % with no failure case (failure = no convergence of the algorithm). The quality of the odometry makes it possible to consider this localization as "ground truth".

The 360° camera was synchronized and calibrated with the LiDAR. The 3D data were colored by projecting on the image (with a timestamp as close as possible to the LiDAR timestamp) each 3D point of the LiDAR.

The final data were split according to the timestamp of points in six files (in binary ply format) with 10 M points in each file. Each point has many attributes stored: *x*, *y*, *z*, *x_lidar_position*, *y_lidar_position*, *z_lidar_position*, *intensity*, *timestamp*, *scan_index*, *scan_angle*, *vertical_laser_angle*, *laser_index*, *red*, *green*, *blue*, *semantic*, *instance*.

For the data annotation, this was done entirely manually with 3 people involved in 3 phases. In phase 1, the dataset was divided into two parts, with one person annotating each part (approximately 100 h of labeling per person). In phase 2, a verification of the annotations was performed by the other person on the part that he did not annotate with feedback and corrections. In phase 3, a third person outside the annotation carried out the verification of the labels on the entire dataset and a consistency check with the annotation in CARLA. The software used for annotation and checks was CloudCompare. The total time in human effort was approximately 300 h to obtain very high quality, as visible in Figure 2. The annotation of the data consisted of adding the semantic information (23 classes) and instance information for the *vehicle* class. The classes are the same as those defined in the CARLA simulator, making it possible to test transfer methods from synthetic to real data.

Figure 2. Paris-CARLA-3D dataset: (**left**) Paris point clouds with color information on LiDAR points; (**right**) manual semantic annotation of the LiDAR points (using the same tags from the CARLA simulator). We can see the large number of details in the manual annotation.

3.2. CARLA (Synthetic Data)

The open source CARLA simulator [27] allows for the simulation of the LiDAR and camera sensors in virtual outdoor environments. Starting from our mobile system (with Velodyne HDL32 and Ladybug5 360° camera), we created a virtual vehicle with the same sensors positioned in the same way as on our real platform. We then launched simulations to generate point clouds in the seven maps of CARLA v0.9.10 (called "Town01" to "Town07"). We finally assembled the scans using the ground truth trajectory and then kept one point cloud with 100 million points per town.

The 3D data were colored by projecting on the image (with a timestamp as close as possible to the LiDAR timestamp) each 3D point of the LiDAR. We used the same colorization process used with the real data from Paris.

The final data were stored in seven files (in binary ply format) with 100 M points in each file (one file = one town = one map in CARLA). We kept the following attributes per point: *x, y, z, x_lidar_position, y_lidar_position, z_lidar_position, timestamp, scan_index, cos_angle_lidar_surface, red, green, blue, semantic, instance, semantic_image*.

The annotation of CARLA data was automatic, thanks to the simulator with semantic information (23 classes) and instances (for the *vehicle* and *pedestrian* classes). We also kept during the colorization process the semantic information available in images in the attribute *semantic_image*.

3.3. Interest in Having Both Synthetic and Real Data

One of the interests of the Paris-CARLA-3D dataset is to have both synthetic and real data. The synthetic data are built with a virtual platform as close as possible to the real platform, allowing us to reproduce certain classic acquisition system issues (such as the difference in point of view of LiDAR and cameras sensors, creating color artifacts on the point cloud). Synthetic data are relatively easy to produce in large quantities (here 700 M points) and with ground truth without additional work for various 3D vision tasks such as classes or instances. It is thus of increasing interest to develop new methods on synthetic data but there is no evidence that they work on real data. With Paris-CARLA-3D and therefore with particular attention to having the same annotations between synthetic and real data, a method can be learned on synthetic data and tested on real data (which we will do in Section 5.2.6). However, we will see that the results remain limited. An interesting and promising approach will be to learn on synthetic data and to develop methods of performing unsupervised adaptation on real data. In this way, the methods will be able to learn from the large amount of data available in synthetic and, even better, from classes or objects that do not frequently meet in reality.

4. Dataset Properties

Paris-CARLA-3D has a linear distance of 550 m in Paris and approximately 5.8 km in CARLA (the same order of magnitude as the number of points ($\times 10$) between synthetic and real). For the real part, this represents three streets in the center of Paris. The area coverage is not large but the number and variety of urban objects, pedestrian movements, and vehicles is important: it is precisely this type of dense urban environment that is challenging to analyze.

4.1. Statistics of Classes

Paris-CARLA-3D is split into seven point clouds for the synthetic CARLA data, Town1 (T_1) to Town7 (T_7), and six point clouds for the real data of Paris, Soufflot0 (S_0) to Soufflot5 (S_5).

For CARLA data, cities can be divided into two groups: urban (T_1, T_2, T_3, and T_5) and rural (T_4, T_6, T_7).

For the Paris data, the point clouds can be divided into two groups: those near the Luxembourg Garden with vegetation and wide roads (S_0 and S_1), and those in a more dense urban configuration with buildings on both sides (S_2, S_3, S_4 and S_5).

The detailed distribution of the classes is presented in Appendix A.

4.2. Color

The point clouds are all colored (RGB information per point coming from cameras synchronized with the LiDAR), making it possible to test methods using geometric and/or appearance modalities.

4.3. Split for Training

For the different tasks presented in this article, according to the distribution of the classes, we chose to split the dataset into the following Train/Val/Test sets:

- Training data: S_1, S_2 (Paris); T_2, T_3, T_4, T_5 (CARLA);
- Validation data: S_4, S_5 (Paris); T_6 (CARLA);
- Test data: S_0, S_3 (Paris); T_1, T_7 (CARLA).

4.4. Transfer Learning

Paris-CARLA-3D is the first mapping dataset that is based on both synthetic and real data (with the same "platform" and the same data annotation). Indeed, simulators are becoming more and more reliable, and the fact of being able to transfer a method from a synthetic dataset created by a simulator to a real dataset is a line of research that could be important in the future.

We will now describe three 3D vision tasks using this new Paris-CARLA-3D dataset.

5. Semantic Segmentation (SS) Task

Semantic segmentation of point clouds is a task of increasing interest over the last several years [4,28]. This is an important step in the analysis of dense data from mobile LiDAR mapping systems. In Paris-CARLA-3D, the points are annotated point-wise with 23 classes whose tags are those defined in the CARLA simulator [27]. Figure 2 shows an example of semantic annotation in the Paris data.

5.1. Task Protocol

We introduce the task protocol to perform semantic segmentation in our dataset, allowing future work to build on the initial results presented here. We have many different objects belonging to the same class, as it the case in towns in the real world. This increases the complexity of the semantic segmentation task.

The evaluation of the performance in semantic segmentation tasks relies on True Positives (TP), False Positives (FP), True Negatives (TN), and False Negatives (FN) for each class c. These values are used to calculate the following metrics by class c: precision P_c, recall R_c, and Intersection over Union IoU_c. To describe the performance of methods, we usually report mean IoU as $mIoU$ Equation (1) and Overall Accuracy as OA.

$$mIoU = \frac{1}{C} \sum_{c=1}^{C} \frac{TP_c}{TP_c + FN_c + FP_c} \tag{1}$$

where C is the number of classes.

5.2. Experiments: Setting a Baseline

In this section, we present experiments performed under different configurations in order to demonstrate the relevance and high complexity of PC3D. We provide two baselines for all experiments with PointNet++ [29] and KPConv [30] architectures, two models widely used in semantic segmentation and which have demonstrated good performance on different datasets [4]. A recent survey with a detailed explanation of the different approaches to performing semantic segmentation on point clouds from urban scenes can be found at [28,31].

One of the challenges of dense outdoor point clouds is that they cannot be kept in memory, due to the high number of points. In both baselines, we used a subsampling strategy based on sphere selection. The spheres were selected using a weighted random, with the class rates of the dataset as probability distributions. This technique permits us to choose spheres centered in less populated classes. We evaluated different radius spheres ($r = 2, 5, 10, 20$ m) and we evidenced that using $r = 10$ m was a good compromise between computational cost and performance. On the one hand, small spheres are fast to process but provide poor information about the environment. On the other hand, large spheres provide richer information about the environment but are too expensive to process.

5.2.1. Baseline Parameters

The first baseline is based on the Pointnet++ architecture, commonly used in deep learning applications. We selected the architecture provided by the authors [29]. It is composed of three abstraction layers as the feature extractor and three MLP as the last part of the model. The number of points and neighborhood radius by layer were taken from the PointNet++ authors for outdoor and dense environments using MSG passing.

The second baseline is based on the KPConv architecture. We selected the KP-FCNN architecture provided by the authors for outdoor scenes [30]. It is composed of a five-layer network, where each layer contains two convolutional blocks, as originally proposed by the ResNet authors [32]. We used $dl_0 = 6$ cm, inspired by the value used by the authors for the Semantic3D dataset.

5.2.2. Implementation Details

We fixed similar training parameters between both baselines (Pointnet++ [29] and KPConv [30]) in order to compare their performance. As pre-processing, point clouds are sub-sampled on a grid, keeping one point per voxel (voxel size of 6 cm). Models learn, validate, and test with these data. Then, when testing, we perform inference with the under-sampled point clouds and then give the labels in "full resolution" with a KNN of the probabilities (not the labels). Spheres were computed in pre-processing (before the training stage) in order to reduce the computational cost. During training, we selected the spheres by class (class of the center point of the sphere) so that the network considered all the classes at each epoch, which greatly reduces the problem of class imbalance of the dataset. At each epoch, we took one point cloud from the dataset (T_i for CARLA and S_i for Paris) and set the number of spheres seen in this point cloud to 100.

Two features were included as input: RGB color information and height of points (z). In order to prevent overfitting, we included geometric data augmentation techniques: elastic distortion, random Gaussian noise with $\sigma = 0.1$ m and clip at 0.05 m, random rotations around z, anisotropic random scale between 0.8 and 1.2, and random symmetry around the x and y axes. We included the following transformations to prevent overfitting due to color information: chromatic jitter with $\sigma = 0.05$, and random dropout of RGB features with a probability of 20%.

For training, we selected the loss function as the sum of Cross Entropy and Power Jaccard with $p = 2$ [33]. We used a patience of 50 epochs (no progress in the validation set) and the optimizer ADAM with a default learning rate of 0.001. Both experiments were implemented using the Torch Points3D library [34] using a GPU NVIDIA Titan X with 12Go RAM.

Parameters presented in this section were chosen from a set of experiments varying the loss function (Cross Entropy, Focal Loss, Jaccard, and Power Jaccard) and input features (RGB and z coordinate or only z coordinate or only RGB or only 1 as input feature). The best results were obtained with the reported parameters.

5.2.3. Quantitative Results

Prediction of test point clouds was performed by using a sphere-based approach using a regular grid and maximum voting scheme. In this case, the spheres' centers were calculated to keep the intersections of spheres at $1/3$ of their radius ($r = 10$ m as done during the training).

We report the obtained results in Table 2. We obtained an overall 13.9 % mIoU for Point-Net++ and 37.5 % mIoU for KPConv. This remains low for state-of-the-art architectures. This shows the difficulty and the wide variety of classes present in this Paris-CARLA-3D dataset. We can also see poorer results on synthetic data due to the greater variety of objects between CARLA cities, while, for the real data, the test data are very close to the training data.

Table 2. Results in semantic segmentation task using PointNet++ and KPConv architectures on our dataset, Paris-CARLA-3D. Results are mIoU in %. For S_0 and S_3, training set is S_1, S_2. For T_1 and T_7, training set is T_2, T_3, T_4 and T_5. Overall mIoU is the mean IoU on the whole test sets (real and synthetic).

Model	Paris		CARLA		Overall
	S_0	S_3	T_1	T_7	mIoU
PointNet++ [29]	13.9	25.8	4.0	12.0	13.9
KPConv [30]	45.2	62.9	16.7	25.3	**37.5**

5.2.4. Qualitative Results

Semantic segmentation of point clouds is better on Paris than on CARLA in all evaluated scenarios. This is an expected behavior because class variability and scene configurations are much more complex in the synthetic dataset. By way of an example, Figures 3–6 display the predicted labels and ground truth from the test sets of Paris and CARLA data. These images were obtained from the KPConv architecture.

From the qualitative results of semantic segmentation, it is evidenced the complexity of our proposed dataset. In the case of Paris data, color information is discriminant enough to separate sidewalks, roads, and road-lines. This is an expected behavior because the point clouds were from the same town and were acquired the same day. However, in CARLA point clouds, color information in ground-like classes changed between different towns. Additionally, in some towns, such as T_2, we included rain during simulations, visible in the color of the road. This characteristic makes the learning stage even more difficult.

Figure 3. Left, prediction in S_0 test set of Paris data using KPConv model. **Right**, ground truth.

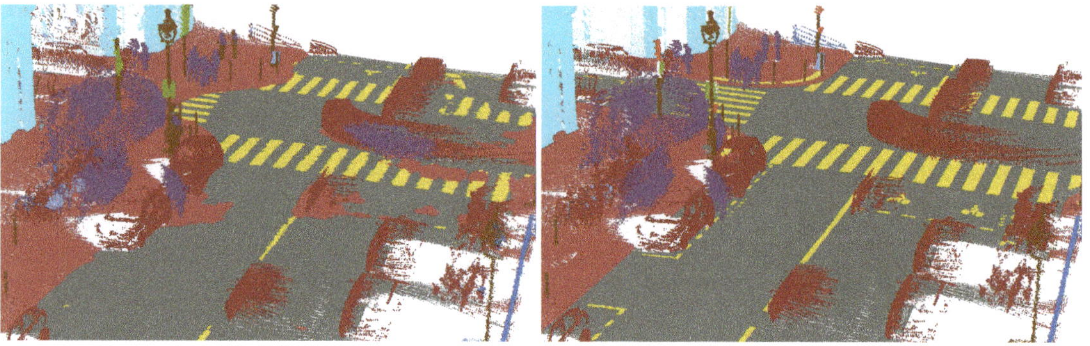

Figure 4. Left, prediction in S_3 test set of Paris data using KPConv model. **Right**, ground truth.

Figure 5. Left, prediction in T_1 test set of CARLA data using KPConv model. **Right**, ground truth.

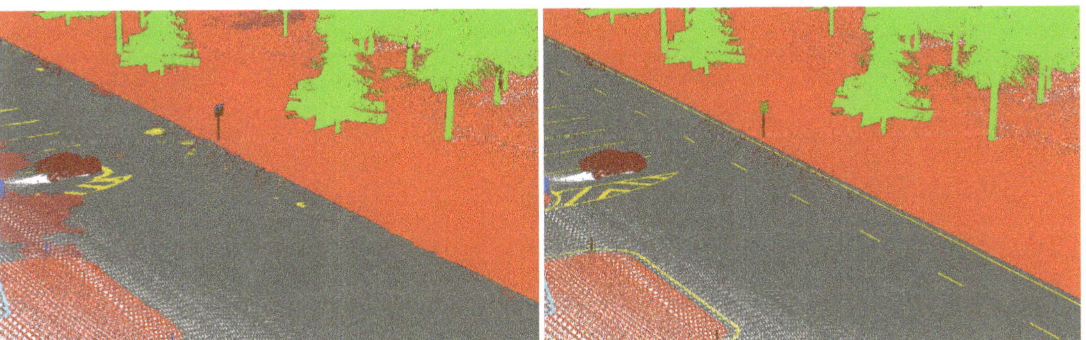

Figure 6. Left, prediction in T_7 test set of CARLA data using KPConv model. **Right**, ground truth.

5.2.5. Influence of Color

We studied the influence of color information during training in the PC3D dataset. In Table 3, we report the obtained results on the test set of semantic segmentation using the KPConv architecture without RGB features. The rest of the training parameters were the same as in the previous experiment. We can see that even if the colorization of the point cloud can create artifacts during the projection step (from the difference in point of view between the LiDAR sensor and the cameras or from the presence of moving objects), the use of the color modality in addition to geometry clearly improved the segmentation results.

Table 3. Results in semantic segmentation task using KPConv [30] architecture on our PC3D dataset with and without RGB colors on LiDAR points. Results are mIoU in %. For S_0 and S_3, training set is S_1, S_2. For T_1 and T_7, training set is T_2, T_3, T_4 and T_5. Overall mIoU is the mean IoU on the whole test sets (real and synthetic).

Model	Paris		CARLA		Overall
	S_0	S_3	T_1	T_7	mIoU
KPConv w/o color	39.4	41.5	35.3	17.0	33.3
KPConv with color	45.2	62.9	16.7	25.3	**37.5**

5.2.6. Transfer Learning

Transfer learning (TL) was performed with the aim of demonstrating the use of synthetic point clouds generated by CARLA to perform semantic segmentation on real-world point clouds. We selected the model with the best performance in the point clouds of the test set from CARLA data, i.e., taking the KPConv architecture (pre-trained on urban towns T_2, T_3, and T_5 since real data are urban data). Then, we took it as a pre-training stage with Paris data.

We carried out different types of experiments as follows: (1) Predict test point clouds of Paris data using the best model obtained in urban towns from CARLA without training in Paris data (*no fine-tuning*); (2) Freeze the whole model except the last layer; (3) Freeze the feature extractor of the network; (4) No frozen parameters; (5) Training a model from scratch using only Paris training data. These scenarios were selected to evaluate the relevance of learned features in CARLA and their capacity to discriminate classes in Paris data. Results are presented in Table 4. The best results using TL were obtained in scenario 4: the model pre-trained in CARLA without frozen parameters during fine-tuning on Paris data. However, scenario 5 (i.e., *no transfer*) ultimately showed superior results.

Table 4. Results in transfer learning for the semantic segmentation task using KPConv architecture on our PC3D dataset. Results are mIoU in %. Pre-training was done using urban towns from CARLA (T_2, T_3, and T_5). *No fine-tuning*: the model was pre-trained on CARLA data without fine-tuning on Paris data. *No frozen parameters*: the model was pre-trained on CARLA without frozen parameters during fine-tuning on Paris data. *No transfer*: the model was trained only on the Paris training set.

Transfer Learning Scenarios	Paris		Overall
	S_0	S_3	mIoU
No fine-tuning	20.6	17.7	19.2
Freeze except last layer	24.1	31.0	27.6
Freeze feature extractor	29.0	41.3	35.2
No frozen parameters	42.8	50.0	46.4
No transfer	45.2	62.9	**51.7**

From Table 4, a first finding is that the current model trained on synthetic data cannot be directly applied to real-world data (the *no fine-tuning* row). This is an expected result, because objects and class distributions in CARLA towns are different from real-world ones.

We may also observe that the performance of *no frozen parameters* is lower than that of *no transfer*: pre-training the network on the synthetic and fine-tuning on the real data decreases the performance compared to training directly on the real dataset. Alternatives are now introduced in order to close the existing gap between synthetic and real data, such as domain adaptation methods.

6. Instance Segmentation (IS) Task

The ability to detect instances in dense point clouds of outdoor environments can be useful for cities for urban space management (for example, to have an estimate of the occupancy of parking spaces through fast mobile mapping) or for building the prior map layer for HD maps in autonomous driving.

We provide instance annotations as follows: in Paris data, instances of *vehicle* class were manually point-wise annotated; in CARLA data, *vehicle* and *pedestrian* instances were automatically obtained by the CARLA simulator. Figure 7 illustrates the instance annotation of vehicles in S_3 Paris data. We found that pedestrians in Paris data were too close to each other to be recognized as separate instances (Figure 8).

6.1. Task Protocol

We introduce the task protocol to evaluate the instance segmentation methods in our dataset.

Evaluation of the performance in the instance segmentation task is different to that in the semantic segmentation task. Inspired by [35] on *things*, we report Segment Matching (SM) and Panoptic Quality (PQ), with $IoU = 0.5$ as the threshold to determine well-predicted instances. We also report the mean IoU, based on IoU by instance i (IoU_i), calculated as follows:

$$IoU_i = \begin{cases} IoU & IoU \geq 0.5 \\ 0, & \text{otherwise} \end{cases} \tag{2}$$

A common issue in LiDAR scanning is the presence of far objects that are unrecognizable due to the small number of points. In the semantic segmentation task, such objects do not affect evaluation metrics, due to their low rate. However, in the instance segmentation task, they may considerably affect the evaluation of the algorithms. In order to provide an evaluation metric having relevance, metrics are computed only with instances closer than $d = 20\,\text{m}$ to the mobile system.

Figure 7. Instances of vehicles in S_3 test set (Paris data).

Figure 8. Pedestrians in Paris data: we can see inside the red circle the difficulty of differentiating the instances of pedestrians.

6.2. Experiments: Setting a Baseline

In this section, we present a baseline for the instance segmentation task and its evaluation with the introduced metrics. We propose a hybrid approach, combining deep learning and mathematical morphology, to predict instance labels. We report the obtained results for each point cloud of the test sets.

As presented by [36], urban objects can be classified using geometrical and contextual features. In our case, we start from already predicted *things* classes (vehicles and pedestrians, in this case) with the best model introduced in Section 5.2.3, i.e., using the KPConv architecture. Then, instances are detected by using Bird's Eye View (BEV) projections and mathematical morphology.

We computed the following BEV projections (with a pixel resolution of 10 cm) for each class:

- Occupancy image (I_b)—binary image with presence or not of *things* class;
- Elevation image (I_h)—stores the maximal elevation among all projected points on the same pixel;
- Accumulation image (I_{acc})—stores the number of points projected on the same pixel.

At this point, three BEV projections were computed for each class: occupancy (I_b), elevation (I_h), and accumulation (I_{acc}). In the following sections, we describe the proposed algorithms to separate the vehicle and pedestrian instances. We highlight that these methods rely on the labels predicted in the semantic segmentation task (Section 5.2.1) using the KPConv architecture.

6.2.1. Vehicles in Paris and CARLA Data

One of the main challenges of this class is the high variability due to the different types of objects that it contains: cars, motorbikes, bikes, and scooters. Additionally, it also includes moving and parked vehicles, which makes it challenging to determine object boundaries.

From BEV projections, vehicle detection is performed as follows:

1. Discard the predicted points of the vehicle if the z coordinate is greater than 4 m in I_h;
2. Connect close components with two consecutive morphological dilations of I_b by a square of 3-pixel size;
3. Fill holes smaller than ten pixels inside each connected component; this is performed with a morphological area closing;
4. Discard instances with less than 500 points in I_{acc};
5. Discard instances not surrounded by ground-like classes in I_b.

6.2.2. Pedestrians in CARLA Data

As mentioned earlier for vehicles, the pedestrian class may contain moving objects. This implies that object boundaries are not always well-defined.

We followed a similar approach as described previously for vehicle instances based on the semantic segmentation results and BEV projections. We first discarded pedestrian points if the z coordinate was greater than 3 m in I_h, and then connected close components and filled small holes, as described for the vehicle class; we then discarded instances with less than 100 points in I_{acc} and, finally, discarded instances not surrounded by ground-like classes in I_b.

6.2.3. Quantitative Results

For vehicles and pedestrians, instance labels of BEV images were back-projected to 3D data in order to provide point-wise predictions. In Table 5, we report the obtained results in instance segmentation using the proposed approach. These results are the first of a method allowing instance segmentation on dense points clouds from 3D mapping, and we hope that it will inspire future methods.

Table 5. Results on test sets of Paris-CARLA-3D for the instance segmentation task. SM: Segment Matching. PQ: Panoptic Quality. *mIoU*: mean IoU. All results are in %.

	# Instances	SM	PQ	mIoU
S_0—Vehicles	10	90.0	70.9	81.6
S_3—Vehicles	86	32.6	40.5	28.0
T_1—Vehicles	41	17.1	20.4	14.2
T_7—Vehicles	27	74.1	72.6	61.2
T_1—Pedestrians	49	18.4	17.0	13.9
T_7—Pedestrians	3	100.0	9.0	66.0
Mean	216	55.3	38.4	44.2

6.2.4. Qualitative Results

In our proposed baseline, instances are separated using BEV projections and geometrical features based on semantic segmentation labels. In some cases, as presented in Figure 9, 2D projections can merge objects in the same instance label if they are too close.

Figure 9. Top, vehicle instances from our proposed baseline using BEV projections and geometrical features in S_3 Paris data. **Bottom**, ground truth.

Close objects and instance intersections are challenging for the instance segmentation task. The former can be tackled by using approaches based directly on 3D data. For the latter, we provide timestamp information by point in each PLY file. The availability of this feature may be useful for future approaches.

Semantic segmentation and instance segmentation could be unified in one task, Panoptic Segmentation (PS): this is a task that has recently emerged in the context of scene understanding [35]. We leave this for future works.

7. Scene Completion (SC) Task

The scene completion (SC) task consists of predicting the missing parts of a scene (which can be in the form of a depth image, a point cloud, or a mesh). This is an important problem in 3D mapping due to holes from occlusions and holes after the removal of unwanted objects, such as vehicles or pedestrians (see Figure 10). It can be solved in the form of 3D reconstruction [37], scan completion [38], or, more specifically, methods to fill holes in a 3D model [39].

Figure 10. Paris data after removal of vehicles and pedestrians. Zones in red circles show the interest in conducting scene completion for 3D mapping, in order to fill holes from removed pedestrians, parked cars, and from the occlusion of other objects, and also to improve the sampling of points in areas far from the LiDAR.

Semantic scene completion (SSC) is the task of filling the geometry as well as predicting the semantics of the points, with the aim that the two tasks carried out simultaneously benefit each other (survey of SSC in [40]). It is also possible to jointly predict the geometry and color during scene completion, as in SPSG [41]. For now, we only evaluate the geometry prediction, as we leave the prediction of simultaneous geometry, semantics, and color for future work.

The vast majority of the existing methods of scene completion (SC) work focus on small indoor scenes, while, in our case, we have a dense outdoor environment with our Paris-CARLA-3D dataset. Completing outdoor LiDAR point clouds is more challenging than data obtained from RGB-D images acquired in indoor environments, due to the sparsity of points obtained using LiDAR sensors. Moreover, larger occluded areas are present in outdoor scenes, caused by static and temporary foreground objects, such as trees, parked vehicles, bus stops, and benches. SemanticKITTI [7] is a dataset conducting scene completion (SC) and semantic scene completion (SSC) on LiDAR data, but they use only one single scan as input, with a target (ground truth) being the accumulation of all LiDAR scans. In our dataset, we seek to complete the "holes" from the accumulation of all LiDAR scans.

7.1. Task Protocol

We introduce the task protocol to perform scene completion on PC3D. Our goal is to predict a more complete point cloud. First, we extract random small chunks from the original point cloud that we transform into a discretized regular 3D grid representation containing the Truncated Signed Distance Function (TSDF) values, which expresses the distance from each voxel to the surface represented by the point cloud. Then, we use a neural network to predict a new TSDF and finally, we extract a point cloud from that TSDF that should be more complete that the input. We used as TSDF the classical signed point-to-plane distance to the closest point of the point cloud as in [42]. Our original point cloud is already incomplete due to the occlusions caused by static objects and the sparsity of the scans. To overcome this incompleteness, we make the point cloud more incomplete by removing 90% of the points (by *scan_index*), and use the incomplete data to compute the TSDF input of the neural network. Moreover, we use the original point cloud containing all of the points as the ground truth and compute the target TSDF. Our approach is inspired by the work done by SG-NN [43] and we do this in order to learn to complete the scene in a self-supervised way. Removing points according to their *scan_index* allows us to create larger "holes" than by removing points at random. For the chunks, we used a grid size of $128 \times 128 \times 128$ and a voxel size of 5 cm (compared to the voxel size of 2 cm used for indoor scenes in SG-NN [43]). Dynamic objects, pedestrians, vehicles, and unlabeled points are first removed from the data using the ground truth semantic information.

To evaluate the completed scene, we use the Chamfer Distance (CD) between the original P_1 and predicted P_2 point clouds:

$$CD = \frac{1}{|P_1|} \sum_{x \in P_1} \min_{y \in P_2} ||x - y||_2 + \frac{1}{|P_2|} \sum_{y \in P_2} \min_{x \in P_1} ||y - x||_2 \tag{3}$$

In a self-supervised context, not having the ground truth and having the predicted point cloud more complete than the target places some limitations on using the CD metric. For this, we introduce a mask that needs to be used to compute the CD only on the points that were originally available. The mask is simply a binary occupancy grid on the original point cloud.

We extract the random chunks as explained previously for Paris (1000 chunks per point cloud) and CARLA (3000 chunks per town) and provide them along with the dataset for future research on scene completion.

7.2. Experiments: Setting a Baseline

In this section, we present a baseline for scene completion using the SG-NN network [43] to predict the missing points (SG-NN predicts only the geometry and not the semantics nor the color). In SG-NN, they use volumetric fusion [44] to compute a TSDF from range images, which cannot be used on LiDAR point clouds. For this, we compute a different TSDF from the point clouds.

Using the cropped chunks, we estimate the normal at each point using PCA as in [42] with $n = 30$ neighbors and obtain a consistent orientation using the LiDAR sensor position provided with the points. Using the normal information, we use the SDF introduced in [42], due to its simplicity and the ease of vectorizing, which reduces the data generation complexity. After obtaining the SDF volumetric representation, we convert the values to voxel units and truncate the function at three voxels, which results in a 3D sparse TSDF volumetric representation that is similar to the input of SG-NN [43]. For the target, we use all the points available in the original point cloud, and for the input, we keep 10% of points (by the scan indices) in each chunk, in order to obtain the "incomplete" point cloud representation.

The resulting sparse tensors are then used for training and the network is trained for 20 epochs with ADAM and a learning rate of 0.001. The loss is a combination of Binary Cross Entropy (BCE) on occupancy and L1 Loss on TSDF prediction. The training was carried out on a GPU NVIDIA RTX 2070 SUPER with 8Go RAM.

In order to increase the number of samples and prevent overfitting, we perform data augmentation on the extracted chunks: random rotation around z, random scaling between 0.8 and 1.2, and local noise addition with $\sigma = 0.05$.

Finally, we extract a point cloud from the TSDF predicted by the network following an approach that is similar to the marching cubes algorithm [45], where we interpolate 1 point per voxel. Finally, we compute the CD (see Equation (3)) between the point cloud extracted from the predicted TSDF and the original point cloud (without dynamic objects) and use the introduced mask to limit the CD computation to known regions (voxels where we have points in the original point cloud).

7.2.1. Quantitative Results

Table 6 shows the results of our experiment on Paris-CARLA-3D data. We can see that the network makes it possible to create point clouds whose distance to the original cloud is clearly smaller.

For further metric evaluation, we provide the mean IoU and ℓ_1 distance between the target and predicted TSDF values on the 2000 and 6000 chunks for Paris and CARLA test sets, respectively. The results are also reported in Table 6.

Table 6. Scene completion results on Paris-CARLA-3D. CD is the mean Chamfer Distance over 2000 chunks for the Paris test set (S_0 and S_3) and 6000 chunks for the CARLA test set (T_1 and T_7). ℓ_1 is the mean ℓ_1 distance between predicted and target TSDF measured in voxel units for 5 cm voxels and *mIoU* is the mean Intersection over Union of TSDF occupancy. Both metrics are computed on known voxel areas. *ori* means original point cloud, *in* is input point cloud (10% of the original), *pred* is the predicted point cloud (computed from predicted TSDF).

Test Set	$CD_{in\leftrightarrow ori}$	$CD_{pred\leftrightarrow ori}$	$\ell_{1pred\leftrightarrow tar}$	$mIoU_{pred\leftrightarrow tar}$
S_0 and S_3 (Paris)	16.6 cm	10.7 cm	0.40	85.3%
T_1 and T_7 (CARLA)	13.3 cm	10.2 cm	0.49	80.3%

7.2.2. Qualitative Results

Figure 11 shows the scene completion result on one point cloud chunk from the CARLA T_1 test set. Figure 12 shows the scene completion result on one chunk from the Paris S_0 test set. We can see that the network manages to produce point clouds quite close visually to the original, despite having as input a sparse point cloud with only 10% of the points of the original.

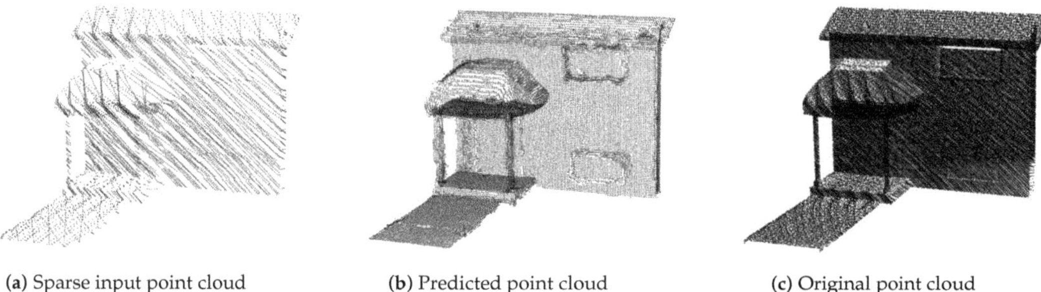

(**a**) Sparse input point cloud   (**b**) Predicted point cloud   (**c**) Original point cloud

Figure 11. Scene completion task for one chunk point cloud in Town1 (T_1) of CARLA test data (training on CARLA data).

(**a**) Sparse input point cloud   (**b**) Predicted point cloud   (**c**) Original point cloud

Figure 12. Scene completion task for one chunk point cloud in Soufflot0 (S_0) of Paris test data (training on Paris data).

7.2.3. Transfer Learning with Scene Completion

Using both synthetic and real data of Paris-CARLA-3D, we tested the training of a scene completion model on CARLA synthetic data to test it on Paris data. With the objective of scene completion on real data chunks (Paris S_0 and S_3), we tested three training scenarios: (1) Training only on real data with Paris training set; (2) Training only on synthetic data with CARLA training set; (3) Pre-train on synthetic data then fine-tune on real data. The results are shown in Table 7. We can see that the Chamfer Distance (CD) is better for the model trained only on synthetic CARLA data: the network is attempting to fill a local plane in large missing regions and smoothing the rest of the geometry. This is an expected behavior, because of the handcrafted geometry present in CARLA, where planar geometric features are predominantly present. Point clouds of real outdoor scenes are not easily obtained and the need to complete missing geometry is becoming increasingly important in vision-related tasks; here, we can see the value of leveraging the large amount of synthetic data present in CARLA to pre-train the network and fine-tune it on other smaller datasets such as Paris when not enough data are available. As we can see in Table 7, pre-training on CARLA and then fine-tuning on Paris allows us to obtain the best predicted TSDF (ℓ_1 and $mIoU$) and point cloud (Chamfer Distance).

Table 7. Results of transfer learning for the scene completion task. CD is the mean Chamfer Distance between point clouds. ℓ_1 is the mean ℓ_1 distance between predicted and target TSDF measured in voxel units for 5 cm voxels and $mIoU$ is the mean Intersection over Union of TSDF occupancy. The mean is over 2000 chunks for Paris data. *ori* means original point cloud, *in* is input point cloud (10 % of the original), *pred* for CD is the predicted point cloud (computed from predicted TSDF), *pred* for IoU, ℓ_1 is the predicted TSDF, and *tar* is the target TSDF.

Test Set: S_0 and S_3 Paris data	$CD_{in\leftrightarrow ori}$	$CD_{pred\leftrightarrow ori}$	$\ell_{1pred\leftrightarrow tar}$	$mIoU_{pred\leftrightarrow tar}$
Trained only on Paris	16.6 cm	10.7 cm	0.40	85.3%
Trained only on CARLA	16.6 cm	8.0 cm	0.48	84.0%
Pre-trained CARLA, fine-tuned on Paris	16.6 cm	7.5 cm	0.35	88.7%

8. Conclusions

We presented a new dataset called Paris-CARLA-3D. This dataset is made up of both synthetic data (700 M points) and real data (60 M points) from the same LiDAR and camera mobile platform. Based on this dataset, we presented three classical tasks in 3D computer vision (semantic segmentation, instance segmentation, and scene completion) with their evaluation protocol as well as a baseline, which will serve as starting points for future work using this dataset.

On semantic segmentation (the most common task in 3D vision), we tested two state-of-the-art methods, PointNet++ and KPConv, and showed that KPConv obtains the best results (37.5% overall mIoU). We also presented a first instance detection method on dense point clouds from mapping systems (with vehicle and pedestrian instances for synthetic data and vehicle instances for real data). For the scene completion task, we were able to adapt a method used for indoor data with RGB-D sensors to outdoor LiDAR data. Even with a simple formulation of the surface, the network manages to learn complex geometries, and, moreover, by using the synthetic data as pre-training, the method obtains better results on the real data.

Author Contributions: Methodology and writing J.-E.D.; data annotation, methodology and writing D.D. and J.P.R.; supervision and reviews S.V.-F., B.M. and F.G. All authors have read and agreed to the published version of the manuscript.

Funding: This research was partially funded by REPLICA FUI 24 project.

Data Availability Statement: The dataset is available at the following URL: https://npm3d.fr/paris-carla-3d, accessed on 15 October 2021.

Appendix A. Complementary on Paris-CARLA-3D Dataset

Appendix A.1. Class Statistics

From CARLA data, as occurs in real-world scenarios, not every class is present in every town: eleven classes are present in all towns (*road, building, sidewalk, vegetation, vehicles, road-line, fence, pole, static, dynamic, traffic sign*), three classes in six towns (*unlabeled, wall, pedestrian*), three classes in five towns (*terrain, guard-rail, ground*), two classes in four towns (*bridge, other*), one class in three towns (*water*), and two classes in two towns (*traffic light, rail-track*).

In Paris data, class variability is smaller than in CARLA data. This is a desired (and expected) feature of these point clouds because they correspond to the same town. However, as is the case with CARLA towns, not every class is present in every point cloud: twelve classes are present in all point clouds (*road, building, sidewalk, road-line, vehicles, other, unlabeled, static, pole, dynamic, pedestrian, traffic sign*), three classes in five point clouds (*vegetation, fence, traffic light*), one class in two point clouds (*terrain*), and seven classes in any point cloud (*wall, sky, ground, bridge, rail-track, guard-rail, water*).

Table A1 shows the detailed statistics of the classes in the Paris-CARLA-3D dataset.

Table A1. Class distribution in Paris-CARLA-3D dataset (in %). Columns headed by S_i are Soufflot from Paris data and T_j are towns from CARLA data.

	Paris						CARLA						
Class	S_0	S_1	S_2	S_3	S_4	S_5	T_1	T_2	T_3	T_4	T_5	T_6	T_7
unlabeled	0.9	1.5	3.9	3.2	1.9	0.9	5.8	2.9	-	7.6	0.0	6.4	1.8
building	14.9	18.9	34.2	36.6	33.1	32.9	6.8	22.6	15.3	4.5	16.1	2.6	3.3
fence	2.3	0.6	0.7	0.8	-	0.4	1.0	0.6	0.0	0.5	3.8	1.5	0.6
other	2.1	3.4	6.7	2.2	2.5	0.4	-	-	-	-	0.1	0.1	0.1
pedestrian	0.2	1.0	0.6	1.0	0.7	0.7	0.1	0.2	0.1	0.0	-	0.1	0.0
pole	0.6	0.9	0.6	0.8	0.7	1.1	0.6	0.6	4.2	0.8	0.8	0.4	0.3
road-line	3.8	3.7	2.4	4.1	3.5	3.4	0.2	0.2	2.9	1.6	2.2	1.3	1.7
road	41.0	49.7	35.0	37.6	40.6	27.5	47.8	37.2	53.1	52.8	44.7	58.0	42.8
sidewalk	10.1	4.2	7.3	6.7	11.9	29.4	22.5	17.5	10.3	1.7	10.5	3.1	0.4
vegetation	18.5	9.0	0.1	0.3	0.1	-	8.7	10.8	2.7	12.8	4.6	8.1	23.1
vehicles	1.3	1.8	6.5	6.5	3.3	1.6	1.7	3.1	0.9	0.5	3.1	4.2	0.9
wall	-	-	-	-	-	-	1.9	3.6	1.4	5.4	5.3	3.4	-
traffic sign	0.1	0.4	0.1	0.1	0.3	0.1	-	0.0	-	0.1	0.0	0.0	0.1
sky	-	-	-	-	-	-	-	-	-	-	-	-	-
ground	-	-	-	-	-	-	-	0.0	0.2	1.4	0.3	0.1	-
bridge	-	-	-	-	-	-	1.7	-	-	0.7	6.6	-	-
rail-track	-	-	-	-	-	-	-	-	7.6	-	0.5	-	-
guard-rail	-	-	-	-	-	-	0.0	-	-	4.3	-	1.2	0.5
static	2.6	2.3	0.3	0.1	0.7	1.5	0.1	0.1	-	-	-	-	-
traffic light	0.1	0.2	0.1	0.1	0.1	-	0.8	0.5	0.3	0.3	0.3	-	-
dynamic	0.3	1.6	1.5	0.2	0.7	0.0	0.1	0.1	0.1	0.3	0.1	0.1	0.1
water	-	-	-	-	-	-	0.4	-	0.0	-	-	-	0.6
terrain	1.4	0.8	-	-	-	-	-	-	0.9	4.8	1.1	9.6	23.8
# Points	**60 M**						**700 M**						

Appendix A.2. Instances

The number of instances in ground truth varies over the point clouds. In the test set from Paris data, Soufflot0 (S_0) has 10 vehicles while Soufflot3 (S_3) has 86. This large difference occurs due to the presence of parked motorbikes and bikes.

With respect to CARLA data, it was observed that in urban towns such as Town1 (T_1), vehicle and pedestrian instances are mainly moving objects. This implies that during simulations, instances can have intersections between them, making their separation challenging.

In the CARLA simulator, the instances of the objects are given by their IDs. If a vehicle/pedestrian is seen several times, the same instance_id is used at different places. This is a problem in the evaluation capacity of detecting correctly the instances. This is why we have divided the CARLA instances using the timestamp of points: separate instances based on a timestamp gap with a threshold of 10 s for vehicles and 5 s for pedestrians.

Appendix B. Images of the Dataset

Figures A1–A6 show top-view images of the different point clouds of the Paris-CARLA-3D dataset.

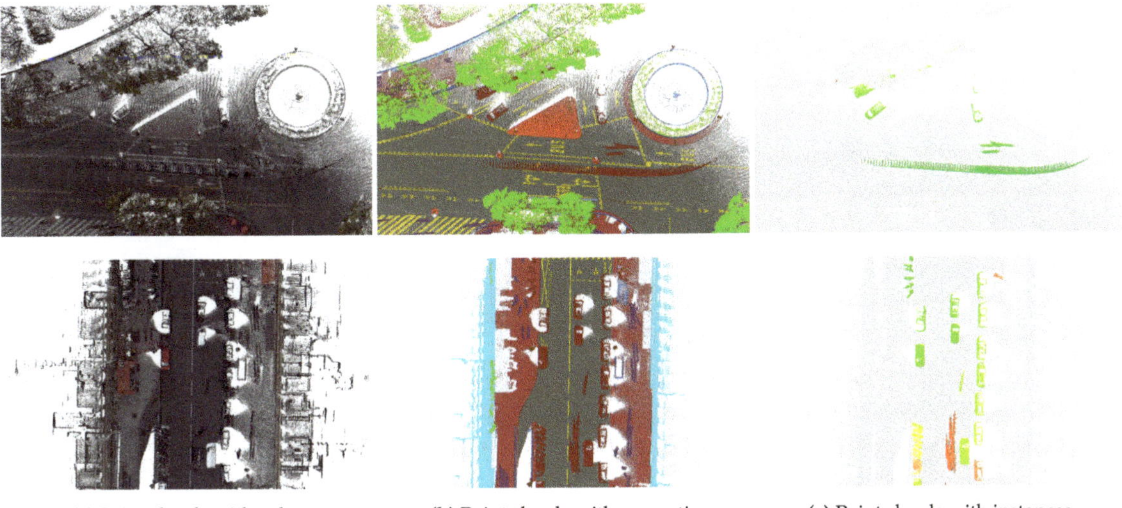

(a) Point clouds with color (b) Point clouds with semantic (c) Point clouds with instances

Figure A1. Paris training set. From **top** to **bottom**: S_1, S_2 (real data).

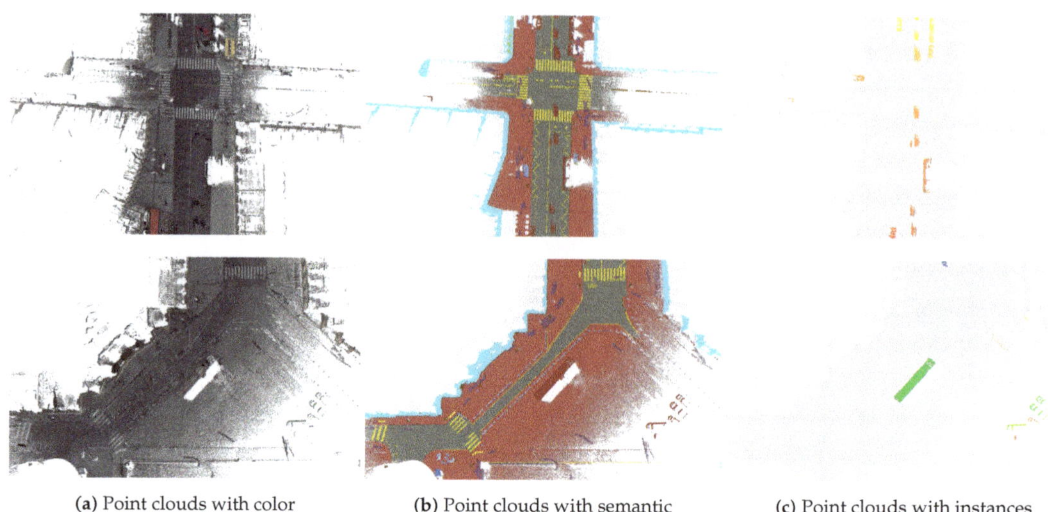

(a) Point clouds with color (b) Point clouds with semantic (c) Point clouds with instances

Figure A2. Paris validation set. From **top** to **bottom**: S_4, S_5 (real data).

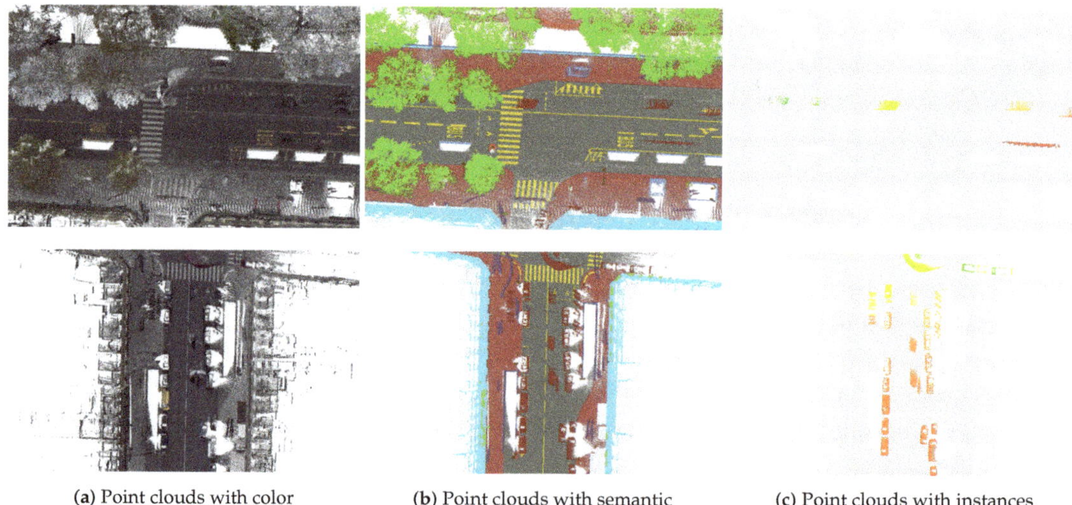

(a) Point clouds with color (b) Point clouds with semantic (c) Point clouds with instances

Figure A3. Paris test set. From **top** to **bottom**: S_0, S_3 (real data).

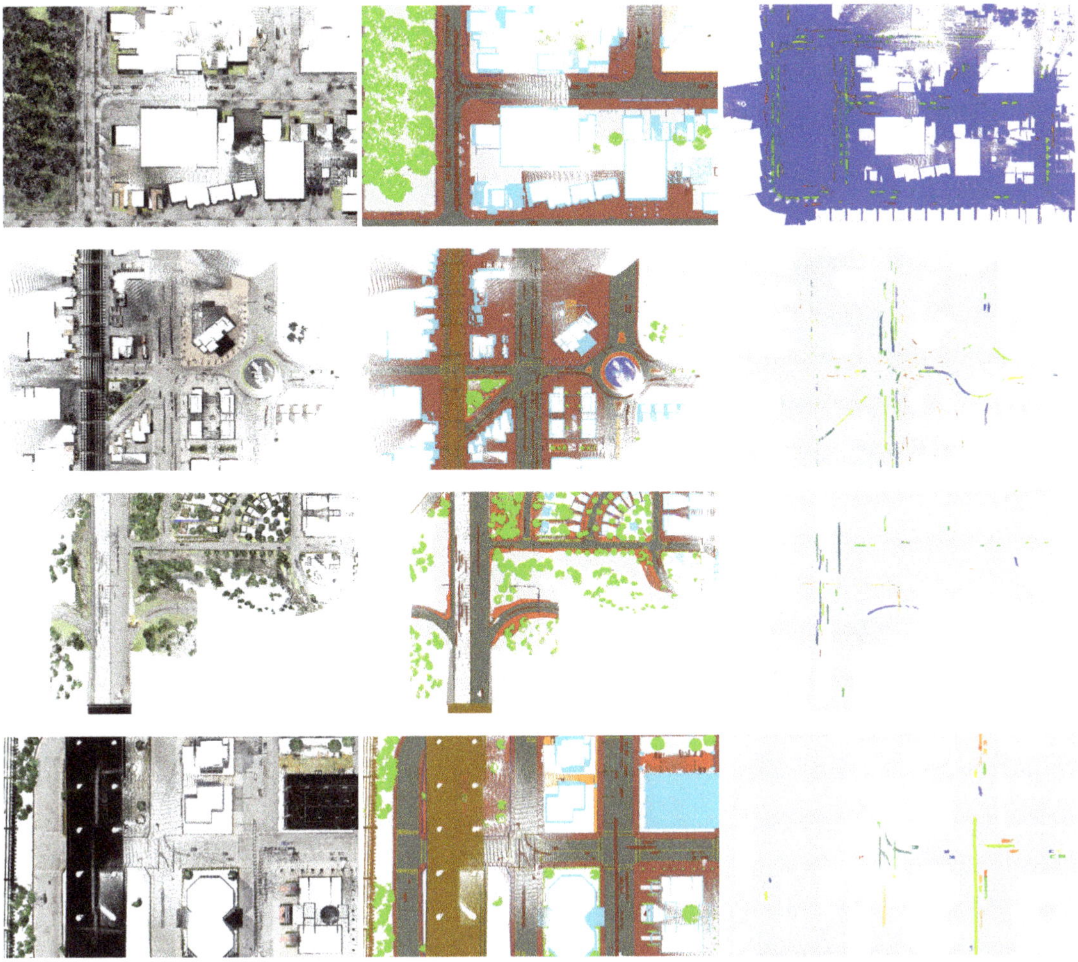

(**a**) Point clouds with color   (**b**) Point clouds with semantic   (**c**) Point clouds with instances

Figure A4. CARLA training set. From **top** to **bottom**: T_2, T_3, T_4, T_5 (synthetic data).

(**a**) Point cloud with color   (**b**) Point cloud with semantic   (**c**) Point cloud with instances

Figure A5. CARLA validation set. T_6 (synthetic data).

(**a**) Point clouds with color　　(**b**) Point clouds with semantic　　(**c**) Point clouds with instances

Figure A6. CARLA test set. From **top** to **bottom**: T_1, T_7 (synthetic data).

References

1. Silberman, N.; Hoiem, D.; Kohli, P.; Fergus, R. Indoor Segmentation and Support Inference from RGBD Images. In Proceedings of the Computer Vision—ECCV 2012, Florence, Italy, 7–13 October 2012; Fitzgibbon, A., Lazebnik, S., Perona, P., Sato, Y., Schmid, C., Eds.; Springer: Berlin/Heidelberg, Germany, 2012; pp. 746–760.
2. Varney, N.; Asari, V.K.; Graehling, Q. DALES: A Large-scale Aerial LiDAR Data Set for Semantic Segmentation. In Proceedings of the 2020 IEEE/CVF Conference on Computer Vision and Pattern Recognition Workshops (CVPRW), Seattle, WA, USA, 14–19 June 2020; pp. 717–726. [CrossRef]
3. Li, X.; Li, C.; Tong, Z.; Lim, A.; Yuan, J.; Wu, Y.; Tang, J.; Huang, R. Campus3D: A Photogrammetry Point Cloud Benchmark for Hierarchical Understanding of Outdoor Scene. In Proceedings of the 28th ACM International Conference on Multimedia, Seattle, WA, USA, 12–16 October 2020; Association for Computing Machinery: New York, NY, USA, 2020; pp. 238–246. [CrossRef]
4. Hu, Q.; Yang, B.; Khalid, S.; Xiao, W.; Trigoni, N.; Markham, A. Towards Semantic Segmentation of Urban-Scale 3D Point Clouds: A Dataset, Benchmarks and Challenges. In Proceedings of the IEEE/CVF Conference on Computer Vision and Pattern Recognition (CVPR), Nashville, TN, USA, 19–25 June 2021; pp. 4977–4987.
5. Behley, J.; Garbade, M.; Milioto, A.; Quenzel, J.; Behnke, S.; Stachniss, C.; Gall, J. SemanticKITTI: A Dataset for Semantic Scene Understanding of LiDAR Sequences. In Proceedings of the 2019 IEEE/CVF International Conference on Computer Vision (ICCV), Seoul, Korea, 27–28 October 2019; pp. 9296–9306. [CrossRef]
6. Tan, W.; Qin, N.; Ma, L.; Li, Y.; Du, J.; Cai, G.; Yang, K.; Li, J. Toronto-3D: A Large-Scale Mobile LiDAR Dataset for Semantic Segmentation of Urban Roadways. In Proceedings of the IEEE/CVF Conference on Computer Vision and Pattern Recognition (CVPR) Workshops, Seattle, WA, USA, 14–19 June 2020.
7. Hackel, T.; Savinov, N.; Ladicky, L.; Wegner, J.D.; Schindler, K.; Pollefeys, M. SEMANTIC3D.NET: A new large-scale point cloud classification benchmark. *arXiv* **2017**, arXiv:1704.03847.
8. Xiao, J.; Owens, A.; Torralba, A. SUN3D: A Database of Big Spaces Reconstructed Using SfM and Object Labels. In Proceedings of the 2013 IEEE International Conference on Computer Vision (ICCV), Sydney, Australia, 1–8 December 2013; pp. 1625–1632. [CrossRef]
9. McCormac, J.; Handa, A.; Leutenegger, S.; Davison, A.J. SceneNet RGB-D: Can 5M Synthetic Images Beat Generic ImageNet Pre-training on Indoor Segmentation? In Proceedings of the 2017 IEEE International Conference on Computer Vision (ICCV), Venice, Italy, 22–29 October 2017; pp. 2697–2706. [CrossRef]
10. Armeni, I.; Sener, O.; Zamir, A.R.; Jiang, H.; Brilakis, I.; Fischer, M.; Savarese, S. 3D Semantic Parsing of Large-Scale Indoor Spaces. In Proceedings of the 2016 IEEE Conference on Computer Vision and Pattern Recognition (CVPR), Las Vegas, NV, USA, 27–30 June 2016; pp. 1534–1543. [CrossRef]
11. Dai, A.; Chang, A.X.; Savva, M.; Halber, M.; Funkhouser, T.; Nießner, M. ScanNet: Richly-Annotated 3D Reconstructions of Indoor Scenes. In Proceedings of the 2017 IEEE Conference on Computer Vision and Pattern Recognition (CVPR), Honolulu, HI, USA, 21–26 July 2017; pp. 2432–2443. [CrossRef]
12. Chang, A.; Dai, A.; Funkhouser, T.; Halber, M.; Niebner, M.; Savva, M.; Song, S.; Zeng, A.; Zhang, Y. Matterport3D: Learning from RGB-D Data in Indoor Environments. In Proceedings of the 2017 International Conference on 3D Vision (3DV), Qingdao, China, 10–12 October 2017; pp. 667–676. [CrossRef]

13. Hurl, B.; Czarnecki, K.; Waslander, S. Precise Synthetic Image and LiDAR (PreSIL) Dataset for Autonomous Vehicle Perception. In Proceedings of the 2019 IEEE Intelligent Vehicles Symposium (IV), Paris, France, 9–12 June 2019; pp. 2522–2529. [CrossRef]
14. Caesar, H.; Bankiti, V.; Lang, A.H.; Vora, S.; Liong, V.E.; Xu, Q.; Krishnan, A.; Pan, Y.; Baldan, G.; Beijbom, O. nuScenes: A Multimodal Dataset for Autonomous Driving. In Proceedings of the 2020 IEEE/CVF Conference on Computer Vision and Pattern Recognition (CVPR), Seattle, WA, USA, 14–19 June 2020; pp. 11618–11628. [CrossRef]
15. Geyer, J.; Kassahun, Y.; Mahmudi, M.; Ricou, X.; Durgesh, R.; Chung, A.S.; Hauswald, L.; Pham, V.H.; Mühlegg, M.; Dorn, S.; et al. A2D2: Audi Autonomous Driving Dataset. *arXiv* **2020**, arXiv:2004.06320.
16. Pan, Y.; Gao, B.; Mei, J.; Geng, S.; Li, C.; Zhao, H. SemanticPOSS: A Point Cloud Dataset with Large Quantity of Dynamic Instances. In Proceedings of the 2020 IEEE Intelligent Vehicles Symposium (IV), Las Vegas, NV, USA, 23 June 2020; pp. 687–693. [CrossRef]
17. Xiao, A.; Huang, J.; Guan, D.; Zhan, F.; Lu, S. SynLiDAR: Learning From Synthetic LiDAR Sequential Point Cloud for Semantic Segmentation. *arXiv* **2021**, arXiv:2107.05399.
18. Deschaud, J.E. KITTI-CARLA: A KITTI-like dataset generated by CARLA Simulator. *arXiv* **2021**, arXiv:2109.00892.
19. Munoz, D.; Bagnell, J.A.; Vandapel, N.; Hebert, M. Contextual classification with functional Max-Margin Markov Networks. In Proceedings of the 2009 IEEE Conference on Computer Vision and Pattern Recognition, Miami, FL, USA, 20–25 June 2009; pp. 975–982. [CrossRef]
20. Serna, A.; Marcotegui, B.; Goulette, F.; Deschaud, J.E. Paris-rue-Madame database: A 3D mobile laser scanner dataset for benchmarking urban detection, segmentation and classification methods. In Proceedings of the 4th International Conference on Pattern Recognition, Applications and Methods (ICPRAM 2014), Loire Valley, France, 6–8 March 2014.
21. Vallet, B.; Brédif, M.; Serna, A.; Marcotegui, B.; Paparoditis, N. TerraMobilita/iQmulus urban point cloud analysis benchmark. *Comput. Graph.* **2015**, *49*, 126–133. [CrossRef]
22. Roynard, X.; Deschaud, J.E.; Goulette, F. Paris-Lille-3D: A large and high-quality ground-truth urban point cloud dataset for automatic segmentation and classification. *Int. J. Robot. Res.* **2018**, *37*, 545–557. [CrossRef]
23. Griffiths, D.; Boehm, J. SynthCity: A large scale synthetic point cloud. *arXiv* **2019**, arXiv:1907.04758.
24. Zhu, J.; Gehrung, J.; Huang, R.; Borgmann, B.; Sun, Z.; Hoegner, L.; Hebel, M.; Xu, Y.; Stilla, U. TUM-MLS-2016: An Annotated Mobile LiDAR Dataset of the TUM City Campus for Semantic Point Cloud Interpretation in Urban Areas. *Remote Sens.* **2020**, *12*, 1875. [CrossRef]
25. Geiger, A.; Lenz, P.; Urtasun, R. Are we ready for Autonomous Driving? The KITTI Vision Benchmark Suite. In Proceedings of the Conference on Computer Vision and Pattern Recognition (CVPR), Providence, RI, USA, 16–21 June 2012.
26. Deschaud, J.E. IMLS-SLAM: Scan-to-Model Matching Based on 3D Data. In Proceedings of the 2018 IEEE International Conference on Robotics and Automation (ICRA), Brisbane, Australia, 21–25 May 2018; pp. 2480–2485. [CrossRef]
27. Dosovitskiy, A.; Ros, G.; Codevilla, F.; Lopez, A.; Koltun, V. CARLA: An Open Urban Driving Simulator. In Proceedings of the 1st Annual Conference on Robot Learning, Mountain View, CA, USA, 13–15 November 2017; pp. 1–16.
28. Bello, S.A.; Yu, S.; Wang, C.; Adam, J.M.; Li, J. Review: Deep Learning on 3D Point Clouds. *Remote Sens.* **2020**, *12*, 1729. [CrossRef]
29. Qi, C.R.; Yi, L.; Su, H.; Guibas, L.J. PointNet++: Deep Hierarchical Feature Learning on Point Sets in a Metric Space. In Proceedings of the 31st International Conference on Neural Information Processing Systems, Long Beach, CA, USA, 4–9 December 2017; Curran Associates Inc.: Red Hook, NY, USA, 2017; pp. 5105–5114.
30. Thomas, H.; Qi, C.R.; Deschaud, J.E.; Marcotegui, B.; Goulette, F.; Guibas, L.J. Kpconv: Flexible and deformable convolution for point clouds. In Proceedings of the IEEE/CVF International Conference on Computer Vision (ICCV), Seoul, Korea, 27–28 October 2019; pp. 6411–6420.
31. Guo, Y.; Wang, H.; Hu, Q.; Liu, H.; Liu, L.; Bennamoun, M. Deep Learning for 3D Point Clouds: A Survey. *IEEE Trans. Pattern Anal. Mach. Intell.* **2020**, *43*, 4338–4364. [CrossRef] [PubMed]
32. He, K.; Zhang, X.; Ren, S.; Sun, J. Deep Residual Learning for Image Recognition. In Proceedings of the 2016 IEEE Conference on Computer Vision and Pattern Recognition (CVPR), Las Vegas, NV, USA, 27–30 June 2016; pp. 770–778. [CrossRef]
33. Duque-Arias, D.; Velasco-Forero, S.; Deschaud, J.E.; Goulette, F.; Serna, A.; Decencière, E.; Marcotegui, B. On power Jaccard losses for semantic segmentation. In Proceedings of the VISAPP 2021: 16th International Conference on Computer Vision Theory and Applications, Vienna, Austria, 8–10 March 2021.
34. Chaton, T.; Chaulet, N.; Horache, S.; Landrieu, L. Torch-Points3D: A Modular Multi-Task Framework for Reproducible Deep Learning on 3D Point Clouds. In Proceedings of the 2020 International Conference on 3D Vision (3DV), Fukuoka, Japan, 25–28 November 2020; pp. 1–10. [CrossRef]
35. Kirillov, A.; He, K.; Girshick, R.; Rother, C.; Dollár, P. Panoptic segmentation. In Proceedings of the IEEE/CVF Conference on Computer Vision and Pattern Recognition (CVPR), Long Beach, CA, USA, 15–20 June 2019; pp. 9404–9413.
36. Serna, A.; Marcotegui, B. Detection, segmentation and classification of 3D urban objects using mathematical morphology and supervised learning. *ISPRS J. Photogramm. Remote Sens.* **2014**, *93*, 243–255. [CrossRef]
37. Gomes, L.; Regina Pereira Bellon, O.; Silva, L. 3D reconstruction methods for digital preservation of cultural heritage: A survey. *Pattern Recognit. Lett.* **2014**, *50*, 3–14. [CrossRef]
38. Xu, Y.; Zhu, X.; Shi, J.; Zhang, G.; Bao, H.; Li, H. Depth Completion From Sparse LiDAR Data With Depth-Normal Constraints. In Proceedings of the IEEE/CVF International Conference on Computer Vision (ICCV), Seoul, Korea, 27 October–2 November 2019.

39. Guo, X.; Xiao, J.; Wang, Y. A Survey on Algorithms of Hole Filling in 3D Surface Reconstruction. *Vis. Comput.* **2018**, *34*, 93–103. [CrossRef]
40. Roldao, L.; de Charette, R.; Verroust-Blondet, A. 3D Semantic Scene Completion: a Survey. *arXiv* **2021**, arXiv:2103.07466.
41. Dai, A.; Siddiqui, Y.; Thies, J.; Valentin, J.; Niessner, M. SPSG: Self-Supervised Photometric Scene Generation From RGB-D Scans. In Proceedings of the IEEE/CVF Conference on Computer Vision and Pattern Recognition (CVPR), Nashville, TN, USA, 21–24 June 2021; pp. 1747–1756.
42. Hoppe, H.; DeRose, T.; Duchamp, T.; McDonald, J.; Stuetzle, W. Surface Reconstruction from Unorganized Points. *SIGGRAPH Comput. Graph.* **1992**, *26*, 71–78. [CrossRef]
43. Dai, A.; Diller, C.; Niessner, M. SG-NN: Sparse Generative Neural Networks for Self-Supervised Scene Completion of RGB-D Scans. In Proceedings of the 2020 IEEE/CVF Conference on Computer Vision and Pattern Recognition (CVPR), Seattle, WA, USA, 14–19 June 2020; pp. 846–855. [CrossRef]
44. Curless, B.; Levoy, M. A Volumetric Method for Building Complex Models from Range Images. In Proceedings of the 23rd Annual Conference on Computer Graphics and Interactive Techniques, New Orleans, LA, USA, 4–9 August 1996; Association for Computing Machinery: New York, NY, USA, 1996; pp. 303–312. [CrossRef]
45. Lorensen W.E.; Cline H. E. Marching cubes: A high resolution 3D surface construction algorithm. In Proceedings of the 14th Annual Conference on Computer Graphics and Interactive Techniques, New York, NY, USA, July 1987; Association for Computing Machinery: New York, NY, USA, 1987; pp. 163–169. [CrossRef]

Article

Design and Evaluation of a Permanently Installed Plane-Based Calibration Field for Mobile Laser Scanning Systems

Erik Heinz *, **Christoph Holst**, **Heiner Kuhlmann and Lasse Klingbeil**

Institute of Geodesy and Geoinformation, University of Bonn, Nussallee 17, 53115 Bonn, Germany; c.holst@igg.uni-bonn.de (C.H.); heiner.kuhlmann@uni-bonn.de (H.K.); l.klingbeil@igg.uni-bonn.de (L.K.)
* Correspondence: e.heinz@igg.uni-bonn.de; Tel.: +49-228-73-3574

Received: 27 December 2019; Accepted: 4 February 2020; Published: 7 February 2020

Abstract: Mobile laser scanning has become an established measuring technique that is used for many applications in the fields of mapping, inventory, and monitoring. Due to the increasing operationality of such systems, quality control w.r.t. calibration and evaluation of the systems becomes more and more important and is subject to on-going research. This paper contributes to this topic by using tools from geodetic configuration analysis in order to design and evaluate a plane-based calibration field for determining the lever arm and boresight angles of a 2D laser scanner w.r.t. a GNSS/IMU unit (Global Navigation Satellite System, Inertial Measurement Unit). In this regard, the impact of random, systematic, and gross observation errors on the calibration is analyzed leading to a plane setup that provides accurate and controlled calibration parameters. The designed plane setup is realized in the form of a permanently installed calibration field. The applicability of the calibration field is tested with a real mobile laser scanning system by frequently repeating the calibration. Empirical standard deviations of <1 ... 1.5 mm for the lever arm and <0.005° for the boresight angles are obtained, which was priorly defined to be the goal of the calibration. In order to independently evaluate the mobile laser scanning system after calibration, an evaluation environment is realized consisting of a network of control points as well as TLS (Terrestrial Laser Scanning) reference point clouds. Based on the control points, both the horizontal and vertical accuracy of the system is found to be < 10 mm (root mean square error). This is confirmed by comparisons to the TLS reference point clouds indicating a well calibrated system. Both the calibration field and the evaluation environment are permanently installed and can be used for arbitrary mobile laser scanning systems.

Keywords: mobile laser scanning; lever arm; boresight angles; plane-based calibration field; configuration analysis; accuracy; controllability; evaluation; control points; TLS reference point clouds

1. Introduction

In recent years, the use of mobile mapping systems has gained increasing acceptance. Clearly, this trend is confirmed by the great number of applications already addressed with this measuring technology, e.g., mapping of road corridors and city areas, determination of clearances, monitoring of infrastructures, or the extraction of geometric road parameters, road markings, and road furniture. A lot of publications and reviews about this topic can be found [1–5].

In the case of mobile mapping, laser scanners or cameras are mounted on a moving platform and sense the environment. Simultaneously, the position and orientation of the platform is determined by georeferencing sensors like GNSS (Global Navigation Satellite System), IMU (Inertial Measurement Unit), or odometry [6]. By fusing the observations of all sensors, a 3D point cloud of the environment is obtained. While mobile mapping systems have reached an operational status, challenges exist in the context of quality assessment and quality control of the measured 3D point clouds. This is due to the fact that the data acquisition with mobile mapping systems comprises a complex processing chain. In this processing chain, many sources of errors exist that affect the quality of the 3D point cloud:

(a)   Observation errors of the georeferencing sensors as well as model errors in the fusion of the georeferencing sensors in order to estimate the position and orientation of the system,

(b)   Observation errors of the mapping sensors due to the instrument, environmental conditions, measuring configuration, and object properties,

(c)   Errors w.r.t. the intrinsic sensor calibration (i.e., instrumental errors) as well as the extrinsic calibration between different sensors (i.e., lever arms and boresight angles),

(d)   Errors w.r.t. the time synchronization between different sensors.

Generally speaking, all sources of errors must be subject to quality assessment and quality control. In this paper, however, we mainly focus on the calibration. In this regard, we address the extrinsic calibration of mobile laser scanning systems, i.e., the determination of the lever arm and boresight angles of a 2D laser scanner w.r.t. to a GNSS/IMU unit. The literature review in Section 2.1 shows that different methods exist for the extrinsic calibration of mobile laser scanning systems. We decided to implement a plane-based approach, since this type of calibration method has successfully been used in many fields of application. The idea of the calibration approach is oriented towards the research in Strübing and Neumann [7] and is based on matching the scanned points of the mobile laser scanning system to a setup of known reference planes. In the course of this, we can estimate corrections for the extrinsic calibration parameters within a Least Squares adjustment. Similar approaches have been used, e.g., in airborne [8], shipborne [9], ground-based [10–12], and indoor laser scanning [7,13].

The plane-based calibration is based on a Least Squares adjustment, which can be analyzed with tools from geodetic configuration analysis [14,15]. Such tools can be used to analyze the impact of random, systematic, and gross observation errors on the calibration process. Such an analysis makes it possible to assess and improve the reference plane setup and, thus, the accuracy and controllability of the estimated calibration parameters. In this way, the calibration is subject to thorough quality control and causes no additional uncertainty in the 3D point cloud. So far, the use of geodetic configuration analysis in the context of calibrating mobile laser scanning systems has only been studied to a limited extent and is the major scientific contribution of this paper. Based on the results of the configuration analysis, we installed a deliberately designed plane-based calibration field for mobile laser scanning systems that provides accurate and controlled calibration parameters. The calibration field was realized outdoors being permanent, stable, weather-resistant, and cost-effective. The calibration procedure in the calibration field takes less than one minute and, thus, can be repeated frequently. In addition to the configuration analysis, repetitive calibrations also increase the controllability of the calibration parameters and allow for a realistic empirical quantification of their accuracy and stability.

Before moving on to application, the quality of the calibration parameters has to be evaluated independently. However, in the case of mobile laser scanning systems, we are faced with a complex processing chain. The functional model of this processing chain is usually not fully known to users. Moreover, the stochastic distribution functions of the input variables are often unknown, necessarily not Gaussian as well as site- and time-dependent. This makes it challenging or even impossible to model the accuracy of 3D point clouds straightforward using error propagation [16,17] or to analyze individual system components like the calibration separately. Therefore, this paper focuses on an empirical evaluation of mobile laser scanning systems as a whole.

The literature review in Section 2.2 outlines common methods for the empirical evaluation of mobile laser scanning systems. Yet, these methods are not standardized [18] and facilities for the evaluation of mobile laser scanning systems are not available on a large scale [19–21]. This is why we decided to build up our own evaluation environment that implements existing evaluation strategies and combines them to a holistic approach. The evaluation environment consists of a dense network of control points as well as accurate reference point clouds of diverse building structures generated with TLS (Terrestrial Laser Scanning). Our evaluation environment allows for a point-based as well as area-based evaluation and can be used for arbitrary systems, independent of the specific setup. Beside the configuration analysis for the plane-based calibration approach and the realization of the calibration field, the installation of the evaluation environment is the second important contribution of this paper. Both the plane-based calibration field and the evaluation environment are permanently installed, readily accessible, and were utilized by our own mobile laser scanning system proving the operationality of our facilities.

This paper is structured as follows: Section 2 surveys the literature. Section 3 introduces our mobile laser scanning system. Section 4 describes the plane-based calibration approach. Section 5 addresses the design of the calibration field using tools from geodetic configuration analysis. Section 6 presents real calibration results of our mobile laser scanning system. In Section 7, the mobile laser scanning system is evaluated in our evaluation environment. Section 8 concludes.

2. Calibration and Evaluation of Mobile Laser Scanning Systems

This section surveys the literature about the calibration (Section 2.1) and evaluation (Section 2.2) of mobile laser scanning systems. The focus regarding calibration is on the lever arm and boresight angles of laser scanners w.r.t. to the georeferencing sensors.

2.1. Calibration of Mobile Laser Scanning Systems

In the case of mobile mapping, a distinction is made between intrinsic and extrinsic calibration. Intrinsic calibration deals with systematic errors of individual sensors, e.g., the phase center of GNSS antennas [22]; axial misalignments, biases, and scale factors of inertial sensors or odometers [23]; or intrinsic corrections for laser scanners [24,25] and cameras [15]. Intrinsic calibration is beyond the scope of this paper. In contrast, extrinsic calibration addresses the determination of lever arms and boresight angles between different sensors or groups of sensors.

This work focuses on the extrinsic calibration of laser scanners. In this context, sensor-based, entropy-based, and geometry-based approaches exist. Sensor-based approaches make use of external sensors, e.g., total stations and laser trackers [26,27], theodolite measuring systems [28], close range photogrammetry [29], or measuring arms [30] in order to directly or indirectly measure lever arm components and boresight angles. Alternatively, in the case of laser scanners with 3D scanning mode, reference points that are known w.r.t. the georeferencing sensors can statically be scanned with the laser scanner in order to solve the extrinsic calibration problem [28,29,31]. Often, sensor-based approaches are elaborate, require additional instruments, or impose specific requirements on the calibration, e.g., clearly defined reference points at the casing of the sensors, a mechanical realization of the sensor frames, or a 3D scanning mode of the laser scanner.

In addition to sensor-based methods, entropy-based approaches can be used, which originate from robotics. Such approaches set up a cost function that describes the consistency within the 3D point cloud. This cost function is derived from scans of the environment with different position and orientation of the platform. The idea is that calibration errors lead to distortions in the 3D point cloud. By adjusting the calibration parameters, the cost function and the distortions can be minimized. This kind of self-calibration has been used for both indoor [32–37] and outdoor systems [33,34,37–40]. In addition to entropy, also other cost functions can be used, e.g., based on geometrical constraints for point neighborhoods such as planarity, curvature, or omnivariance [37,41]. However, the sensitivity towards the calibration parameters strongly depends on the environment and the trajectory, for which variation in all six degrees of freedom is required [40]. Especially for ground-based systems variation in roll, pitch, and height is difficult, which limits the accuracy and controllability of the calibration. Furthermore, the approaches do not provide direct quality measures for the calibration parameters. The accuracy of the calibration is usually evaluated based on repeated calibrations and the consistency of the 3D point clouds. Thus, thorough quality assessment and quality control is difficult.

Another way of self-calibration is to perform special driving maneuvers past building facades or cylindrical objects [34,42,43]. Potential tilts or displacements of the objects in the 3D point clouds are used to correct errors in the boresight angles or the time synchronization. The problem with these methods is that they calibrate single system components one after another. However, errors in a 3D point cloud typically result from the interaction of multiple system components. This means that errors in the 3D point cloud cannot be traced back unambiguously to errors in the calibration. Similar methods are also used for the extrinsic calibration of multi beam echo sounders adapted to ships using patch-test procedures [27] as well as for strip adjustment in airborne and UAV-based laser scanning (Unmanned Aerial Vehicle) [24,44].

The last category are geometry-based approaches, where constraints between the scan points of the mobile laser scanning system and geometric primitives are used for determining the extrinsic calibration parameters within a Least Squares adjustment. In this context plane-based methods are most common and have successfully been applied in airborne laser scanning [8,24,45–47], but also for systems mounted on UAVs [48,49], on ships [9], as well as for ground-based [10–12,18,26,34,50–52] and indoor systems [7,13,34,53]. While based on the same idea, plane-based calibration approaches vary in implementation. The most important differences are the number of calibration parameters (some only calibrate the boresight angles), the usage of natural or artificial planes, and the general handling of the planes. The latter means that the plane normals can either be determined before the calibration by independent means or the plane normals can be estimated as unknowns in the adjustment leading to a self-calibration approach. For the sake of completeness, please note that also other geometric forms can be used, e.g., spheres [54] or catenaries [52].

Clearly, there are many different approaches for the extrinsic calibration of laser scanners. We implemented a plane-based approach, which is based on minimizing the differences between scan points of the mobile laser scanning system and known reference plane equations by adjusting the lever arm and the boresight angles between the laser scanner and the GNSS/IMU unit. The basic idea of this approach is oriented towards the research in Strübing and Neumann [7].

Most publications highlight the importance of the plane configuration for the calibration, i.e., setup, number, and size of the planes. The principal measures for assessing and improving the quality of the configuration are the sensitivity of the plane configuration towards changes in the calibration parameters as well as the variances and correlations of the calibration parameters. Other quality criteria are the radius of convergence of the adjustment and the reduction of plane residuals after calibration. The advantage of plane-based approaches is that they are based on a Least Squares adjustment, which can be interpreted as a geodetic network. This means that we can use tools from geodetic configuration analysis in order to assess the quality of the estimated calibration parameters not only in terms of sensitivity and accuracy, but also in terms of controllability by investigating the robustness w.r.t. to gross observation errors [14,15]. So far, controllability in the context of calibrating mobile laser scanning systems has only been addressed to a limited extent [8]. The major scientific contribution of this paper is that we use tools from configuration analysis to investigate the impact of random, systematic, and gross observation errors on the calibration. In the course of this, we derive a plane setup that provides both accurate and controlled calibration parameters. This paper is connected to our previous publication in Heinz et al. [11].

2.2. Evaluation of Mobile Laser Scanning Systems

The evaluation strategies for analyzing the accuracy of mobile laser scanning systems can be classified into point-based, area-based, and parameter-based methods. Point-based methods use either natural control points (e.g., building corners, manholes, poles, or road markings) or artificial control points (e.g., targets or markers), which are extracted from the 3D point clouds and compared to reference values that were determined by other surveying methods like total stations, leveling, GNSS, or TLS (e.g., [18–21,26,28,29,31,33,55–59]). Such methods assess the absolute accuracy of a system as a whole, instead of analyzing individual components. Moreover, the precision can be examined by measuring control points multiple times or by analyzing relative point distances. In this regard, the horizontal and vertical components are often investigated separately.

In addition to control points, area-based evaluation methods are widespread. In this respect, existing 3D city models can be utilized as a reference for evaluation [60–62]. Furthermore, reference point clouds (e.g., from TLS) are commonly used for evaluation (e.g., [10,11,18,21,33,63–65]), since software packages provide algorithms for the comparison of two point clouds. In this regard, it is also possible to mesh patches of the road surface that were measured multiple times with the mobile laser scanning system and compare these patches to each other [19,59]. Area-based evaluation methods supplement the point-based strategies, as mobile laser scanning is an area-based measuring technology. However, the interpretation of the differences of a mobile point cloud to a model or a reference point cloud is difficult due to the dependency on the structure of the measured object. For instance, height errors cannot be detected when driving parallel to a vertical wall.

A third way of evaluation emerges in the context of application. Geometric parameters of road surfaces, building structures, or objects can repeatedly be extracted from the point clouds of multiple passes. The repeatability of such parameters also indicates the quality of a system (e.g., [18,59,66]).

The review indicates that point-based as well as area-based approaches using control points and reference point clouds are common strategies for the evaluation of mobile laser scanning systems. Basically, such strategies provide an empirical evaluation of the system as a whole. In our view, this is the most effective evaluation strategy due to the complex and partially unknown processing chain. Yet, the associated methods are not standardized [18] and facilities for the evaluation of mobile laser scanning systems are not available on a large scale [19–21]. Therefore, we realized an own permanent evaluation environment with control points and TLS reference point clouds (cf. Section 7) that can be used for arbitrary mobile laser scanning systems.

3. Mobile Laser Scanning System

The mobile laser scanning system that is utilized within this work is shown in Figure 1. The core component is a navigation-grade inertial navigation system iMAR iNAV-FJI-LSURV [67] with fiber-optic gyroscopes, servo accelerometers, and RTK-GNSS (Real Time Kinematic). An odometer is optional. For the trajectory estimation, we use the software Waypoint Inertial Explorer 8.80 [68]. Depending on GNSS quality, accuracies of centimeters and centidegrees or better can be reached for the position and orientation. For mapping, a 2D laser scanner Z+F Profiler 9012A is used [69]. This instrument is specified with an accuracy of millimeters and comes with a maximum profile rate of 200 Hz, a maximum scan rate of 1 MHz and a special hardware optimization that decreases the measurement noise in the close-range [70]. The system can be adapted to a trolley or a van (Figure 1). A more detailed view of the system is given in Figure 2. A case study on a motorway showed that the accuracy of the system is about millimeter to centimeter under good GNSS conditions [59].

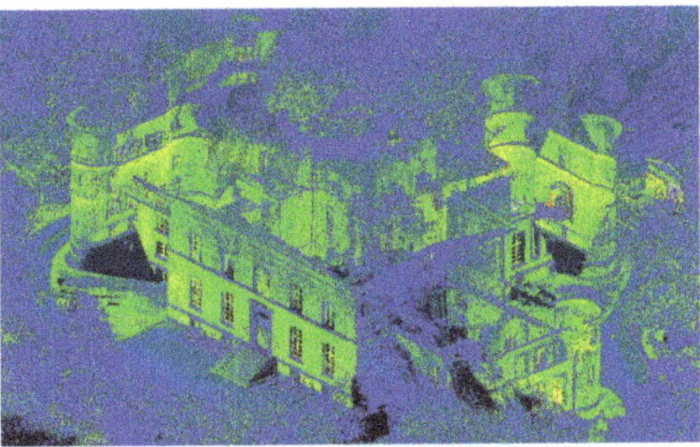

Figure 1. Mobile laser scanning system of the University of Bonn adapted to a trolley (**left**) or a van (**middle**). The 3D point cloud (**right**) was recorded during a mapping campaign in Bonn, Germany.

4. Calibration Approach

This section addresses the plane-based calibration approach. Firstly, the calibration parameters and the georeferencing equation for the generation of a 3D point cloud are introduced (Section 4.1). Following this, the calibration procedure is described (Section 4.2). Finally, we discuss details about the estimation and the quality assessment of the calibration parameters (Section 4.3).

4.1. Calibration Parameters and Georeferencing Equation

The measuring of a 3D point cloud using mobile laser scanning is based on the data fusion of multiple sensors. The goal of the extrinsic calibration is to determine the mutual installation position and orientation of these sensors on the platform. In this paper, we address the determination of the lever arm $[\Delta x, \Delta y, \Delta z]^T$ as well as the boresight angles α, β, and γ of a 2D laser scanner w.r.t. a GNSS/IMU unit (Figure 2, left). The calibration approach in Section 4.2 aims at determining these parameters, which are assumed to be temporally constant. However, this has to be checked regularly.

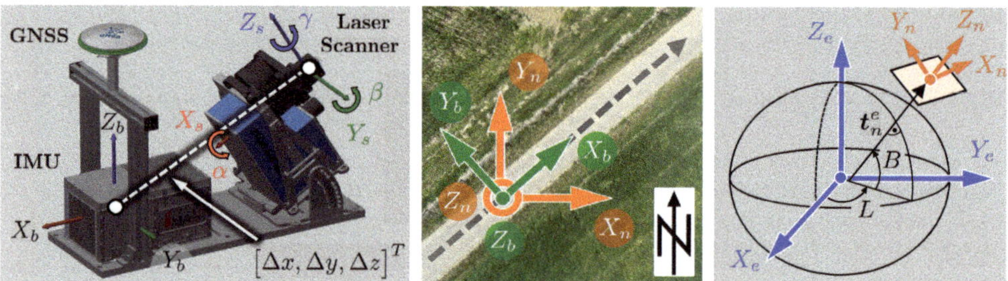

Figure 2. Different coordinate frames (i.e., s(canner)-frame, b(ody)-frame, n(avigation)-frame, and e(arth)-frame) and transformations for generating a georeferenced 3D point cloud.

In the context of mobile laser scanning, an object point $[x_s, y_s, z_s]^T$ in the local sensor frame of the 2D laser scanner (s-frame) can be written as $[0, d_s \cdot \sin(b_s), d_s \cdot \cos(b_s)]^T$, where d_s is the measured distance and b_s the scanning angle. In order to obtain a georeferenced point cloud, the object points of the 2D laser scanner must be transformed into the body-frame of the platform (b-frame), which often coincides with the coordinate system of the IMU. For this transformation, the extrinsic calibration parameters are needed, i.e., the lever arm $[\Delta x, \Delta y, \Delta z]^T$ as well as the boresight angles α, β, and γ (Figure 2, left). Subsequently, the orientation angles of the platform, i.e., roll ϕ, pitch θ, and yaw ψ, are used to transform the object points into the navigation-frame (n-frame), which is a north-oriented local level frame, whose origin coincides with that of the b-frame (Figure 2, middle). Finally, the object points are transformed to a superordinate earth-fixed and earth-centered coordinate frame, which we denote as e-frame (Figure 2, right), by using ellipsoidal longitude L and latitude B, as well as the translation vector $t_n^e = [t_x, t_y, t_z]^T$. All these transformations can be combined to the georeferencing equation, which provides georeferenced scan points $[x_e, y_e, z_e]_{MLS}^T$ (MLS: Mobile Laser Scanning):

$$\begin{bmatrix} x_e \\ y_e \\ z_e \end{bmatrix}_{MLS} = \begin{bmatrix} t_x \\ t_y \\ t_z \end{bmatrix} + \boldsymbol{R}_n^e(L, B) \cdot \boldsymbol{R}_b^n(\phi, \theta, \psi) \cdot \left(\boldsymbol{R}_s^b(\alpha, \beta, \gamma) \begin{bmatrix} x_s = 0 \\ y_s = d_s \cdot \sin(b_s) \\ z_s = d_s \cdot \cos(b_s) \end{bmatrix} + \begin{bmatrix} \Delta x \\ \Delta y \\ \Delta z \end{bmatrix} \right), \quad (1)$$

where $\boldsymbol{R}_i^j(\cdot)$ denotes an Euler rotation matrix between two frames i and j. Afterwards, the global Cartesian coordinates $[x_e, y_e, z_e]_{MLS}^T$ are often transformed to application-related coordinate frames, e.g., UTM (Universal Transverse Mercator) and ellipsoidal heights. In the case of small areas, it is also possible to transform the object points from the n-frame to a local frame (l-frame) by omitting the rotation matrix $\mathbf{R}_n^e(L, B)$ with ellipsoidal longitude L and latitude B, and just using the translation vector $\mathbf{t}_n^l = [t_e, t_n, t_h]^T$. This leads to the simplified equation:

$$
\begin{bmatrix} x_l \\ y_l \\ z_l \end{bmatrix}_{MLS} = \begin{bmatrix} t_e \\ t_n \\ t_h \end{bmatrix} + \mathbf{R}_b^n(\phi, \theta, \psi) \cdot \left(\mathbf{R}_s^b(\alpha, \beta, \gamma) \begin{bmatrix} x_s = 0 \\ y_s = d_s \cdot \sin(b_s) \\ z_s = d_s \cdot \cos(b_s) \end{bmatrix} + \begin{bmatrix} \Delta x \\ \Delta y \\ \Delta z \end{bmatrix} \right).
\tag{2}
$$

4.2. Calibration Procedure

The calibration approach is based on the use of reference planes [7]. In this section, we describe the calibration procedure with a given plane setup. Yet, we are not interested in how this plane setup was derived. The design of the plane setup and the influence of its configuration on the quality of the estimated calibration parameters are discussed in Section 5. The general workflow of the calibration procedure is illustrated in Figure 3.

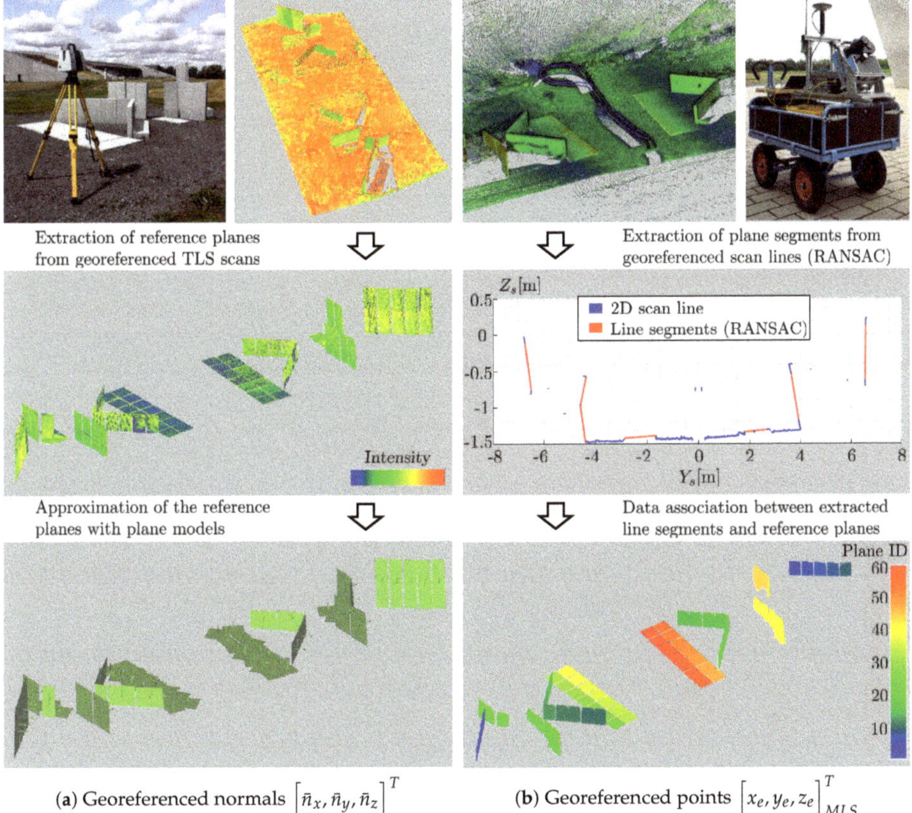

Figure 3. (a) Generating reference normals for the plane setup using TLS (Terrestrial Laser Scanning), (b) Generating a point cloud of the plane setup using the mobile laser scanning system. The scan points of the mobile laser scanning system are automatically extracted from the scan lines using RANSAC (Random Sample Consensus) and assigned to the correct reference plane.

The first step of the calibration comprises the determination of reference values for the plane setup (Figure 3, left column). Therefore, the plane setup is scanned with TLS from several stations. The resulting TLS point cloud is georeferenced in the e-frame or l-frame using control points. As a result, each plane is represented by a number of georeferenced points $[x_e, y_e, z_e]_{TLS}^T$. Subsequently, the georeferenced TLS points are approximated with a plane model, which is parameterized by the normal vector $[n_x, n_y, n_z]^T$ and the distance parameter d_n. By normalizing the normal vector with d_n, the plane equation is described by only three parameters $[\bar{n}_x, \bar{n}_y, \bar{n}_z]^T$ [71]:

$$[x_e, y_e, z_e]_{TLS} \cdot \begin{bmatrix} n_x \\ n_y \\ n_z \end{bmatrix} - d_n = [x_e, y_e, z_e]_{TLS} \cdot \begin{bmatrix} \bar{n}_x \\ \bar{n}_y \\ \bar{n}_z \end{bmatrix} - 1 \overset{!}{=} 0. \tag{3}$$

The estimated normal vectors $[\bar{n}_x, \bar{n}_y, \bar{n}_z]^T$ serve as reference information for the calibration of the mobile laser scanning system.

Following this, the plane setup is scanned with the mobile laser scanning system (Figure 3, right column). By using the georeferencing equation and approximate calibration parameters, a mobile point cloud of the plane setup in the e-frame or l-frame is calculated (cf. Equation (1) or Equation (2)). The calibration approach is based on the constraint that the georeferenced points $[x_e, y_e, z_e]_{MLS}$ of the mobile laser scanning system must fulfill the plane equations as given from the TLS survey in Equation (3). This leads to the following observation equation for the calibration:

$$\underbrace{\left[\begin{bmatrix} t_x \\ t_y \\ t_z \end{bmatrix} + R_n^e(L, B) \cdot R_b^n(\phi, \theta, \psi) \cdot \left(R_s^b(\alpha, \beta, \gamma) \begin{bmatrix} 0 \\ d_s \cdot \sin b_s \\ d_s \cdot \cos b_s \end{bmatrix} + \begin{bmatrix} \Delta x \\ \Delta y \\ \Delta z \end{bmatrix} \right) \right]^T}_{[x_e, y_e, z_e]_{MLS}^T} \cdot \begin{bmatrix} \bar{n}_x \\ \bar{n}_y \\ \bar{n}_z \end{bmatrix} - 1 \overset{!}{=} 0. \tag{4}$$

Equation (4) leads to a system of equations that can be solved using a Least Squares adjustment within the Gauß-Helmert Model [15]. In the adjustment, the plane normals $[\bar{n}_x, \bar{n}_y, \bar{n}_z]^T$ are considered as an error-free reference. However, it is also possible to use them as observations when introducing an adequate variance information. The observations in the adjustment are the positions $[t_x, t_y, t_z]^T$ and the orientation angles ϕ, θ, and ψ of the mobile laser scanning system, as well as the distances d_s and the angular observations b_s of the 2D laser scanner. The lever arm $[\Delta x, \Delta y, \Delta z]^T$ and the boresight angles α, β, and γ are the parameters to be estimated. For small areas the ellipsoidal longitude L and latitude B can be considered as error-free constant values. In the case of performing the parameter estimation in the l-frame, the leveled positions $[t_e, t_n, t_h]^T$ are used as observations.

Please note that not each scan point of the mobile laser scanning system is assigned an own position and orientation, but all scan points within the same scanning profile are assigned the mean position and orientation of that scanning profile. This is reasonable for several reasons:

(i)   Significant reduction of the normal equation system enhancing the computational efficiency,

(ii) Position and orientation of adjacent scan points are highly correlated and, thus, do not provide a new and independent information,

(iii) Errors that might be induced by this simplification are negligible due to the high profile rate of the 2D laser scanner, the low speed of the platform, and the accuracy of the trajectory (e.g., at a speed of $0.75\ \frac{m}{s}$ and a profile rate of 200 Hz, the maximum position error is < 1 mm).

In order to assign each scan point of the mobile laser scanning to the correct reference plane, the scan lines of the mobile laser scanning system are separately pre-processed. Please note that the pre-processing is not done on the full 3D point cloud of the mobile laser scanning system, but on individual scan lines. This is illustrated in Figure 3 (right column). Initially, the scan lines are given in the s-frame of the 2D laser scanner. By ignoring the X_s-component, a 2D scan is obtained, where the planes are represented as line segments. These line segments are extracted from the scan lines by using a RANSAC (Random Sample Consensus) approach [72]. In the RANSAC approach, thresholds for the maximum distance of a scanned point to a line segment as well as for the minimum size of a line segment are used. The thresholds were empirically determined based on the noise of the 2D laser scanner and the known size of the reference planes. Subsequently, the extracted line segments are transformed to the e-frame or l-frame by using the georeferencing equation and approximate calibration parameters (cf. Equation (1) or Equation (2)). Finally, the extracted line segments are inserted into the reference plane equations from TLS and assigned to the plane that gives the minimum distance error. This results in a segmented point cloud of the mobile laser scanning system (Figure 3, bottom right).

4.3. Quality Criteria for the Estimation of the Calibration Parameters

The basic equation of the calibration approach is Equation (4), which is used as an observation equation for a Least Squares adjustment within the Gauß-Helmert model [15]. By using Equation (5) and Equation (6), the calibration parameters $\hat{p} = [\Delta x, \Delta y, \Delta z, \alpha, \beta, \gamma]^T$ and their covariance matrix $\Sigma_{\hat{p}\hat{p}}$ can be estimated, respectively:

$$\hat{p} = p_0 + \left[A^T \left(B\Sigma_{ll}B^T \right)^{-1} A \right]^{-1} A^T \left(B\Sigma_{ll}B^T \right)^{-1} w, \tag{5}$$

$$\Sigma_{\hat{p}\hat{p}} = \left[A^T \left(B\Sigma_{ll}B^T \right)^{-1} A \right]^{-1}, \tag{6}$$

where p_0 denotes approximate calibration parameters, the matrizes A and B contain the partial derivatives of Equation (4) w.r.t. the parameters and observations, respectively, Σ_{ll} is the covariance matrix of the observations and w is the vector of discrepancies. The covariance matrix $\Sigma_{\hat{p}\hat{p}}$ indicates the accuracy of the estimated calibration parameters $\hat{p}$ and is an important criterion for analyzing the quality of the calibration parameters. The main goal of the calibration is that the uncertainty of the 3D point cloud, which is principally caused by the observation errors of the GNSS/IMU unit and the 2D laser scanner, does not increase significantly by the uncertainty $\Sigma_{\hat{p}\hat{p}}$ of the calibration.

In order to define a target accuracy for the estimated calibration parameters, we performed an error propagation of Equation (2) with optimistic assumptions for the accuracy of the observations. These accuracies were oriented towards empirical values and manufacturer specifications and are summarized in Table 1. Two different scenarios were simulated, i.e., with and without an uncertainty for the calibration parameters. In Table 1, these two scenarios are denoted with cases (i) and (ii). The error propagation was executed in a local frame (l-frame) with east, north, and height component in order to distinguish between horizontal and vertical accuracy. A point grid was generated for the scan profile of the 2D laser scanner with a typical operating range of 50 m. For each grid point, a covariance matrix was calculated via error propagation. Based on this, a 3D point error $\sigma_{3D} = \sqrt{\sigma_{x_l}^2 + \sigma_{y_l}^2 + \sigma_{z_l}^2}$ was derived from the trace of each covariance matrix.

Figure 4 (left) visualizes the 3D point errors σ_{3D}^{obs} based on observation errors only (cf. Table 1, case (i)). In the close range, position errors are dominant with about 20 mm. Because of orientation errors, the uncertainty increases at higher distances. Figure 4 (right) shows the additional uncertainty of the point cloud when an uncertainty for the calibration parameters is added, i.e., $\sigma_{3D}^{add} = \sigma_{3D}^{obs+cal} - \sigma_{3D}^{obs}$ (cf. Table 1, case (ii) − case (i)). A radially symmetrical pattern is visible. However, the additional uncertainty is $\sigma_{3D}^{add} < 1$ mm within a radius of 50 m. We define this to be the goal of the calibration. Thus, given optimistic assumptions for the accuracy of the observations, the calibration parameters must be determined with a standard deviation of $\leq 1 \dots 1.5$ mm for the lever arm and $\leq 0.005°$ for the boresight angles (cf. Table 1, case (ii)).

Table 1. Standard deviations for the error propagation of the georeferencing equation (Equation (2)).

	Position		Orientation		2D Laser Scanner		Calibration Parameters		
	t_e, t_n	t_h	ϕ, θ	ψ	d_s	b_s	$\Delta x, \Delta y$	Δz	α, β, γ
(i) Obs	0.01 m	0.015 m	0.005°	0.010°	0.001 m	0.005°	–	–	–
(ii) Obs + Cal	0.01 m	0.015 m	0.005°	0.010°	0.001 m	0.005°	0.001 m	0.0015 m	0.005°

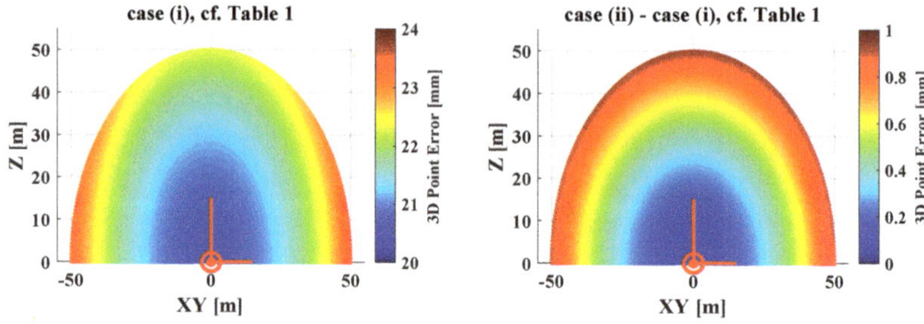

Figure 4. (**Left**) 3D point error of the mobile point cloud with observation errors only (Table 1, case (i)), (**Right**) Additional uncertainty of the point cloud due to calibration errors (Table 1, case (ii) – case (i)).

Beside the accuracy of the calibration parameters, the controllability of the estimation process is the second important quality criterion. Controllability addresses, if gross errors in the observations can be detected in an outlier test (i.e., internal controllability), and if not, how much undetected gross errors affect the parameter estimation (i.e., external controllability) [14,15]. The internal controllability can be analyzed using partial redundancies $r_i \in [0, 1]$, which can be calculated for each observation l_i according to [73,74]:

$$r_i = \left(\Sigma_{vv} \Sigma_{ll}^{-1} \right)_{ii} = \left(\Sigma_{ll} B^T \Sigma_{\overline{ll}}^{-1} \left[I - A \left(A^T \Sigma_{\overline{ll}}^{-1} A \right)^{-1} A^T \Sigma_{\overline{ll}}^{-1} \right] B \right)_{ii}, \tag{7}$$

where Σ_{vv} is the covariance matrix of the residuals and $\Sigma_{\overline{ll}} = B \Sigma_{ll} B^T$. Partial redundancies r_i describe the contribution of an observation to the redundancy r of an adjustment, i.e., $\sum r_i = r$. They further indicate the amount of an observation error that is transferred to the own residual v_i. Thus, high partial redundancies increase the probability that gross errors are detected. In geodetic network adjustment, values of $r_i > 0.3$ are recommended [14]. When testing normalized residuals, it is even possible to calculate the minimum detectable outlier ∇l_i [14,75,76]:

$$\nabla l_i = \delta_0(\alpha_T, \beta_T) \cdot \frac{\sigma_{l_i}}{\sqrt{r_i}}, \quad \sigma_{l_i} \in \Sigma_{ll}, \tag{8}$$

where δ_0 is called non-centrality parameter, which is a function of the type I error probability α_T and the type II error probability β_T (e.g., $\delta_0(0.001, 0.20) = 4.13$). Based on the minimum detectable outlier ∇l_i, we can calculate the impact ∇p_i of an undetected outlier on the parameter estimation:

$$\nabla p_i = \left(A^T \Sigma_{\overline{ll}}^{-1} A \right)^{-1} A^T \Sigma_{\overline{ll}}^{-1} B \left[0, \dots, \nabla l_i, \dots, 0 \right]^T. \tag{9}$$

Both r_i and ∇l_i in the observation space as well as ∇p_i in the parameter space allow for a rigorous analysis of the controllability of the parameter estimation in terms of gross errors [14,15].

5. Simulation of the Calibration

The configuration of the plane setup has a substantial impact on the quality of the calibration parameters. Therefore, we developed a simulation environment in order to find a plane setup that meets priorly defined quality criteria. For this purpose, we interpret the plane-based calibration as a geodetic network and investigate the parameter estimation process with geodetic quality criteria for both accuracy and controllability (cf. Section 4.3). This strategy is typically applied in geodetic network adjustment [14], but also for, e.g., surface approximation [71], calibration of TLS [77] or measuring arms [78], and VLBI scheduling (Very Long Baseline Interferometry) [79]. Hereinafter, Section 5.1 describes the simulation environment and Section 5.2 analyzes the realized plane setup.

5.1. Simulation Environment

Our simulation environment for the extrinsic calibration of mobile laser scanning systems makes use of the robotic simulation toolbox V-REP (Virtual Robot Experimentation Platform [80]). V-REP enables users to design mobile platforms, which can be equipped with different sensors like GNSS, IMU, laser scanners, or cameras. We utilized V-REP to rebuild the mobile laser scanning system that was introduced in Section 3 (Figure 5, left). In addition to this, V-REP provides the possibility to design environments, in which mobile platforms can operate. Thus, we can also use V-REP to create specific plane setups (Figure 5, middle). Given the mobile laser scanning system and the plane setup, we can simulate the calibration process (Figure 5, right). The plane setup and the simulated sensor observations are then exported to MATLAB (Matrix Laboratory, MathWorks), where we implemented the plane-based calibration approach. In MATLAB, we can analyze and improve the plane setup with quality criteria for both accuracy and controllability (cf. Section 5.2).

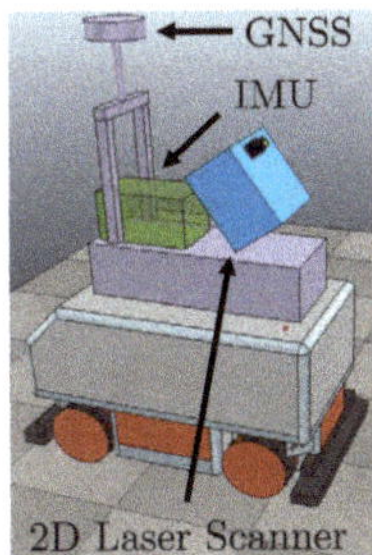

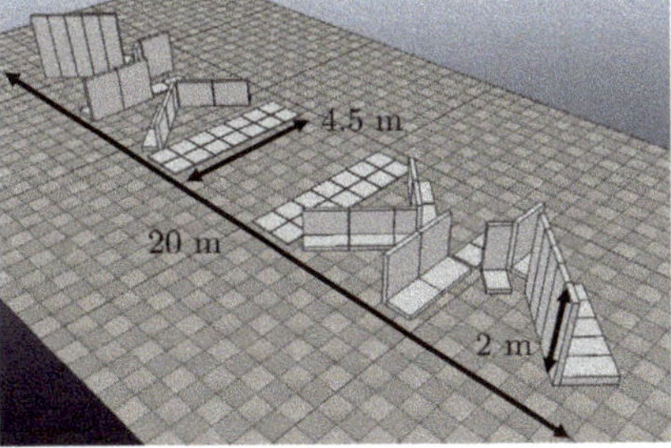

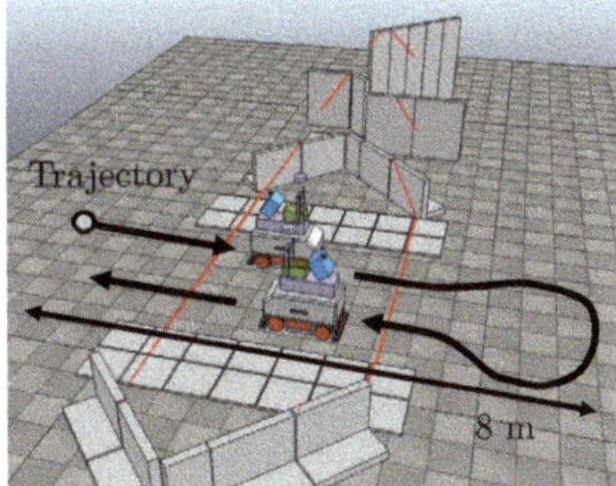

Figure 5. Using V-REP (Virtual Robot Experimentation Platform) for the simulation of the plane-based calibration approach: Mobile laser scanning system (**left**), plane setup (**middle**), kinematic data acquisition in the calibration field (**right**).

A rigorous optimization of the plane setup based on an all-embracing target function is hardly feasible [14,77]. Recently, Hartmann et al. [53] published an approach based on a genetic algorithm in order to optimize the plane configuration for the calibration of a mobile indoor laser scanning system. However, their target function for the optimization of the plane setup is based on the accuracy of the calibration parameters and does not include criteria for the controllability of the parameter estimation. Moreover, boundary conditions for permissible plane setups are required.

In our case, the determination of an appropriate plane setup is based on an iterative refinement of the plane setup. This strategy starts with an educated guess for the plane setup. This educated guess is iteratively refined based on expert knowledge and the quality criteria for both accuracy and controllability of the calibration parameters as introduced in Section 4.3. The iteration is stopped when the calibration parameters meet priorly defined requirements. In addition, we also specify boundary conditions for the plane setup in the calibration field that result from practical reasons:

- available area for the calibration field is 10 m × 20 m (cf. Figure 5, middle and right),

- fast and simple calibration procedure for mobile laser scanning systems (cf. Figure 5, right),
- for the reference planes we want to use standard face concrete elements as used for retaining walls in civil engineering (cf. Figure 5 and Section 6.1); such face concrete elements are highly planar, cost-effective, stable, and sufficiently robust for a permanent outdoor installation.

5.2. Simulation Results

This section presents the simulation results of the plane setup in Figure 5, which was finally realized in the calibration field (Section 6.1). This plane setup is scanned with the mobile laser scanning system in two passes. The difference between first and second pass is that the orientation of the system has been turned by 180° (Figure 5, right). The investigation of the plane setup and the calibration procedure can be split into three parts. We start with analyzing the sensitivity of the plane setup w.r.t. the calibration parameters, i.e., how changes of the lever arm and boresight angles affect the consistency between the reference planes and the point cloud of the mobile laser scanning system. This is connected to the separability of the calibration parameters (Section 5.2.1). Following this, we examine the impact of random and systematic observation errors on the precision and accuracy of the estimated calibration parameters (Section 5.2.2). Finally, we investigate the controllability of the parameter estimation, i.e., its robustness w.r.t. gross observation errors (Section 5.2.3).

5.2.1. Sensitivity of the Plane Setup and Separability of the Calibration Parameters

In this section, we examine the sensitivity of the plane setup w.r.t. changes in the calibration parameters. Figure 6a depicts the point-to-plane distances of the plane setup when the calibration parameters deviate from their true values by either 5 mm (lever arm) or 0.05° (boresight angles).

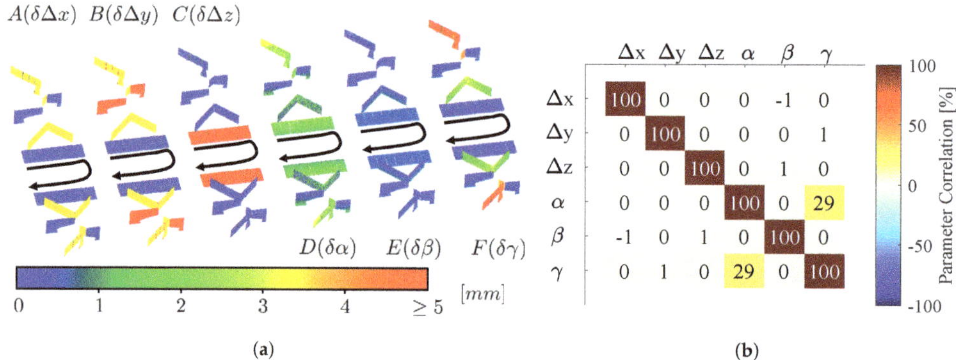

	Δx	Δy	Δz	α	β	γ
Δx	100	0	0	0	-1	0
Δy	0	100	0	0	0	1
Δz	0	0	100	0	1	0
α	0	0	0	100	0	29
β	-1	0	1	0	100	0
γ	0	1	0	29	0	100

(a) (b)

Figure 6. (a) Sensitivity of the plane setup towards deviation of the calibration parameters. The plots show the point-to-plane distance when the lever arm or the boresight angles deviate from their true values: A ($\delta\Delta x = 5$ mm), B ($\delta\Delta y = 5$ mm), C ($\delta\Delta z = 5$ mm), D ($\delta\alpha = 0.05°$), E ($\delta\beta = 0.05°$), and F ($\delta\gamma = 0.05°$) (b) The calibration parameters are basically uncorrelated and, thus, clearly separable. Only the boresight angles α and γ show a small correlation of 29 % (explanation in the text).

For the lever arm components (A, B, C), there is always a subset of planes that is sensitive towards the parameter deviation. For instance, the planes that are inclined by 45° w.r.t. the direction of motion of the platform have a sensitivity of 70% towards changes in Δx and Δy (yellow color in A and B). In addition, the reference planes that are parallel to the direction of motion are fully sensitive towards changes in Δy (red color in B). The planes on the ground are fully sensitive towards changes in Δz (red color in C). Regarding the boresight angles, the plane setup has a sufficient sensitivity towards changes in α (D) and γ (F), but is less sensitive towards changes in β (E). The lower sensitivity towards β means that it is more difficult to estimate this parameter based on the minimization of point-to-plane distances like it is done in the calibration. This is proved by both the simulation results (Figure 7) and the results based on real data (Section 6.2), because in both cases the boresight angle β has the biggest standard deviation of all boresight angles. The sensitivity towards β could be improved by mounting the mobile laser scanning system on a high van. In this case, the distance between the mobile laser scanning system and the ground planes, which are most sensitive towards β, is increased.

In Figure 6a, we can see that different parameter deviations lead to different patterns of the point-to-plane distances. Thus, making the assumptions that we have no other sources of errors, we can separate erroneous calibration parameters by analyzing their unique deviation patterns. This indicates a good separability of the calibration parameters. The good separability of the calibration parameters is verified by their correlation matrix in Figure 6b, which is derived from the covariance matrix $\Sigma_{\hat{p}\hat{p}}$ of the parameter estimation (cf. Equation (6)). According to the correlation matrix, the calibration parameters are basically uncorrelated, except for a small correlation of 29% between the boresight angles α and γ. The reason for this correlation is the sequence of boresight rotations using Euler angles. Due to the $\beta = 30°$ tilt of the 2D laser scanner (cf. Figure 2, left), the horizontal projection of the X_s-axis is smaller than the length of the X_s-axis. In the extreme case of $\beta = 90°$, the projection is zero. In navigation, this phenomenon is known as gimbal lock [24], where the boresight angles α and γ rotate around the same axis and cannot be separated from each other (100 % correlation). Thus, in the case of $\beta = 0°$, the correlation between α and γ is 0 % as is demonstrated in [11]. In our case, the 2D laser scanner is tilted by $\beta = 30°$ and, thus, the correlation is between 0 % and 100 %.

5.2.2. Impact of Random and Systematic Observation Errors

In order to examine how random observation errors of the mobile laser scanning system affect the parameter estimation, we carried out a Monte Carlo simulation with 1000 realizations of the calibration process. Given the plane setup, true calibration parameters, and the observations of the mobile laser scanning system, the noise of the observations was sampled from a Gaussian distribution. The associated standard deviations were oriented towards manufacturer information and empirical values (cf. Table 1). The Monte Carlo simulation was performed in a local frame (l-frame) with east, north, and height component. This allows for a better interpretation of the simulation results. As a result, 1,000 realizations for the estimated calibration parameters $\hat{p} = [\Delta x, \Delta y, \Delta z, \alpha, \beta, \gamma]^T$ and their covariance matrix $\Sigma_{\hat{p}\hat{p}}$ were obtained. According to Förstner and Wrobel (chap. 4.6.8, pp. 139–141) [15], the unbiasedness of $\hat{p}$ and $\Sigma_{\hat{p}\hat{p}}$ within the simulation can be checked by different statistical tests. All simulations passed these tests.

The grey histograms in Figure 7 show the distributions of the calibration parameters after the Monte Carlo simulation based on random observation errors only. The red vertical lines indicate the true calibration parameters, i.e., the expectation values. It can be stated that all parameters are unbiased if only random observation errors are simulated. For the three lever arm components, a standard deviation of $\sigma < 1$ mm is obtained. These results meet the defined target accuracies of 1 mm for Δx and Δy as well as 1.5 mm for Δz (cf. Table 1). The accuracy of the lever arm mainly depends on the accuracy of the position $[t_e, t_n, t_h]^T$ of the mobile laser scanning system. The higher standard deviation of Δz results from the higher standard deviation of the height component t_h, which is assumed to be 1.5 times the standard deviation of the horizontal position t_e and t_n (cf. Table 1).

For the boresight angles α, β, and γ, standard deviations of $\sigma < 0.001°$ are obtained, which also meet the defined target accuracy of $0.005°$ (cf. Table 1). The accuracy of the boresight angles mainly depends on the accuracy of the orientation angles of the mobile laser scanning system. In this regard, α is mostly aligned with the roll angle ϕ, β is mostly aligned with the pitch angle θ, and γ is mostly aligned with the yaw angle ψ. Thus, the smaller standard deviation of α as compared to γ is expected, since the roll angle is assumed to be more accurate than the yaw angle (cf. Table 1). However, the boresight angle β is estimated with the lowest accuracy. At this point, we recall the sensitivity analysis in Section 5.2.1, where we found that the plane setup is less sensitive towards the boresight angle β. Thus, the higher standard deviation of the boresight angle β can be attributed to the lower sensitivity of the plane setup. However, the target accuracy of $0.005°$ is still achieved.

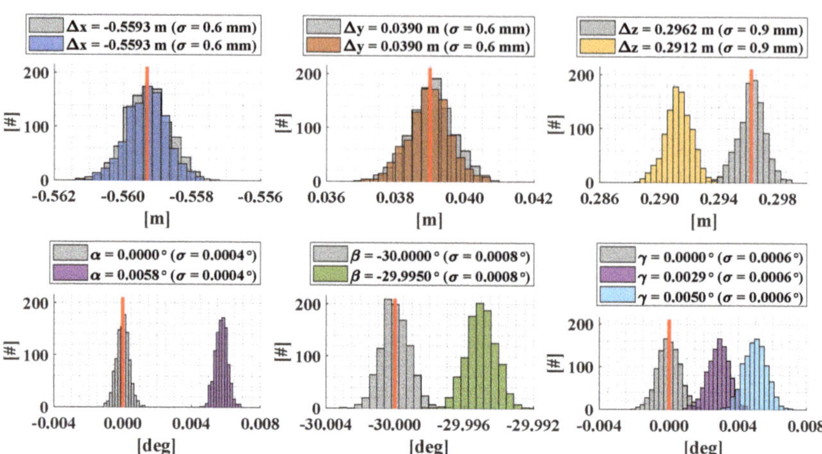

Figure 7. Distributions of the estimated calibration parameters for a Monte Carlo Simulation with 1000 realizations. The grey histograms correspond to random observation errors only (cf. Table 1, case (i)). The colored histograms correspond to random observation errors with additional systematic observation errors: $\delta t_e = 5$ mm (dark blue), $\delta t_n = 5$ mm (orange), $\delta t_h = 5$ mm (yellow), $\delta\phi = 0.005°$ (purple), $\delta\theta = 0.005°$ (green), $\delta\psi = 0.005°$ (light blue). The red vertical lines indicate the true values.

Inherently, the position and orientation observations of a mobile laser scanning system are prone to systematic errors due to the use of GNSS or uncorrected errors of the IMU. Therefore, we repeated the Monte Carlo simulation with additional biases of 5 mm and 0.005° for the position and orientation observations of the mobile laser scanning system. As in the case of random errors, we found that systematic errors of the position mainly affect the lever arm, while systematic errors of the orientation angles mainly affect the boresight angles. The results of the Monte Carlo simulation with both random and systematic observation errors are visualized as colored histograms in Figure 7.

Clearly, the lever arm components Δx and Δy remain unbiased in the case of systematic observation errors (cf. Figure 7, top left, top middle). The lever arm components Δx and Δy mainly depend on the horizontal position t_e and t_n of the mobile laser scanning system. In the simulation, we found that the associated systematic errors δt_e and δt_n are completely transferred to the residuals of t_e and t_n and, thus, do not affect the parameters. This results from the repeated measurements of the plane setup with opposite direction of motion of the platform (cf. Figure 5, right). Basically, this kind of double measurement can be considered as a measurement in two faces, which is an established strategy in geodesy for the elimination of systematic errors [81]. For the height component, however, the systematic error δt_h is not eliminated in this way, which is why the lever arm component Δz is biased by 5 mm (Figure 7, top right). For the elimination of a systematic height error δt_h, the mobile platform would need to measure the plane setup upside down in the second pass.

Systematic orientation errors of the mobile laser scanning system are also not eliminated by the double measurement. A systematic error of the pitch angle θ is completely transferred to the boresight angle β, which is biased by 0.005° (Figure 7, bottom middle, green histogram). The same applies to a systematic error of the yaw angle ψ, which is completely transferred to the boresight angle γ (Figure 7, bottom right, light blue histogram). However, in contrast to this, a systematic error of the roll angle ϕ affects the boresight angles α and γ (Figure 7, bottom left and right, purple histograms). This phenomenon is caused by the correlation between the boresight angles α and γ (cf. Section 5.2.1 and Figure 6b) and the sequence of the Euler rotations from the n-frame back to the s-frame, i.e., $R_b^n(\psi)$ $\rightarrow R_b^n(\theta) \rightarrow R_b^n(\phi) \rightarrow R_s^b(\gamma) \rightarrow R_s^b(\beta) \rightarrow R_s^b(\alpha)$. In the case of a systematic error in the roll angle ϕ, this error is partially compensated by the boresight angle γ before it reaches the boresight angle α

due to the correlation between these two parameters. In contrast to this, a systematic error in the yaw angle ψ can completely be compensated by the boresight angle γ before it reaches the boresight angle α. Similar considerations can be made for a systematic error in the pitch angle θ.

The simulations indicate that systematic errors in the trajectory estimation corrupt the calibration results. However, such systematic errors are often site- and time-dependent, e.g., in the case of GNSS. This means that if we repeat the calibration multiple times over a certain time period, systematic errors might change in magnitude and sign. In this way, systematic errors could get a more random characteristic and, thus, could be reduced by averaging the calibration results of multiple runs.

Another source of systematic errors, which we have not discussed yet, is the 2D laser scanner. In this respect, we simulated a range finder offset of d_0 = 2 mm and analyzed its impact on the calibration parameters. As a result, the range finder offset mainly biases the lever arm components Δx ($\approx$0.4 mm) and Δz ($\approx$1.4 mm), as well as the boresight angle β ($\approx$0.033°). However, the used plane setup is sensitive towards a range finder offset. Thus, this intrinsic parameter can be added to the functional model in Equation (4) and estimated as part of the calibration. This is demonstrated in Section 6.3, where we estimate the range finder offset d_0 for test purposes based on real data.

Finally, we also took errors of the TLS point cloud into consideration, which is the basis for the determination of the reference plane equations (cf. Equation (3)). In this respect, a global registration error of the TLS point clouds seems to be the most serious problem. Hence, we simulated translation errors of up to 3 mm as well as rotation errors of up to 0.02°. We found that errors in the horizontal position of the TLS point cloud as well as tilting errors are filtered out to $\geq$85% in the calibration, i.e., such errors are transferred to the residuals of the adjustment. As in the case of systematic errors of the horizontal position of the mobile laser scanning system, this results from the repeated measurements of the plane setup with opposite direction of motion of the platform. Registration errors in height and azimuth, however, bias Δz and γ, respectively. In practice, the registration of the TLS point cloud is based on a dense network of highly accurate control points (cf. Sections 6.1 and 7.1). This provides a high degree of accuracy and controllability. In the light of this, we expect registration errors in height and azimuth to be smaller than the target accuracy for the calibration parameters, i.e., 1 ... 1.5 mm for the lever arm components and 0.005° for the boresight angles.

5.2.3. Impact of Gross Observation Errors

In addition to the accuracy, the quality of the estimated calibration parameters is also determined by the detectability of gross observation errors and the impact of undetected gross observation errors on the parameters. Therefore, we calculated the partial redundancies r_i (cf. Equation (7)) and the minimum detectable outliers ∇l_i (cf. Equation (8)) of the position and orientation observations of the mobile laser scanning system. The results are visualized in Figure 8.

Figure 8 (left column) shows that the partial redundancies of the position are close to 1 and, thus, well controlled. Gross errors of $\nabla t_{e,n} \geq 48$ mm for the horizontal position and $\nabla t_h \geq 72$ mm for the height component can be detected due to the different horizontal and vertical accuracy of the mobile laser scanning system. Undetected gross errors in the position bias the lever arm by $|\nabla p_{t_{e,n}}| \leq 0.17$ mm and $|\nabla p_{t_h}| \leq 0.26$ mm, respectively (not shown in Figure 8). The boresight angles are not affected by gross errors in the position.

Regarding the orientation (Figure 8, right column), the roll angle ϕ and yaw angle ψ are well controlled with partial redundancies of $r_i \geq 0.85$. However, the pitch angle θ is weakly controlled with partial redundancies of $r_i < 0.3$. This is connected to the lower sensitivity of the plane setup towards the boresight angle β (Figure 6, case E), which is mostly aligned with the pitch angle θ. Accordingly, the detectability of gross errors is worse for pitch ($\nabla\theta \geq 0.05° - 0.4°$) than for roll ($\nabla\phi \geq 0.025°$) and yaw ($\nabla\psi \geq 0.05°$). The impact of undetected gross errors in roll and yaw on the boresight angles is $|\nabla p_\phi| \leq 0.00012°$ and $|\nabla p_\psi| \leq 0.00018°$, respectively. For the pitch angle, the impact is about two to three times larger with $|\nabla p_\theta| \leq 0.00031°$ (not shown in Figure 8). However, despite the small partial redundancies of the pitch angles, their impact values are very small and do not affect the calibration. The impact of gross errors in the orientation on the lever arm is <40 μm and, thus, negligible.

Due to the high redundancy of the adjustment, the partial redundancies of the 2D laser scanner points are nearly $r_i = 1$. Hence, outliers can reliably be detected and do not affect the calibration.

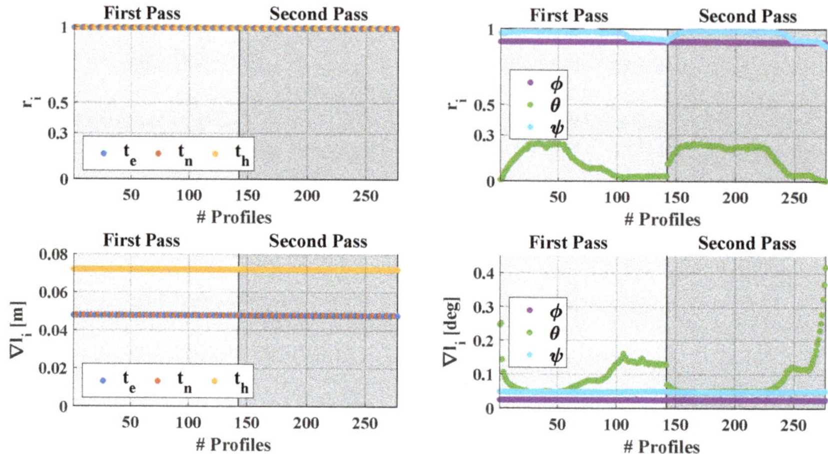

Figure 8. Partial redundancies r_i (**top**) and minimum detectable outliers ∇l_i (**bottom**) of the position $[t_e, t_n, t_h]^T$ and orientation angles ϕ, θ, ψ of the GNSS/IMU unit (Global Navigation Satellite System, Inertial Measurement Unit). The plots show the values for each profile of first and second pass with opposite direction of motion of the platform (Figure 5, right).

Principally, the observations of the mobile laser scanning system are well controlled, which means that undetected gross errors do not considerably affect the calibration. This controllability is especially important for the position and the yaw angle observations of the mobile laser scanning system, because GNSS—which is prone to gross errors—contributes to these observations.

6. Calibration of the Mobile Laser Scanning System

This section presents empirical calibration results of the mobile laser scanning system that was introduced in Section 3. Section 6.1 describes the calibration measurements, Section 6.2 discusses the lever arm and boresight calibration. In Section 6.3, we address the applicability of the calibration field for estimating the range finder offset as additional intrinsic calibration parameter.

6.1. Calibration Measurements

The plane setup as shown in Figure 5 and validated in Section 5.2 was realized in the form of a permanently installed calibration field. The calibration field is shown in Figure 9. As postulated in Section 5.1, the calibration field covers an area of 10 m × 20 m and was realized using cost-effective, stable, and robust face concrete elements from civil engineering. For the calibration, the system was attached to a trolley due to its easy handling in the calibration field (Figure 9, right).

Initially, reference values for the plane setup were determined with TLS. For this purpose, a Leica ScanStation P50 was set up on five stations in order to completely cover all planes without obstructions. The scans were performed in two faces for reducing systematic errors of the TLS [81]. The TLS point clouds were georeferenced using a network of tie points as well as highly accurate georeferenced control points. All tie and control points were signalized with special BOTA8 targets (Bonn Target 8), which have a square stellar black and white pattern with a size of 0.3 m × 0.3 m (cf. Section 7). The BOTA8 targets have been developed at the University of Bonn and allow for an accurate registration of TLS scans [82]. The mean absolute error after georeferencing was 1.2 mm. Following this, plane models were fitted to the TLS point clouds, which serve as reference information for the calibration. Please note that the control points for the georeferencing of the TLS point clouds are part of a bigger network of control points that was principally realized for the point-based evaluation of the mobile laser scanning systems (cf. Section 7).

For empirically analyzing the quality of the calibration parameters, the calibration procedure as shown in Figure 5 was repeated 98 times. Each calibration run took less than one minute. The 98 runs were performed in four independent blocks with 14, 21, 31, and 32 runs, respectively. After each block, the system was reinitialized. In addition, the blocks were measured at different days and daytimes. That way, different GNSS constellations were covered. Table 2 gives an overview.

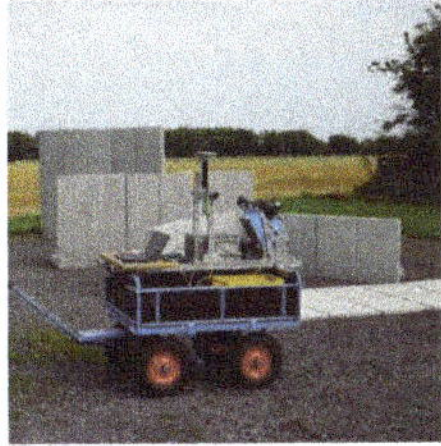

Figure 9. Realization of the plane-based calibration field for the calibration of mobile laser scanning systems (**left**, middle). Mobile laser scanning system during the calibration measurements (**right**).

Table 2. Time schedule of the 98 calibration runs, which were measured in July and August 2019.

Block	Date	GPS Time	Runs	Block	Date	GPS Time	Runs
1	31/07/19	01:40 p.m.–01:55 p.m.	14	3	05/08/19	10:30 a.m.–10:50 a.m.	31
2	05/08/19	09:05 a.m.–09:25 a.m.	21	4	05/08/19	11:45 a.m.–12:10 p.m.	32

6.2. Calibration of Lever Arm and Boresight Angles

The calibration runs were processed independently. A total of 98 realizations of the calibration parameters were obtained that indicate the repeatability of the calibration. The results are visualized in Figure 10. The mean values and standard deviations are stated on top of the histograms. For better visual assessment, an ideal Gaussian distribution is added to the histograms.

The standard deviations of the lever arm components range from 0.9 mm to 4.5 mm. According to the simulation in Section 5.2, the horizontal components Δx and Δy are more precise than the vertical component Δz. This is due to the fact that the quality of the lever arm calibration mainly depends on the accuracy of the position of the GNSS/IMU unit. In the case of GNSS/IMU integration, the height component is normally less accurate than the horizontal component [22,23]. In particular, systematic height errors are critical, since such errors can hardly be detected in the calibration and affect the lever arm component Δz. In contrast, systematic errors in the horizontal position can be eliminated by measuring the plane setup with opposite direction of motion of the platform (cf. Section 5.2.2). Hence, Δz is expected to be worse than Δx and Δy due to potential systematic height errors. These results are in good accordance with the simulations in Section 5.2. According to the simulations, gross errors in position most likely do not affect the lever arm, since they can reliably be detected.

The standard deviations of the boresight angles range from 0.0012° to 0.0188°. According to the simulations, the boresight angle β is estimated with the lowest accuracy. This is due to the fact that the plane setup is less sensitive towards β than towards α and γ (cf. Section 5.2.1). Generally, the accuracy of the boresight angles depends on the accuracy of the orientation angles of the GNSS/IMU unit. While roll and pitch can be estimated from the accelerometers, the yaw determination is more challenging, since it depends on the quality of the initial IMU alignment and the quality of GNSS during motion [23]. In this respect, the yaw angle is more prone to systematic errors than roll and pitch. Basically, systematic errors in the yaw angle affect the boresight angle γ. In Figure 10, the histograms of the parameters α and β appear to be closer to a Gaussian distribution than the histogram of the parameter γ. This may indicate that systematic errors have affected the determination of γ.

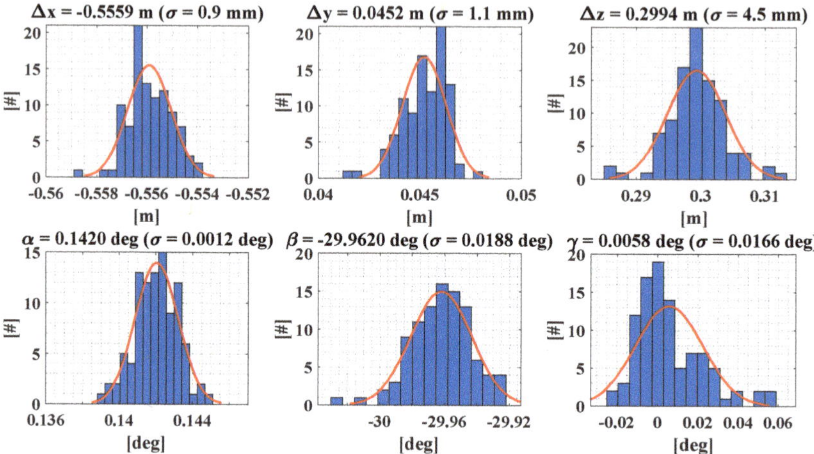

Figure 10. Results of the 98 calibration runs: Lever arm components Δx, Δy, and Δz (**top row**), boresight angles α, β, and γ (**bottom row**). The mean values and standard deviations σ are stated on the top of the histograms. For comparison, an ideal Gaussian distribution is added to the plots.

The standard deviations of the histograms in Figure 10 indicate the precision of one realization of the calibration parameters. Except for the parameter Δx, Δy, and α, the required target accuracy of 1 … 1.5 mm for the lever arm as well as 0.005° for the boresight angles as defined in Section 5.2 is not reached. Generally, the empirical standard deviations are higher than those from simulation. This might be caused by systematic errors or neglected correlations that exist in reality, but were not simulated. For n uncorrelated realizations of the calibration parameters, however, the accuracy of the mean value can be improved by $\sqrt{n}$. In our case, 15 statistically independent calibrations are needed to reach the target accuracies for Δz, β, and γ (Table 3). Obviously, individual calibration runs are not fully uncorrelated. Therefore, we recommend to perform 15 calibration runs in the calibration field and to repeat this after a break of 1–3 h. After some hours the GNSS constellation is changed, which should lead to different systematic errors in the trajectory estimation.

Table 3. Standard deviations for a single calibration run and multiple calibration runs.

Parameter	$\sigma_{\Delta x}$	$\sigma_{\Delta y}$	$\sigma_{\Delta z}$	σ_{α}	σ_{β}	σ_{γ}
Target Accuracy	1.0 mm	1.0 mm	1.5 mm	0.0050°	0.0050°	0.0050°
1 Realization	0.9 mm	1.1 mm	4.5 mm	0.0012°	0.0188°	0.0166°
15 Realizations	0.2 mm	0.3 mm	1.2 mm	0.0003°	0.0049°	0.0043°
98 Realizations	0.1 mm	0.1 mm	0.5 mm	0.0001°	0.0019°	0.0017°

Figure 11 shows the additional uncertainty $\sigma_{3D}^{add} = \sigma_{3D}^{obs+cal} - \sigma_{3D}^{obs}$ of the mobile point cloud that results from the difference between a point cloud that includes an uncertainty of the calibration ($\sigma_{3D}^{obs+cal}$) and a point cloud that does not include an uncertainty of the calibration (σ_{3D}^{obs}); cf. Figure 4 (left). Clearly, the additional uncertainty of the point cloud is $\sigma_{3D}^{add} < 1$ mm within a 50 m radius when using a mean calibration. Please remember that this was the overall goal of the calibration.

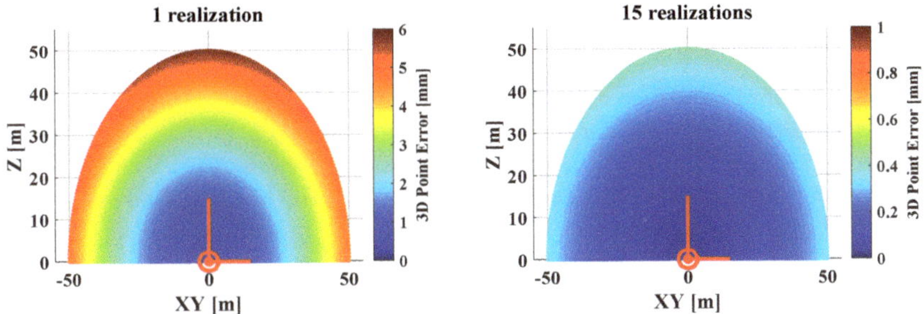

Figure 11. Additional uncertainty σ_{3D}^{add} of the mobile point cloud due to the uncertainty of the calibration parameters for a single calibration run (**left**) and 15 calibration runs (**right**), cf. Table 3.

6.3. Calibration of Range Finder Offset

As mentioned in Section 5.2.2, the calibration field is also sensitive towards the range finder offset d_0 of the 2D laser scanner. The right part of Figure 12 depicts the sensitivity of the plane setup towards an uncorrected range finder offset of $d_0 = 5$ mm (cf. Figure 6). The range finder offset d_0 can simply be added to the calibration model (cf. Equation (4)):

$$\left[\begin{bmatrix} t_x \\ t_y \\ t_z \end{bmatrix} + \boldsymbol{R}_n^e(L,B) \cdot \boldsymbol{R}_b^n(\phi,\theta,\psi) \cdot \left(\boldsymbol{R}_s^b(\alpha,\beta,\gamma) \begin{bmatrix} 0 \\ (d_s + d_0) \cdot \sin b_s \\ (d_s + d_0) \cdot \cos b_s \end{bmatrix} + \begin{bmatrix} \Delta x \\ \Delta y \\ \Delta z \end{bmatrix} \right) \right]^T \cdot \begin{bmatrix} \bar{n}_x \\ \bar{n}_y \\ \bar{n}_z \end{bmatrix} - 1 \overset{!}{=} 0. \quad (10)$$

For test purposes, we estimated the range finder offset d_0 for each of the 98 calibation runs. The result is shown in the right part of Figure 12. The related histogram has a mean value of -0.05 mm and an empirical standard deviation of 0.08 mm. Accordingly, the range finder offset can be estimated in our calibration field. In this case, however, the range finder offset is of negligible magnitude.

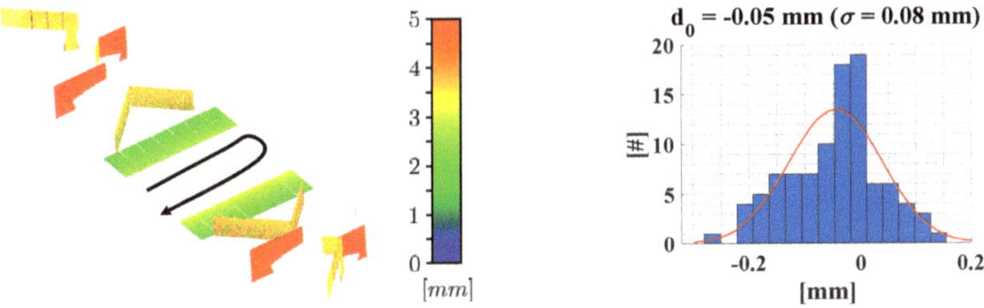

Figure 12. (**Left**) Sensitivity of the plane setup towards a range finder offset. The colored plot shows the point-to-plane distance in the case of an unmodelled range finder offset of $d_0 = 5$ mm. (**Right**) Distribution of the estimated range finder offsets for the 98 calibration runs. The mean value is -0.05 mm and the empirical standard deviation is 0.08 mm. For comparison, an ideal Gaussian distribution is added.

7. Evaluation of the Mobile Laser Scanning System

This section addresses the evaluation of the mobile laser scanning system. In Section 7.1, we introduce the evaluation environment. Section 7.2 describes the evaluation measurements. Based on this, two different evaluation strategies are pursued, i.e., a point-based evaluation using control points (Section 7.3) and an area-based evaluation using TLS reference point clouds (Section 7.4).

7.1. Evaluation Environment

The evaluation environment contains a network of control points (6 pillars, 16 ground points, 15 building points) that covers an area of about 1.5 km^2 with variable GNSS conditions (Figure 13). The control point coordinates were determined in a network adjustment using measurements from total stations, GNSS baselines, and levelings. The network is linked to the official German coordinate systems, i.e., ETRS89/UTM32 (European Terrestrial Reference System 1989, Universal Transverse Mercator, Zone 32) for position, and DHHN16 (Deutsches Haupthöhennetz 2016) for physical heights. In addition, the ellipsoidal heights w.r.t. GRS80 (Geodetic Reference System 1980) were determined. The connection between the ellipsoidal heights and the physical heights was realized by using the German Combined Quasi Geoid Model (GCG16) [83]. The network was surveyed in June 2017 and July 2018 for checking its stability [84]. The final control point coordinates are based on a combined network adjustment of both epochs and have estimated standard deviations of <1 mm.

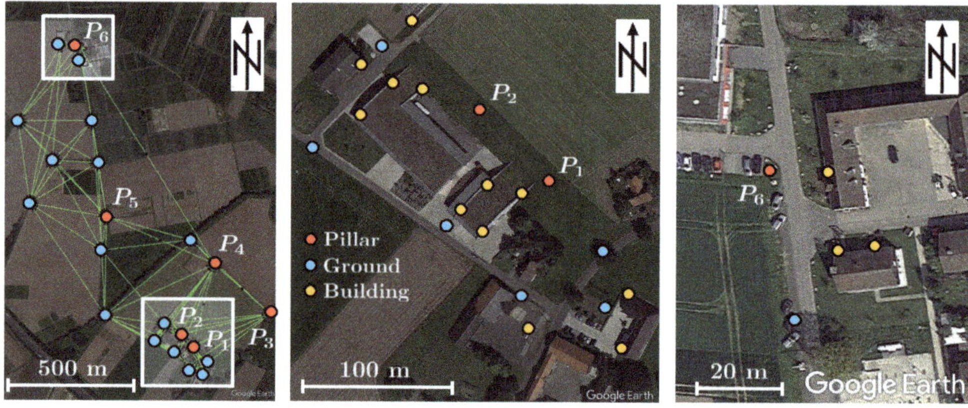

Figure 13. Network of control points with six pillars, 15 building points, and 16 ground points. The control points can be signalized with BOTA8 (Bonn Target 8) laser scanning targets (Source: Google Earth, modified).

In addition to the control points, TLS reference point clouds of all buildings in the area of the control points were measured (Figure 14). The TLS scans were captured from multiple stations in two faces and georeferenced using the network of control points like it was done for the calibration field in Section 6.1. The mean absolute error after georeferencing was 1.2 mm. The TLS reference point clouds can be used for an area-based evaluation of mobile laser scanning systems. In addition, a point-based evaluation is possible based on the control points. For this purpose, the pillars and building points can be signalized with special BOTA8 laser scanning targets [82] (cf. Figure 15), which are applicable to both TLS registration and evaluation of mobile laser scanning systems.

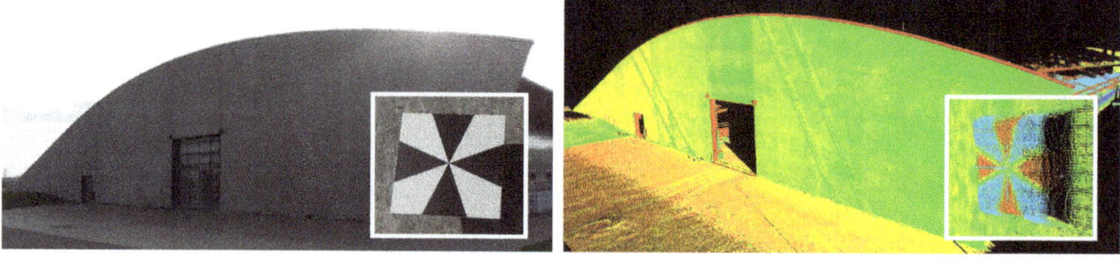

Figure 14. TLS (Terrestrial Laser Scanning) reference point clouds georeferenced using the control points and BOTA8 (Bonn Target 8) laser scanning targets.

7.2. Evaluation Measurements

For the evaluation of the mobile laser scanning system, five test runs were carried out in November 2018 and August 2019 using both a trolley and a van as carrier platform. Table 4 gives an overview.

Table 4. Time schedule of the evaluation measurements in November 2018 and August 2019.

Run	Date	GPS Time	Duration	Platform	Covered Distance	Targets
1	13/11/18	09:26 a.m.–09:45 a.m.	18:37 min	Trolley	1250 m	24
2	05/08/19	09:25 a.m.–09:52 a.m.	26:24 min	Trolley	1570 m	31
3	05/08/19	10:50 a.m.–11:15 a.m.	24:28 min	Trolley	1370 m	29
4	05/08/19	12:10 p.m.–12:35 p.m.	24:22 min	Trolley	1330 m	27
5	05/08/19	01:10 p.m.–01:54 p.m.	43:17 min	Van	7990 m	37

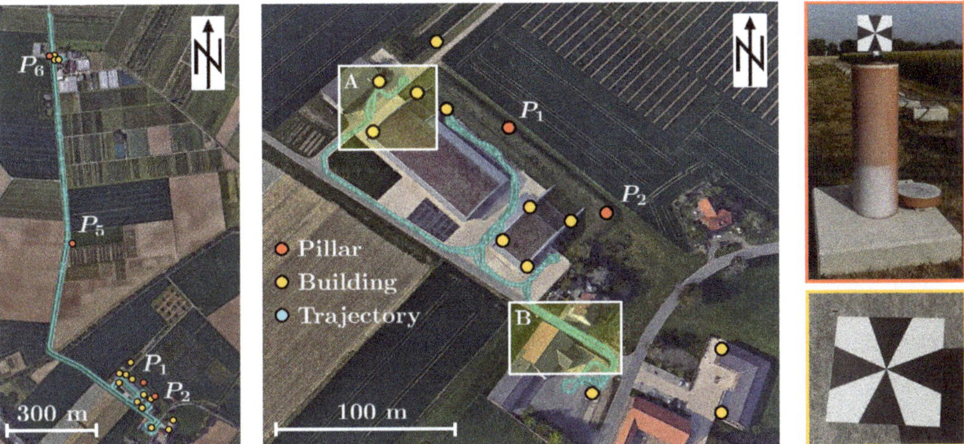

Figure 15. Evaluation of the mobile laser scanning system: Van trajectory (**left**), trolley trajectory (**middle**), control points with BOTA8 targets (**right**). The yellow boxes indicate the test sites A and B for an area-based evaluation of the mobile laser scanning system (Source: Google Earth, modified).

During the test runs, a total distance of 13.5 km was covered. The trajectory was different for each run and lasted between 18 and 43 min. As an example, the cyan dots in Figure 15 indicate the trajectories of run 4 with a trolley and run 5 with a van (cf. Table 4). For each test run, the mobile laser scanning system was reinitialized. For a point-based evaluation, BOTA8 targets were adapted to the control points and scanned with the mobile laser scanning system. Each control point was scanned multiples times leading to a total number of 148 target scans that can be used for a point-based evaluation of the system. For an area-based evaluation using TLS reference point clouds, two selected test sites A and B in the evaluation environment were repeatedly scanned with the mobile laser scanning system (Figure 15). Test site A contains a face concrete wall with dimensions of 9.5 m × 50 m (cf. Figure 14), and test site B is a road passage with houses. The mobile scans of the test sites A and B were also performed during the five test runs in parallel to the scans of the control points.

7.3. Point-Based Evaluation Using Control Points

For the point-based evaluation, the coordinates of the control points were extracted from the point clouds of the mobile laser scanning system and compared to the reference values. That way, an evaluation of the entire processing chain including the calibration is possible. Figure 16 shows the histograms of the differences for east, north, and height component for all 148 target scans. An ideal Gaussian distribution is added to the histograms for better visual interpretation. In addition, the mean values ΔE, ΔN, and ΔH as well as the empirical standard deviations σ are stated on top of the histograms. The histograms in Figure 16 have the approximate shape of a Gaussian distribution with absolute mean values of <2 mm and standard deviations of σ < 9 mm. The differences between the horizontal and vertical components is marginal. These results indicate a high accuracy of the mobile laser scanning system with a satisfactory extrinsic calibration. Moreover, these results also indicate the high quality of the control points in our evaluation environment.

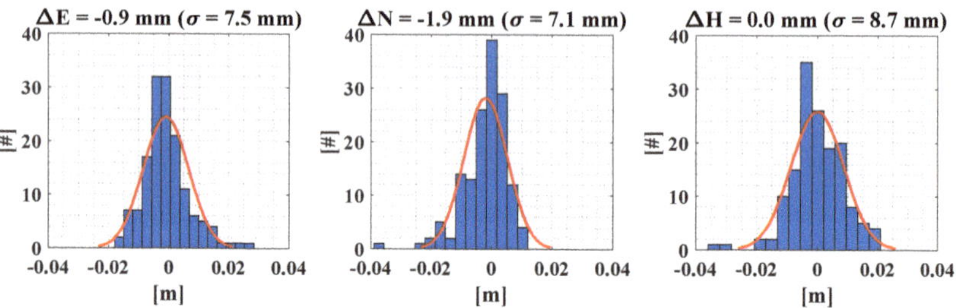

Figure 16. Point-based evaluation of the mobile laser scanning system for east, north, and height component. A total number of 148 target scans were compared to the reference values. The mean values ΔE, ΔN, and ΔH as well as the standard deviations σ are stated on top of the histograms.

7.4. Area-Based Evaluation Using TLS Reference Point Clouds

For an area-based evaluation, the 3D point clouds of the mobile laser scanning system were compared with TLS reference point clouds. For this purpose, we made use of the M3C2 algorithm (Multiscale-Model-to-Model-Cloud) in the software CloudCompare. The M3C2 algorithm allows for a robust distance computation and comparison between two point clouds [85,86]. Both test sites A and B (cf. Figure 15) were scanned five times with the mobile laser scanning system. In order to demonstrate the potential of an area-based evaluation, we calculated the mobile point clouds with two different sets of calibration parameters, i.e., with an approximate calibration that was taken from the construction plan of the system as well as with the calibration that we determined in our plane-based calibration field. For the plane-based calibration, we utilized the mean values of all 98 calibration runs (cf. Figure 10) as these are the best available calibration parameters.

The results of the area-based evaluation are listed in Table 5, which shows the mean, median, standard deviation (STD), and root mean square error (RMS) of the deviations between the mobile point clouds and the TLS reference point clouds that were obtained from the M3C2 comparison. When using the approximate calibration, mean and median values between −18.5 mm and 8.7 mm are obtained (Table 5, left column). In contrast, the plane-based calibration considerably reduces the mean and median values to −6.3 mm to 4.0 mm (Table 5, right column). Thus, the unbiasedness of the point clouds is improved by the plane-based calibration. When using the approximate calibration, we obtain standard deviations and RMS values between 8.5 mm and 33.4 mm. For the plane-based calibration, these values are reduced to 4.1 mm to 9.9 mm. In general, when using the plane-based calibration, the results of the area-based evaluation are in good accordance with the results of the point-based evaluation based on the control points (cf. Figure 16).

Please note that the standard deviations and RMS values are generally smaller in test site A than in test site B. This might be caused by the fact that test site A is not sensitive to all types of errors. In test site A, for example, errors that cause a shift of the mobile point cloud that is parallel to the wall cannot be detected by the M3C2 point cloud comparison. Contrary to this, test site B is more adequate to evaluate the overall system performance, since test site B shows more variation w.r.t. surface position and orientation. This is a drawback of the area-based evaluation strategy using algorithms for the direct comparison of point clouds.

Table 5. Area-based evaluation of the mobile laser scanning system using calibration parameters from the construction plan (**left**) and from the plane-based calibration field (**right**). The table shows the mean, median, standard deviation (STD), and root mean square error (RMS) of an M3C2 comparison to a TLS reference point cloud. The rows marked with * correspond to the histograms in Figure 17.

	Construction Plan				Plane-Based Calibration Field			
Calibration Parameters	$\Delta x = -0.5594$ m $\Delta y = +0.0390$ m $\Delta z = +0.2962$ m	$\alpha = \pm0.0000°$ $\beta = -30.0000°$ $\gamma = \pm0.0000°$			$\Delta x = -0.5559$ m $\Delta y = +0.0452$ m $\Delta z = +0.2994$ m	$\alpha = +0.1420°$ $\beta = -29.9620°$ $\gamma = +0.0058°$		
Test Site	**Mean**	**Median**	**STD**	**RMS**	**Mean**	**Median**	**STD**	**RMS**
A	8.7 mm	8.7 mm	8.5 mm	12.2 mm	−6.3 mm	−6.3 mm	4.1 mm	7.6 mm
A *	−18.4 mm	−18.5 mm	8.8 mm	20.4 mm	−1.1 mm	−0.6 mm	4.4 mm	4.5 mm
A	−16.3 mm	−16.2 mm	10.0 mm	19.1 mm	−3.2 mm	−3.5 mm	5.2 mm	6.1 mm
A	−17.5 mm	−17.4 mm	10.1 mm	20.2 mm	−4.4 mm	−5.2 mm	6.3 mm	7.7 mm
A	7.4 mm	7.6 mm	9.1 mm	11.7 mm	3.6 mm	3.5 mm	4.4 mm	5.7 mm
B	1.5 mm	0.3 mm	21.1 mm	21.1 mm	2.5 mm	4.0 mm	8.7 mm	9.0 mm
B *	−9.7 mm	−1.2 mm	27.0 mm	28.7 mm	−0.3 mm	0.6 mm	7.6 mm	7.6 mm
B	−7.9 mm	−2.2 mm	26.2 mm	27.4 mm	0.6 mm	0.7 mm	6.5 mm	6.5 mm
B	−6.9 mm	−2.2 mm	24.9 mm	25.9 mm	−0.7 mm	−0.8 mm	8.3 mm	8.3 mm
B	−14.6 mm	−8.0 mm	30.0 mm	33.4 mm	−4.3 mm	−3.7 mm	8.9 mm	9.9 mm

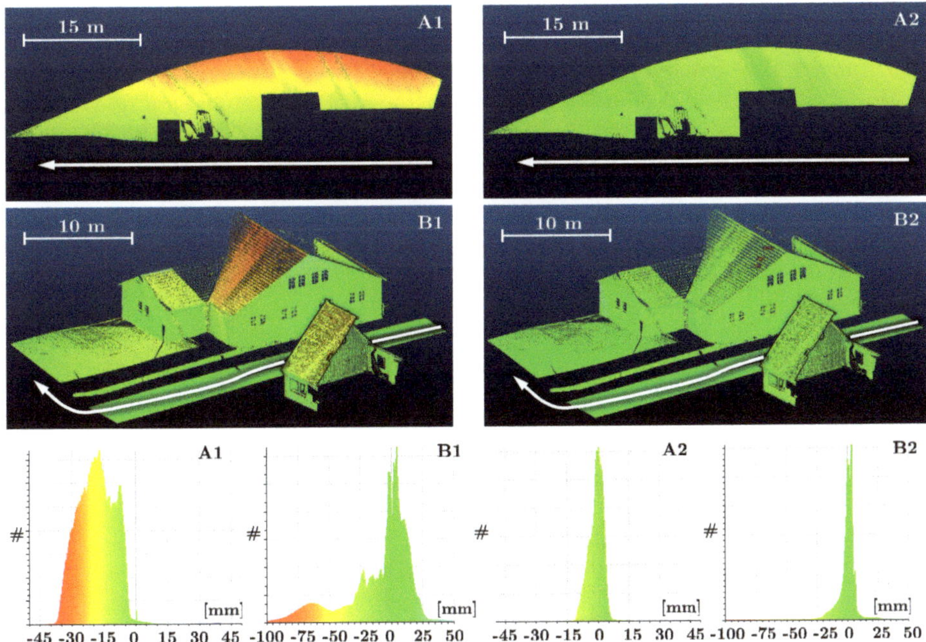

Figure 17. M3C2 comparison between point clouds of the mobile laser scanning system and TLS reference point clouds by using calibration parameters taken from the construction plan (A1, B1, **left column**) and from the plane-based calibration field (A2, B2, **right column**).

However, the area-based evaluation strategy also carries huge potential for an examination of mobile laser scanning systems. In order to demonstrate this, Figure 17 shows the scatter plots and histograms of the M3C2 point cloud comparison at both test sites A and B for the measurements that are marked with an asterisk (*) in Table 5. A1 and B1 correspond to the approximate calibration taken from the construction plan, and A2 and B2 correspond to the plane-based calibration. At test site A, the mobile laser scanning system was moved parallel to a wall (Figure 17, top row). This wall is also shown in Figure 14. When using the approximate calibration parameters (A1), deviations of up to -40 mm are visible increasing from the bottom up. This pattern indicates a tilting error of the point cloud of the mobile laser scanning system. When using the plane-based calibration (A2), such deviations are vanished. The improvement is also visible in the related histograms. The distribution of A2 is considerably smaller and more unbiased in comparison to A1. When having a closer look to the calibration parameters, which are stated at the beginning of Table 5, we can see that there is a big difference of $0.1420°$ for the boresight angle α. This calibration parameter is mostly aligned with the roll axis of the mobile platform. Hence, the tilting error of the wall in A1 can probably be traced back to an erroneous boresight angle α. This example demonstrates that certain errors can be revealed by an area-based evaluation, when using an appropriate object in combination with a specific driving maneuver. This strategy was already proposed in previous publications [34,42–44]. However, it is difficult to unambiguously trace back deviations to its origin. For instance, the pattern in A1 could also be caused by an error in the roll angle. In order to clarify this, it would be conceivable to frequently repeat such measurements within a certain period of time making the assumptions that a rolling error is changing, but the calibration error is constant.

Test site A is not sensitive for all kinds of errors, e.g., height errors or errors parallel to the wall. Thus, we extended our investigations to test site B, where the mobile laser scanning system passes a heterogeneous scene with vertical, horizontal, and inclined surfaces (Figure 17, middle row). Test site B is more sensitive to different types of errors and, thus, more adequate to evaluate the mobile laser scanning system. Clearly, the erroneous boresight angle α can be detected at the rooftop (B1). When using the plane-based calibration (B2), the deviations have a mean value of -0.3 mm and a standard deviation of 7.6 mm. Due to the diversity of the scene, such results indicate a satisfactory quality of the calibration parameters. As indicated by the numbers in Table 5 and the histograms in Figure 17, the deviations are not perfectly Gaussian distributed and not completely unbiased, even after applying the plane-based calibration. This is due to the error characteristics of the mobile laser scanning system, which is mostly affected by systematic errors of the trajectory estimation. However, the 2D laser scanner and the M3C2 algorithm also contribute to the error budget.

8. Conclusions and Outlook

This paper presents the design and evaluation of a plane-based calibration field for determining the lever arm and boresight angles of a 2D laser scanner w.r.t. a GNSS/IMU unit on a mobile platform. In addition to this, the calibration field is sensitive for the estimation of the range finder offset of the 2D laser scanner. The calibration field was designed on the basis of a geodetic configuration analysis and Monte Carlo simulations. In this respect, the impact of random, systematic, and gross observation errors on the calibration was analyzed leading to a plane setup that provides accurate and controlled calibration parameters with a standard deviation of $\leq 1 \dots 1.5$ mm for the lever arm components and $\leq 0.005°$ for the boresight angles. This was empirically verified by calibration measurements with our own mobile laser scanning system. Using tools from geodetic configuration analysis to analyze both the accuracy and the controllability of the parameter estimation process in the context of calibrating mobile laser scanning systems has only been addressed to a limited extent so far and is the major scientific contribution of this work. The plane-based calibration field was realized outdoors being permanent, stable, weather-resistant, and cost-effective. The associated calibration procedure in the calibration field takes less than one minute and, thus, can be repeated frequently. In addition to the configuration analysis, repetitive calibrations also increase the controllability of the calibration parameters and allow for a realistic empirical quantification of their accuracy and stability.

In order to evaluate the mobile laser scanning system and the calibration, a dense network of control points and TLS reference point clouds of diverse building structures was installed. These facilities allow for a point-based as well as an area-based evaluation of the overall performance of mobile laser scanning systems. We could demonstrate that the TLS reference point clouds can also partially be utilized to evaluate individual components of the system. Due to the complexity of the systems, however, the evaluation of individual components is challenging. Moreover, the rigorous modeling of the system accuracy by considering an error budget for each component is hardly feasible. Therefore, we argue that an empirical evaluation of the overall system, as conducted within this work, is the most effective strategy. According to our tests, both the point-based and the area-based evaluation indicate that the accuracy of our mobile laser scanning system can be specified with an RMS of <10 mm for the east, the north, and the height component, separately. This accuracy applies to the measuring conditions in our evaluation environment and might change for other fields of applications. Please note that the obtained evaluation results for our mobile laser scanning system are in good accordance with the results that were obtained in a prior case study on a motorway [59].

In the future, we plan to calibrate and evaluate systems other than the one used within this work using our facilities. Moreover, initial tests have been accomplished regarding the expansion of the plane-based calibration field to extrinsic camera calibration (e.g., [87–89]).

Author Contributions: E.H. and C.H. designed the plane-based calibration field; E.H., L.K., and H.K. installed the plane-based calibration field and the evaluation environment; E.H. performed the experiments; E.H. and L.K. analyzed the data; E.H. wrote the paper; All authors have read and agreed to the published version of the manuscript.

Funding: This research was partially funded by the Deutsche Forschungsgemeinschaft (DFG, German Research Foundation) under Germany's Excellence Strategy - EXC 2070 – 390732324.

Conflicts of Interest: The authors declare no conflict of interest.

References

1. Williams, K.; Olsen, M.J.; Gene, V.R.; Glennie, C. Synthesis of Transportation Applications of Mobile LIDAR. *Remote Sens.* **2013**, *5*, 4652–4692. [CrossRef]
2. Guan, H.; Li, J.; Cao, S.; Yu, Y. Use of mobile LiDAR in road information inventory: A review. *Int. J. Image Data Fusion* **2016**, *7*, 219–242. [CrossRef]
3. Gargoum, S.A.; El Basyouny, K. A literature synthesis of LiDAR applications in transportation: Feature extraction and geometric assessments of highways. *GISci. Remote Sens.* **2019**, *56*, 864–893. [CrossRef]
4. Soilán, M.; Sánchez-Rodríguez, A.; del Río-Barral, P.; Perez-Collazo, C.; Arias, P.; Riveiro, B. Review of Laser Scanning Technologies and Their Applications for Road and Railway Infrastructure Monitoring. *Infrastructures* **2019**, *4*, 58. [CrossRef]
5. Wang, Y.; Chen, Q.; Zhu, Q.; Liu, L.; Li, C.; Zheng, D. A Survey of Mobile Laser Scanning Applications and Key Techniques over Urban Areas. *Remote Sens.* **2019**, *11*, 1540. [CrossRef]
6. Puente, I.; González-Jorge, H.; Martínez-Sánchez, J.; Arias, P. Review of mobile mapping and surveying technologies. *Measurement* **2013**, *46*, 2127–2145. [CrossRef]
7. Strübing, T.; Neumann, I. Positions- und Orientierungsschätzung von LIDAR-Sensoren auf Multisensorplattformen. *Z. Für Geodäsie Geoinf. Und Landmanag. (ZfV)* **2013**, *138*, 210–221.
8. Filin, S. Recovery of Systematic Biases in Laser Altimetry Data Using Natural Surfaces. *Photogramm. Eng. Remote Sens.* **2003**, *69*, 1235–1242. [CrossRef]
9. Lu, X.; Feng, C.; Ma, Y.; Yang, F.; Shi, B.; Su, D. Calibration method of rotation and displacement systematic error for ship-borne mobile surveying systems. *Surv. Rev.* **2017**, *51*, 78–86. [CrossRef]
10. Heinz, E.; Eling, C.; Wieland, M.; Klingbeil, L.; Kuhlmann, H. Development, Calibration and Evaluation of a Portable and Direct Georeferenced Laser Scanning System for Kinematic 3D Mapping. *J. Appl. Geod.* **2015**, *9*, 227–243. [CrossRef]
11. Heinz, E.; Eling, C.; Wieland, M.; Klingbeil, L.; Kuhlmann, H. Analysis of different reference plane setups for the calibration of a mobile laser scanning system. In *Ingenieurvermessung 17, Beiträge zum 18. Internationalen Ingenieurvermessungskurs, Graz, Österreich*; Lienhart, W., Ed.; Wichmann Verlag: Berlin, Germany, 2017; pp. 131–145.
12. Hong, S.; Park, I.; Lee, J.; Lim, K.; Choi, Y.; Sohn, H.G. Utilization of a Terrestrial Laser Scanner for the Calibration of Mobile Mapping Systems. *Sensors* **2017**, *17*, 474. [CrossRef] [PubMed]

13. Hartmann, J.; Paffenholz, J.A.; Strübing, T.; Neumann, I. Determination of Position and Orientation of LiDAR Sensors on Multisensor Platforms. *J. Surv. Eng.* **2017**, *143*. [CrossRef]

14. Niemeier, W. *Ausgleichungsrechnung—Statistische Auswertemethoden (2., überarbeitete und erweiterte Auflage)*; de Gruyter: Berlin, Germany; New York, NY, USA, 2008.

15. Förstner, W.; Wrobel, B. *Photogrammetric Computer Vision—Statistics, Geometry, Orientation and Reconstruction*; Springer International Publishing: Cham, Switzerland, 2016.

16. Glennie, C. Rigorous 3D error analysis of kinematic scanning LIDAR systems. *J. Appl. Geod.* **2007**, *1*, 147–157. [CrossRef]

17. Mezian, M.; Vallet, B.; Soheilian, B.; Paparoditis, N. Uncertainty Propagation For Terrestrial Mobile Laser Scanner. *Int. Arch. Photogramm. Remote Sens. Spat. Inf. Sci.* **2016**, *41*, 331–335. [CrossRef]

18. Hauser, D.; Glennie, C.; Brooks, B. Calibration and Accuracy Analysis of a Low-Cost Mapping-Grade Mobile Laser Scanning System. *J. Surv. Eng.* **2016**, *142*, 04016011. [CrossRef]

19. Barber, D.; Mills, J.; Smith-Voysey, S. Geometric validation of a ground-based mobile laser scanning system. *ISPRS J. Photogramm. Remote Sens.* **2008**, *63*, 128–141. [CrossRef]

20. Kaartinen, H.; Hyyppä, J.; Kukko, A.; Jaakkola, A.; Hyyppä, H. Benchmarking the Performance of Mobile Laser Scanning Systems Using a Permanent Test Field. *Sensors* **2012**, *12*, 12814–12835. [CrossRef]

21. Hofmann, S.; Brenner, C. Accuracy assessment of Mobile Mapping Point Clouds Using the Existing Environment as Terrestrial Reference. *Int. Arch. Photogramm. Remote Sens. Spat. Inf. Sci.-ISPRS Arch.* **2016**, *41*, 601–608. [CrossRef]

22. Teunissen, P.J.G.; Montenbruck, O. (Eds.) *Springer Handbook of Global Navigation Satellite Systems*; Springer International Publishing: Cham, Switzerland, 2017.

23. Groves, P.D. *Principles of GNSS, Inertial, and Multisensor Integrated Navigation Systems*, 2nd ed.; Artech House: Boston, MA, USA; London, UK, 2013.

24. Vosselman, G.; Maas, H.G. *Airborne and Terrestrial Laser Scanning*; CRC Press, Whittles Publishing: Dunbeath, Scotland, 2010; ISBN 9781439827987.

25. Holst, C.; Neuner, H.; Wieser, A.; Wunderlich, T.; Kuhlmann, H. Calibration of Terrestrial Laser Scanners. *Allg. Vermess.-Nachrichten (AVN)* **2016**, *123*, 147–157.

26. Gräfe, G. Kinematische Anwendungen von Laserscannern im Straßenraum. Ph.D. Thesis, Gottfried Wilhelm Leibniz Universität Hannover, Fakultät für Bauingenieurwesen und Geodäsie, Hanover, Germany, 2007.

27. Brüggemann, T.; Artz, T.; Weiß, R. Kalibrierung von Multisensorsystemen. In *Schriftenreihe des DVW, Band 91, Hydrographie 2018 – Trend zu Unbemannten Messsystemen*; Wißner Verlag: Augsburg, Germany, 2018; pp. 29–46.

28. Hesse, C. Hochauflösende kinematische Objekterfassung mit terrestrischen Laserscannern. Ph.D. Thesis, Gottfried Wilhelm Leibniz Universität Hannover, Fakultät für Bauingenieurwesen und Geodäsie, Hanover, Germany, 2007.

29. Vennegeerts, H. Objektraumgestützte kinematische Georeferenzierung für Mobile-Mapping-Systeme. Ph.D. Thesis, Gottfried Wilhelm Leibniz Universität Hannover, Fakultät für Bauingenieurwesen und Geodäsie, Hanover, Germany, 2011.

30. Eling, C.; Klingbeil, L.; Wieland, M.; Kuhlmann, H. Direct Georeferencing of Micro Aerial Vehicles - System Design, System Calibration and First Evaluation Tests. *Photogramm. Fernerkund. Geoinf. (PFG)* **2014**, *2014*, 227–237. [CrossRef]

31. Talaya, J.; Alamus, R.; Bosch, E.; Serra, A.; Kornus, W.; Baron, A. Integration of a terrestrial laser scanner with GPS/IMU orientation sensors. In Proceedings of the XXth ISPRS Congress, Istanbul, Turkey, 12–23 July 2004.

32. Sheehan, M.; Harrison, A.; Newman, P. Self-calibration for a 3D laser. *Int. J. Robot. Res.* **2011**, *31*, 675–687. [CrossRef]

33. Elseberg, J.; Borrmann, D.; Nüchter, A. Algorithmic Solutions for Computing Precise Maximum Likelihood 3D Point Clouds from Mobile Laser Scanning Platforms. *Remote Sens.* **2013**, *5*, 5871–5906. [CrossRef]

34. Keller, F. Entwicklung eines forschungsorientierten Multi-Sensor-Systems zum kinematischen Laserscanning innerhalb von Gebäuden. Ph.D. Thesis, HafenCity Universität Hamburg, Arbeitsgebiet Ingenieurgeodäsie und geodätische Messtechnik, Hamburg, Germany, 2015.

35. Nüchter, A.; Borrmann, D.; Elseberg, J.; Redondo, D. A Backpack-mounted 3D Mobile Scanning System. *Allg. Vermess.-Nachrichten (AVN)* **2015**, *122*, 301–307.

36. Nüchter, A.; Borrmann, D.; Koch, P.; Kühn, M.; May, S. A Man-Portable, IMU-Free Mobile Mapping System. In Proceedings of the ISPRS Annals of the Photogrammetry, Remote Sensing and Spatial Information Sciences (ISPRS Geospatial Week 2015), La Grande Motte, France, 28 September–3 October 2015; Volume II-3/W5.

37. Hillemann, M.; Weinmann, M.; Mueller, M.S.; Jutzi, B. Automatic Extrinsic Self-Calibration of Mobile Mapping Systems Based on Geometric 3D Features. *Remote Sens.* **2019**, *11*, 1955. [CrossRef]

38. Maddern, W.; Harrison, A.; Newman, P. Lost in Translation (and Rotation): Rapid Extrinsic Calibration for 2D and 3D LIDARs. In Proceedings of the IEEE International Conference on Robotics and Automation (ICRA), Saint Paul, MN, USA, 14–19 May 2012 2012.

39. Maddern, W.; Pascoe, G.; Linegar, C.; Newman, P. 1 year, 1000 km: The Oxford RobotCar dataset. *Int. J. Robot. Res.* **2017**, *36*, 3–15. [CrossRef]

40. Hillemann, M.; Meidow, J.; Jutzi, B. Impact of different trajectories on extrinsic self-calibration for vehicle-based mobile laser scanning systems. In Proceedings of the International Archives of the Photogrammetry, Remote Sensing and Spatial Information Sciences, PIA19+MRSS19 - Photogrammetric Image Analysis & Munich Remote Sensing Symposium, Munich, Germany, 18–20 September 2019; Volume XLII-2/W16.

41. Levinson, J.; Thrun, S. Unsupervised Calibration for Multi-beam Lasers. In *Experimental Robotics. Springer Tracts in Advanced Robotics*; Khatib, O., Kumar, V., Sukhatme, G., Eds.; Springer: Berlin/Heidelberg, Germany, 2014; Volume 79, pp. 179–193.

42. Keller, F.; Sternberg, H. Multi-Sensor Platform for Indoor Mobile Mapping: System Calibration and Using a Total Station for Indoor Applications. *Remote Sens.* **2013**, *5*, 5805–5824. [CrossRef]

43. Sternberg, H.; Keller, F.; Willemsen, T. Precise indoor mapping as a basis for coarse indoor navigation. *J. Appl. Geod.* **2013**, *7*, 231–246. [CrossRef]

44. Li, Z.; Tan, J.; Liu, H. Rigorous Boresight Self-Calibration of Mobile and UAV LiDAR Scanning Systems by Strip Adjustment. *Remote Sens.* **2019**, *1*, 442. [CrossRef]

45. Friess, P. Toward a rigorous methodology for airborne laser mapping. In Proceedings of the International Calibration and Validation Workshop EuroCOW, Castelldefels, Spain, 25–27 January 2006.

46. Skaloud, J.; Lichti, D. Rigorous approach to bore-sight self-calibration in airborne laser scanning. *ISPRS J. Photogramm. Remote Sens.* **2006**, *61*, 47–59. [CrossRef]

47. Lindenthal, S.M.; Ussyshkin, V.R.; Wang, J.G.; Pokorny, M. Airborne LIDAR: A fully-automated self-calibration procedure. In Proceedings of the International Archives of the Photogrammetry, Remote Sensing and Spatial Information Sciences (ISPRS Calgary 2011 Workshop), Calgary, AB, Canada, 29–31 August 2011; Volume XXXVIII-5/W12.

48. Ravi, R.; Shamseldin, T.; Elbahnasawy, M.; Lin, Y.J.; Habib, A. Bias Impact Analysis and Calibration of UAV-Based Mobile LiDAR System With Spinning Multi-Beam Laser Scanner. *Appl. Sci.* **2018**, *8*, 297. [CrossRef]

49. Keyetieu, R.; Seube, N. Automatic Data Selection and Boresight Adjustment of LiDAR Systems. *Remote Sens.* **2019**, *11*, 1087. [CrossRef]

50. Rieger, P.; Studnicka, N.; Pfennigbauer, M.; Zach, G. Boresight alignment method for mobile laser scanning systems. *J. Appl. Geod.* **2010**, *4*, 13–21. [CrossRef]

51. Glennie, C. Calibration and Kinematic Analysis of the Velodyne HDL-64E S2 Lidar Sensor. *Photogramm. Eng. Remote Sens.* **2012**, *78*, 339–347. [CrossRef]

52. Chan, T.O.; Licht, D.D.; L., G.C. Multi-feature based boresight self-calibration of a terrestrial mobile mapping system. *ISPRS J. Photogramm. Remote Sens.* **2013**, *82*, 112–124. [CrossRef]

53. Hartmann, J.; von Gösseln, I.; Schild, N.; Dorndorf, A.; Paffenholz, J.A.; Neumann, I. Optimisation of the calibration process of a k-tls based multi-sensor-system by genetic algorithms. In Proceedings of the International Archives of the Photogrammetry, Remote Sensing and Spatial Information Sciences (2019 ISPRS Geospatial Week 2019), Enschede, The Netherlands, 10–14 June 2019; Volume XLII-2/W13, pp. 1655–1662.

54. Chen, S.; Liu, J.; Wu, T.; Huang, W.; Liu, K.; Yin, D.; Liang, X.; Hyyppä, J.; R., C. Extrinsic Calibration of 2D Laser Rangefinders Based on a Mobile Sphere. *Remote Sens.* **2018**, *10*, 1176. [CrossRef]

55. Vennegeerts, H.; Martin, J.; Becker, M.; Kutterer, H. Validation of a kinematic laserscanning system. *J. Appl. Geod.* **2008**, *2*, 79–84. [CrossRef]

56. Kukko, A.; Kaartinen, H.; Hyyppä, J.; Chen, Y. Multiplatform Mobile Laser Scanning: Usability and Performance. *Sensors* **2012**, *12*, 11712–11733. [CrossRef]

57. Schlichting, A.; Brenner, C.; Schön, S. Bewertung von Inertial/GNSS-Modulen mittels Laserscannern und bekannter Landmarken. *Photogramm. Fernerkundung Geoinf. (PFG)* **2014**, *2014*, 5–15. [CrossRef]

58. Mao, Q.; Zhang, L.; Li, Q.; Hu, Q.; Yu, J.; Feng, S.; Ochieng, W.; Gong, H. A Least Squares Collocation Method for Accuracy Improvement of Mobile LiDAR Systems. *Remote Sens.* **2015**, *7*, 7402–7424. [CrossRef]

59. Heinz, E.; Eling, C.; Klingbeil, L.; Kuhlmann, H. On the applicability of a scan-based mobile mapping system for monitoring the planarity and subsidence of road surfaces—Pilot study on the A44n motorway in Germany. *J. Appl. Geod.* **2020**, *14*, 39–54. [CrossRef]

60. Haala, N.; Peter, M.; Kremer, J.; Hunter, G. Mobile LIDAR mapping for 3D point cloud collection in urban areas—A performance test. In Proceedings of the ISPRS Archives—Volume XXXVII Part B5, XXIst ISPRS Congress, Beijing, China, 3–11 July 2008; pp. 1119–1124.

61. Bureick, J.; Vogel, S.; Neumann, I.; Unger, J.; Alkhatib, H. Georeferencing of an Unmanned Aerial System by Means of an Iterated Extended Kalman Filter Using a 3D City Model. *PFG - J. Photogramm. Remote Sens. Geoinf. Sci.* **2019**. [CrossRef]

62. Dehbi, Y.; Lucks, L.; Behmann, J.; L., K.; Plümer, L. Improving GPS Trajectories Using 3D City Models and Kinematic Point Clouds. In Proceedings of the 4th International Conference on Smart Data and Smart Cities, ISPRS Annals of the Photogrammetry, Remote Sensing and Spatial Information Sciences, Kuala Lumpur, Malaysia, 1–3 October 2019; Volume IV-4/W9, pp. 35–42.

63. Toschi, I.; Rodríguez-Gonzálvez, P.; Remondino, F.; Minto, S.; Orlandini, S.; Fuller, A. Accuracy Evaluation of a Mobile Mapping System with Advanced Statistical Methods. In Proceedings of the 3D Virtual Reconstruction and Visualization of Complex Architectures, Avila, Spain, 25–27 February 2015; Volume XL-5/W4, pp. 245–253.

64. Hartmann, J.; Trusheim, P.; Alkhatib, H.; Paffenholz, J.A.; Diener, D.; Neumann, I. High Accurate Pointwise (Geo-)Referencing of a k-TLS Based Multi-Sensor-System. In Proceedings of the 2018 ISPRS TC IV Mid-Term Symposium 3D Spatial Information Science—The Engine of Change, Delft, The Netherlands, 1–5 October 2018.

65. Tucci, G.; Visintini, D.; Bonora, V.; Parisi, E.I. Examination of Indoor Mobile Mapping Systems in a Diversified Internal/External Test Field. *Appl. Sci.* **2018**, *8*, 401. [CrossRef]

66. Kalenjuk, S.; Rebhan, M.J.; Lienhart, W.; Marte, R. Large-scale monitoring of retaining structures: New approaches on the safety assessment of retaining structures using mobile mapping. In *Proceedings SPIE, Sensors and Smart Structures Technologies for Civil, Mechanical and Aerospace Systems 2019*; International Society for Optics and Photonics: Bellingham, WA, USA, 2019; Volume 10970.

67. IMAR Navigation GmbH. Inertial Navigation System iNAV-FJI-LSURV. Technical Report. Available online: http://www.imar.de/index.php/en/products/by-product-names (accessed on 6 February 2020).

68. NovAtel Inc. Waypoint Inertial Explorer 8.80 Post Processing Software. 2019. Available online: http://www2.novatel.com/waypointrelease (accessed on 19 October 2019).

69. Zoller & Fröhlich GmbH. Z+F Profiler 9012A, 2D Laser Scanner. Technical report. Available online: http://www.zf-laser.com (accessed on 6 February 2020).

70. Heinz, E.; Mettenleiter, M.; Kuhlmann, H.; Holst, C. Strategy for Determining the Stochastic Distance Characteristics of the 2D Laser Scanner Z+F Profiler 9012A with Special Focus on the Close Range. *Sensors* **2018**, *18*, 2253. [CrossRef] [PubMed]

71. Holst, C.; Artz, T.; Kuhlmann, H. Biased and unbiased estimates based on laser scans of surfaces with unknown deformations. *J. Appl. Geod.* **2014**, *8*, 169–184. [CrossRef]

72. Fischler, M.A.; Bolles, R.C. Random sample consensus: A paradigm for model fitting with applications to image analysis and automated cartography. *Commun. ACM* **1981**, *24*, 381–395. [CrossRef]

73. Förstner, W. Ein Verfahren zur Schätzung von Varianz- und Kovarianzkomponenten. *Allg. Vermess.-Nachrichten (AVN)* **1979**, *86*, 446–453.

74. Förstner, W. Reliability Analysis of Parameter Estimation in Linear Models with Applications to Mensuration Problems in Computer Vision. *Comput. Vis. Graph. Image Process.* **1987**, *40*, 273–310. [CrossRef]

75. Baarda, W. *Statistical Concepts in Geodesy*; Netherlands Geodetic Commission, Publications on Geodesy, New Series; Netherlands Geodetic Commission: Delft, The Netherlands, 1967; Volume 2, Number 4.

76. Baarda, W. *A Testing Procedure for Use in Geodetic Networks*; Netherlands Geodetic Commission, Publications on Geodesy, New Series; Netherlands Geodetic Commission: Delft, The Netherlands, 1968; Volume 2, Number 5.

77. Medić, T.; Kuhlmann, H.; Holst, C. Designing and Evaluating a User-Oriented Calibration Field for the Target-Based Self-Calibration of Panoramic Terrestrial Laser Scanners. *Remote Sens.* **12**, 2020, 15. [CrossRef]
78. Dupuis, J.; Holst, C.; Kuhlmann, H. Improving the Kinematic Calibration of a Coordinate Measuring Arm using Configuration Analysis. *Precis. Eng.* **2017**, *50*, 171–182. [CrossRef]
79. Leek, J.; Artz, T.; Nothnagel, A. Optimized scheduling of VLBI UT1 intensive sessions for twin telescopes employing impact factor analysis. *J. Geod.* **2015**, *89*, 911–924. [CrossRef]
80. Robotics, C. V-REP—Virtual Robot Experimentation Platform. Technical Report. Available online: http://www.coppeliarobotics.com/ (accessed on 2 October 2019).
81. Holst, C.; Medic, T.; Kuhlmann, H. Dealing with systematic laser scanner errors due to misalignment at area-based deformation analyses. *J. Appl. Geod.* **2018**, *12*, 169–185. [CrossRef]
82. Janßen, J.; Medic, T.; Kuhlmann, H.; Holst, C. Decreasing the Uncertainty of the Target Center Estimation at Terrestrial Laser Scanning by Choosing the Best Algorithm and by Improving the Target Design. *Remote Sens.* **2019**, *2019*, 845; doi:10.3390/rs11070845. [CrossRef]
83. Bundesamt für Kartographie und Geodäsie (BKG). *Quasigeoid der Bundesrepublik Deutschland—GCG2016 (German Combined QuasiGeoid 2016)*; Technical report; Available online: https://sg.geodatenzentrum.de/web_public/gdz/dokumentation/deu/quasigeoid.pdf (accessed on 6 February 2020).
84. Heunecke, O.; Kuhlmann, H.; Welsch, W.; Eichhorn, A.; Neuner, H. *Handbuch Ingenieurgeodäsie: Auswertung geodätischer Überwachungsmessungen (2., neu bearbeitete und erweiterte Auflage)*; Wichmann Verlag: Berlin/Offenbach am Main, Germany, 2013.
85. Lague, D.; Brodu, N.; Leroux, J. Accurate 3D comparison of complex topography with terrestrial laser scanner: Application to the Rangitikei canyon (N-Z). *ISPRS J. Photogramm. Remote Sens.* **2013**, *82*, 10–26. [CrossRef]
86. Cloud Compare. 3D Point Cloud and Mesh Processing Software—Open Source Project. Technical report. Available online: https://www.danielgm.net/cc/ (accessed on 6 February 2020).
87. Zhang, Q.; Pless, R. Extrinsic Calibration of a Camera and Laser Range Finder (improves camera calibration). In Proceedings of the IEEE/RSJ International Conference on Intelligent Robots and Systems (IROS), Sendai, Japan, 28 September–2 October 2004; pp. 2301–2306.
88. Unnikrishnan, R.; Hebert, M. *Fast Extrinsic Calibration of a Laser Rangefinder to a Camera*; Technical report; Robotics Institute, Carnegie Mellon University: Pittsburgh, PA, USA, 2005.
89. Geiger, A.; Moosmann, F.; Car, O.; Schuster, B. Automatic Camera and Range Sensor Calibration using a single Shot. In Proceedings of the IEEE International Conference on Robotics and Automation (ICRA), Saint Paul, MN, USA, 14–18 May 2012; pp. 3936–3943.

Article

Ideal Angular Orientation of Selected 64-Channel Multi Beam Lidars for Mobile Mapping Systems

Bashar Alsadik

Faculty of Geo-Information Science and Earth Observation (ITC), University of Twente, 7500 AE Enschede, The Netherlands; b.s.a.alsadik@utwente.nl

Received: 13 January 2020; Accepted: 4 February 2020; Published: 5 February 2020

Abstract: Lidar technology is thriving nowadays for different applications mainly for autonomous navigation, mapping, and smart city technology. Lidars vary in different aspects and can be: multi beam, single beam, spinning, solid state, full 360 field of view FOV, single or multi pulse returns, and many other geometric and radiometric aspects. Users and developers in the mapping industry are continuously looking for new released Lidars having high properties of output density, coverage, and accuracy while keeping a lower cost. Accordingly, every Lidar type should be well evaluated for the final intended mapping aim. This evaluation is not easy to implement in practice because of the need to have all the investigated Lidars available in hand and integrated into a ready to use mapping system. Furthermore, to have a fair comparison; it is necessary to ensure the test applied in the same environment at the same travelling path among other conditions. In this paper, we are evaluating two state-of-the-art multi beam Lidar types: Ouster OS-1-64 and Hesai Pandar64 for mapping applications. The evaluation of the Lidar types is applied in a simulation environment which approximates reality. The paper shows the determination of the ideal orientation angle for the two Lidars by assessing the density, coverage, and accuracy and presenting clear performance quantifications and conclusions.

Keywords: Lidar; point cloud density; point cloud coverage; mobile mapping systems; 3D simulation; Pandar64; Ouster OS-1-64

1. Introduction

Currently, several Lidar manufacturers are found around the world where their business is growing with the high demand for the Lidar products in different industries. The major industry pushing Lidar companies for continuing development and being competitive is the thriving autonomous navigation/driving industry [1–4]. On the other hand, mapping companies are benefiting from this continuous development to use the Lidars as active scanning devices from the ground based or aerial based systems.

Lidar available types are varied in many geometric and radiometric aspects and this has an impact on the final aim of application and the selling prices as well. The most common Lidar types are multi beam time-of-flight TOF scanners graduating from 8 up to 128 beams, which are spinning devices at a high rotation speed of 5–30 Hz [5–8]. Solid state Lidars are also available with a limited field of view FOV like the Luminar H-series type [9] and Blickfeld [10]. Furthermore, single beam Lidars are available in the market with high geometric properties of range accuracy, beam divergence, angular resolution, etc., such as Riegel VUX H [11].

Since mobile mapping applications have different requirements than autonomous navigation, the orientation of the Lidar is not necessarily posed vertically on its base on the car without tilt. Accordingly, to have a better density and coverage of the scanned features such as road surfaces, it is necessary to tilt the Lidar in a mobile mapping system (MMS).

However, finding the ideal orientation angle of the Lidar device that fits the mapping requirements is not an easy task especially with the variety of the offered Lidar characteristics by the manufacturers [12,13]. Furthermore, the angular orientation determination is vital and more complicated when placing more than one Lidar in a mapping system to increase the production amount of data.

Confirming point clouds coverage and density are common quality assurance QA and quality control QC tasks in workflows that comprise the processing of Lidar data [14,15]. Thus, in this paper, the ideal orientation of the Lidar device is estimated based on the production efficiency in terms of density of points, coverage amount, and the predicted accuracy.

For mobile mapping and geospatial applications, the density of points on road surfaces is of a high demand [16,17] beside the density and coverage on off-ground features like building facades in the urban environments. The major mobile mapping applications for: engineering surveys, digital terrain modeling, clearance computations, road asset inventory, drainage analysis, virtual three-dimensional (3D) design, as-built documentation and structural inspection requires in general high point densities (>100 points in one squared meter pts/m^2), and high accuracies (<5 cm) [16]. Furthermore, the required mapping coverage is governed among many factors by the Lidar scanning range, FOV, and the number of the installed Lidar tools in the mobile mapping system.

It should be noted that the density, accuracy, and coverage characteristics sometimes contradict to each other. As an example, Lidars which have a maximum range up to 250 m can increase the coverage compared to Lidars with a maximum of 100 m scanning range. However, this result in a higher beam divergence and range uncertainty which should be counted.

Accordingly, the best orientation angle of the Lidar is the one producing a compromised higher number of points on the road surface and facades and ensures a high percentage of coverage and accuracy.

An investigation of two newly released state-of-the-art 64 channel Lidar types is applied to quantify the ideal angular orientation that fits for mobile mapping applications.

Both Lidar types are multi beam spinning Lidar types with 64 channels and represent a potential Lidar candidate for MMS especially with their reasonable size, weight, and cost. These two Lidars are lately released from different manufacturers in China and USA. The first Lidar is the Hesai Pandar 64 (Figure 1a) and the second Lidar is the Ouster OS-1-64 (Figure 1b).

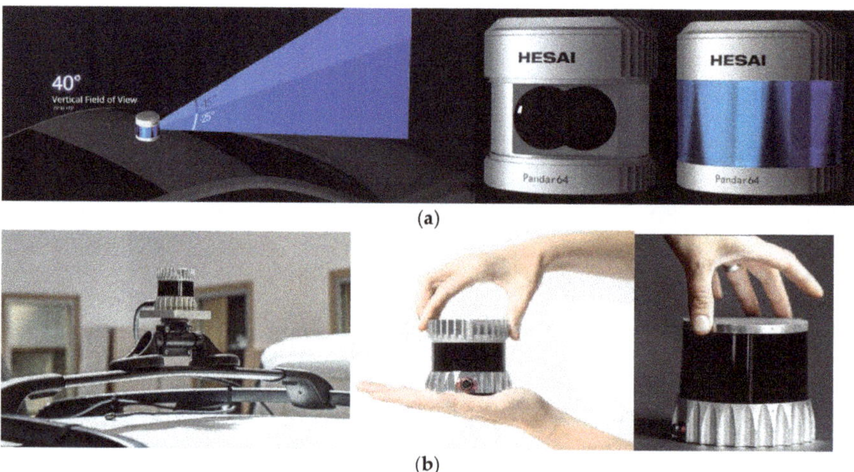

Figure 1. (**a**) Pandar64 [11]. (**b**) Ouster Lidar types [13,15].

Pandar64 has a longer maximum range of 200 m at 10% target reflectivity while having an irregular beams distribution or what may be called gradient distribution. On the other hand, Ouster OS-1-64 has a shorter range of 120 m @ 80% target reflectivity with linear distribution of beams and smaller beam divergence. The specifications of every Lidar type are summarized in Table 1, which shows that every type has its own advantages. Initially and based on the specifications, they are both suitable for high productivity mapping applications. A thorough investigation of their efficiency will be presented in Section 4.

Table 1. Specifications of the investigated Lidar types.

Lidar Sensor/System	Ouster	HESAI
Type/version	OS-1 -64	Pandar64
Max. Range	≤120 m@80%	≤200 m@10%
Range Accuracy 1σ	±3 cm	±2 cm
Beam divergence	0.13° (2.2 mrad)	0.177° (3 mrad)
Output rate pts/sec.	≈1.3 million	≈1.15 million
FOV—Vertical	≈33.2° (±16.6°)	≈40° (+15°: −25°)
Rotation rate	10–20 Hz	10–20 Hz
Vertical resolution	0.53°	0.167°
Horizontal resolution	0.35° at 20 Hz	0.4° at 20 Hz
Number of beams	64	64
Lidar mechanism	Spinning	Spinning
Distribution of beams	Linear	Gradient
Weight	396 g	1.5 kg
Power consumption	20 W	22 W
Estimated cost price	≅$12 k	≅$30 k

In Figure 2, the beam distribution pattern is given for both Lidars as published by their manufacturers [18,19]. Furthermore, to give the reader a visual impression about how the scans are expressed for each Lidar type, a simulation scanning space surrounded by walls is applied. Every Lidar type is placed at the center of the space, while the Lidar is placed on its base without a tilt in a stationary scan of one revolution.

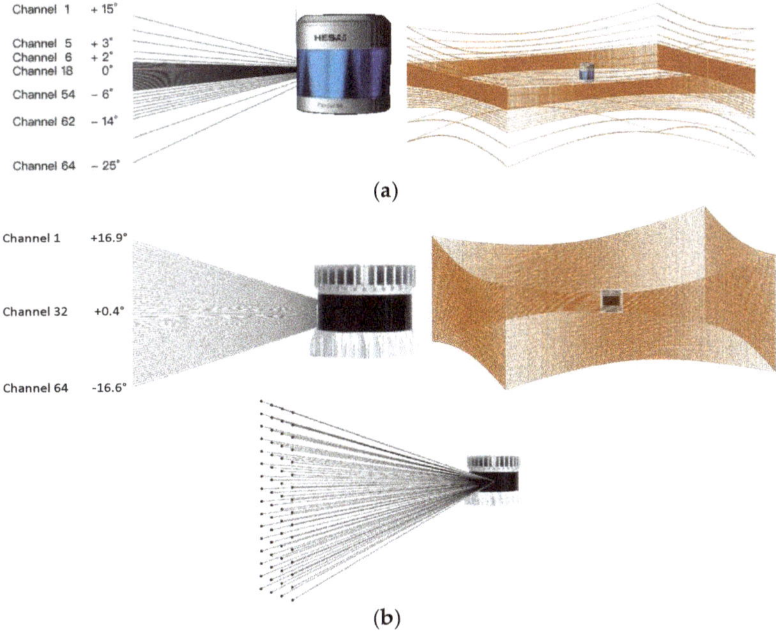

Figure 2. Scanning patterns of the two Lidar types. (**a**) Pandar64 Lidar. (**b**) OS-1-64 Lidar.

2. Methods

In this section, the method including evaluation parameters and simulation assumptions about the Lidar performance will be given. Namely, the angular orientations, scanning patterns, and the assessment of density, accuracy, and coverage. Figure 3 illustrates the methodology workflow. For both investigated Lidar types, the scanning routine is built based on the scanning properties given by the manufacturers. Then, a trajectory is defined in a 3D simulated urban model where a virtual mobile mapping system equipped with a single Lidar is mounted in a vehicle driving along the trajectory. It should be noted that the simulations are applied using the Blender tool [20].

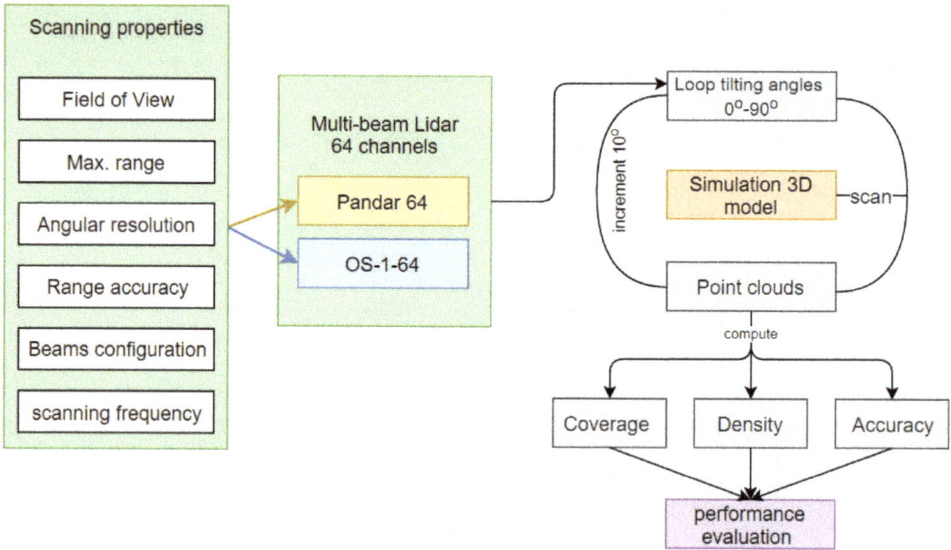

Figure 3. Methodology workflow.

In every run, a new tilting angle is initiated to orient the Lidar device and then a simulated scanning is applied. The produced point clouds in every angle set up is analyzed in terms of coverage, density, and accuracy and then a final performance evaluation is concluded.

2.1. Lidar Angular Set Up

In this research paper, the ideal angular orientation of the two Lidar types is investigated to determine their suitability for the mobile mapping applications. The investigated angular range of the Lidar is defined as shown in Figure 4.

As shown in Figure 4a, the 0° tilting angle (blue color) is defined when the Lidar is placed on its base, and then the Lidar scanning lines will constitute concentric circles on the ground (Figure 4b). This 0° Lidar orientation is the mostly used in autonomous driving. On the other hand, when the Lidar is rotated 90° (red in Figure 4a), the scans will hit the ground surface close to right angles and result in a parallel scan line pattern (Figure 4b).

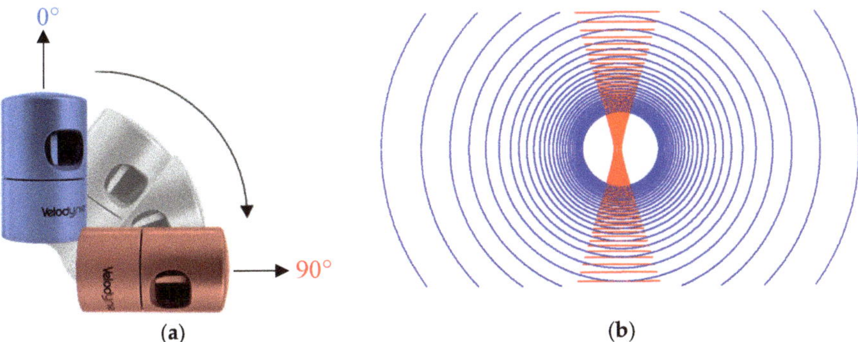

Figure 4. (**a**) Lidar tilt angles illustration. (**b**) Patterns of the Lidar beams at 0° (blue) and 90° (red) angles on a road surface from a top view.

2.2. Accuracy Assessment

Accuracy assessment is an important measure to check the quality of the scanned objects at the different tilting angles of the Lidar. This estimation of accuracy can be achieved by applying the Lidar error model of Equation (1) [21]:

$$P_i^w(t) = M_{INS}^w(t)P_i^{INS}(t) + T_{INS}^w(t) \tag{1}$$

where $P_i^w(t)$ represents the 3D coordinate of a point (X_i, Y_i, Z_i) in the world coordinate system at time t, $T_{INS}^w(t)$ is the position of the inertial navigation system INS in the world coordinate system at time t, $M_{INS}^w(t)$ is the rotation matrix between the INS body frame and the world coordinate system at time t, and $P_i^{INS}(t)$ is the position of the target point in the INS body frame at time t.

However, a pre-calibrated Lidar device is assumed in this study and as a result, errors in the lever arm offset, and boresight angles are assumed insignificant. Accordingly, the error propagation equation model is simplified where the position of every scanned point X_i, Y_i, Z_i at time t can be formulated as in Equation (2):

$$\begin{aligned}
X_i &= X_1 + R\,cos(Az)\,cos(V) \\
Y_i &= Y_1 + R\,sin(Az)\,cos(V) \\
Z_i &= Z_1 + R\,sin(V)
\end{aligned} \tag{2}$$

where R is the measured range distance from the Lidar to the object point P, Az is the measured azimuth angle of the laser beam, V is the vertical angle of the laser beam measured from the horizon, and X_1, Y_1, and Z_1 are the coordinates of the Lidar sensor at t.

It should be noted that the errors of the Lidar coordinates X_1, Y_1, Z_1 are predicted in normal cases from an integrated INS. If the accuracy in the navigated Lidar coordinates is not included in the model of Equation (2), then the estimated errors of the scanned points will represent the relative accuracy.

An error propagation can be applied using Jacobian Matrix J to estimate the errors of the scanned points in a point cloud as follows in Equations (3) and (4) [22]:

$$J = \begin{bmatrix} \frac{\partial Xi}{\partial Az} & \frac{\partial Xi}{\partial V} & \frac{\partial Xi}{\partial R} \\ \frac{\partial Yi}{\partial Az} & \frac{\partial Yi}{\partial V} & \frac{\partial Yi}{\partial R} \\ \frac{\partial Zi}{\partial Az} & \frac{\partial Zi}{\partial V} & \frac{\partial Zi}{\partial R} \end{bmatrix} = \begin{bmatrix} -R\,sin(A)cos(V) & -Rcos(A)sin(V) & cos(A)cos(V) \\ R\,cos(A)cos(V) & -Rsin(A)sin(V) & sin(A)cos(V) \\ 0 & Rcos(V) & sin(V) \end{bmatrix} \tag{3}$$

$$\sum_{XYZ} = J \sum_{R,Az,V} J^t = \begin{bmatrix} \sigma_{Xi}^2 & & \\ & \sigma_{Yi}^2 & \\ & & \sigma_{Zi}^2 \end{bmatrix} \tag{4}$$

where $\Sigma_{R,Az,V}$ is the variance-covariance matrix of the scanned range and angles to the point cloud, Σ_{XYZ} is the variance-covariance matrix of the derived coordinates of the point cloud, $\sigma_{Xi}, \sigma_{Yi}, \sigma_{Zi}$ are the standard deviations of point i coordinates, and t is the matrix transpose.

For illustration, exaggerated ellipsoids of errors are shown in Figure 5 for the scan sweep at 30° of a multi beam Lidar on the ground. Figure 5 illustrates how errors are larger for scanned ground points at longer scanning ranges away from the Lidar.

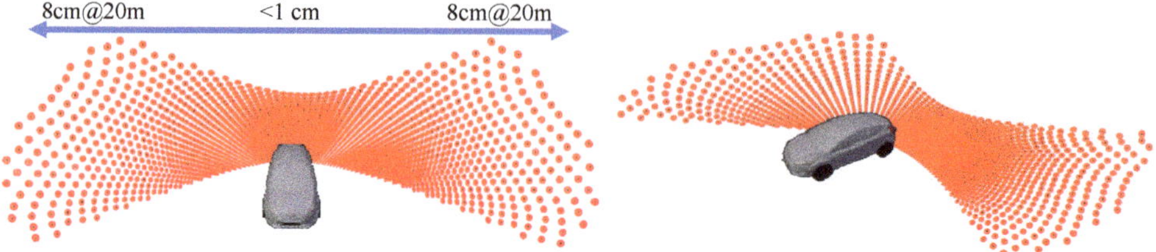

Figure 5. Errors representation of a scanned road surface by a single multi beam Lidar mounted at the rear part of a car roof.

2.3. Density Assessment

Computing the point density is one of the performance evaluation categories in this paper and it describes the number of points in one squared meter (pts/m^2). The density computation will be applied simply by finding the nearest neighbor points of every scanned point within a searching radius and sort their number as density.

As there will be a definite opportunity for having empty space between measurements, it is also possible to describe density by the point spacing [23]. Spacing between points can be calculated from the density as in Equation (5) [16]:

$$sample\ spacing = \sqrt{\frac{1}{point\ density}} \tag{5}$$

However, this simple equation assumes a uniform distribution of points which is not the case with the two Lidar types under investigation. Worth to mention is that in this paper, point clouds were simulated assuming one pulse return and this also complied with the US Geological Survey standards.

2.4. Coverage Assessment

Coverage or completeness is one of the important quality measures of point clouds. Although this is a difficult to define task for ground-based mapping systems, it is possible to assess in this paper, since the reference simulated urban 3D model is well defined.

The coverage percentage is computed by subtracting the scanned point cloud (blue color) from the reference model point cloud (gray color) as shown in Figure 6. First, the proximity between the scanned points and the reference evenly spaced points were computed. Then, only assigned points in the reference point cloud were considered for assessing the coverage against the whole reference point cloud.

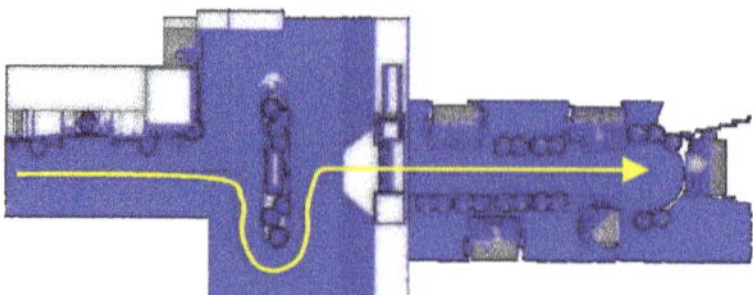

Figure 6. *Cont.*

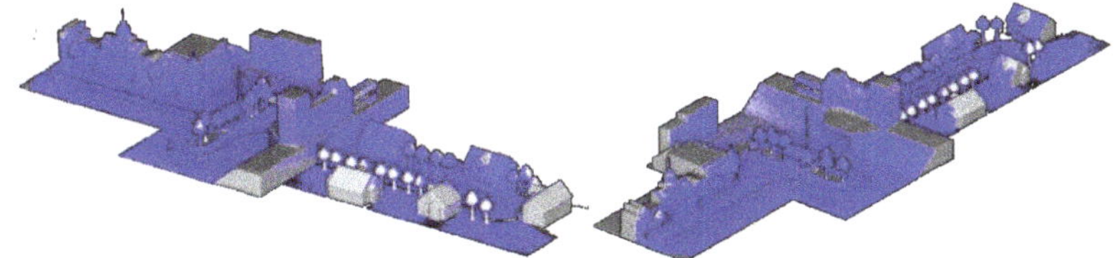

Figure 6. Coverage evaluation where blue color represents the scanned point cloud subtracted out of the simulated reference gray point cloud.

2.5. Scanning Pattern Assessment

To decide which orientation angels can be investigated in the experiment of the two Lidar types, we applied a scanning simulation at an angular range of 0°–90° at a 10° interval. This is simply simulated by placing every Lidar at 2 m height in an MMS where a revolution scan is applied to a planar ground and a façade (Figure 7).

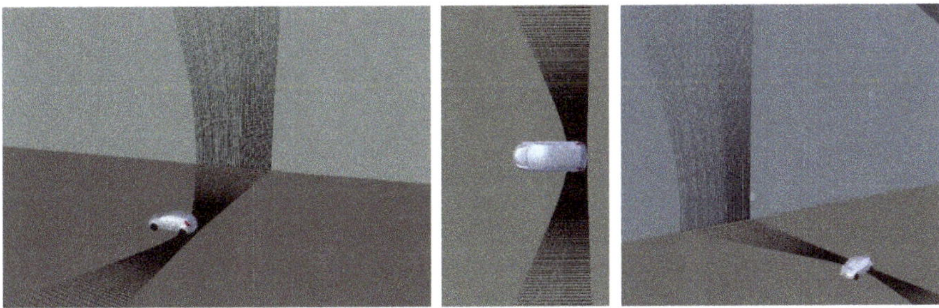

Figure 7. A simulated scan on the ground and aside facade.

The pattern of a stationary scanning sweep on the ground for the two Lidar types of the selected tilt angles from 0°–90° is shown in (Figure 8) with a scanning range of 20 m.

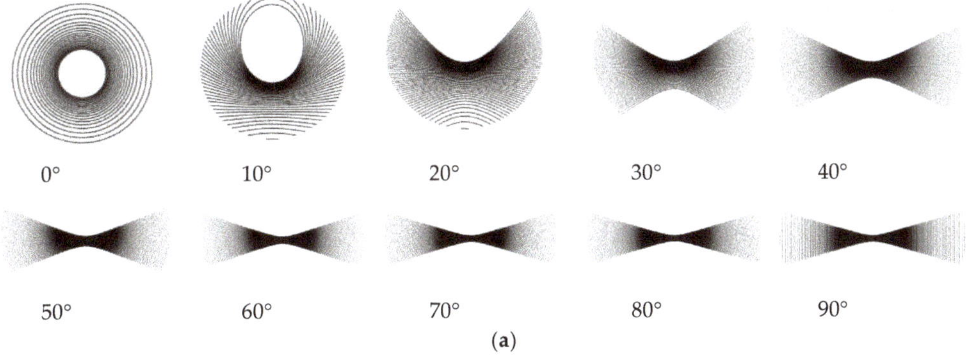

Figure 8. *Cont.*

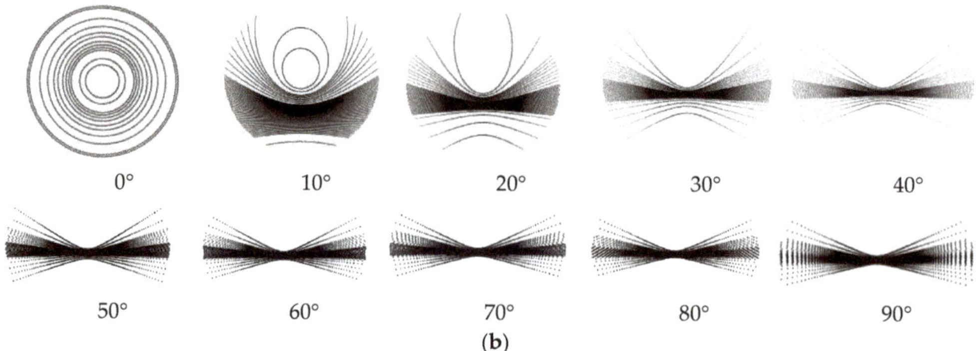

Figure 8. Stationary scanning patterns at a scanning range of 20 m on the ground. (**a**) OS-1-64. (**b**) Pandar64.

Both Lidar types scanning patterns shown in Figure 8 are analyzed in terms of density, coverage, and accuracy to get the performance histograms shown in Figure 9.

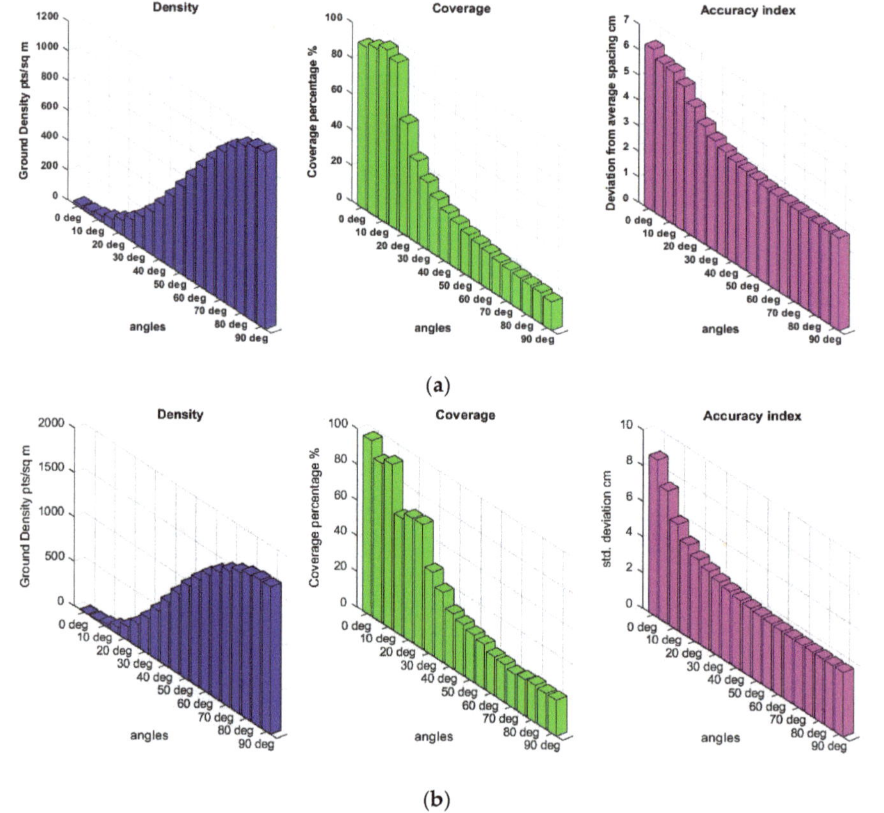

Figure 9. Performance test of the individual scans. (**a**) OS-1-64. (**b**) Pandar64.

A slight difference between OS-1-64 and Pandar64 is found in the coverage aspect. Whenever the tilt angle increases greater than 10°, OS-1-64 coverage scanning footprint decreases. On the other hand, for Pandar64, the coverage area decreases constantly when the tilting angle is greater than 25°. This might be related to: (a) the symmetry of the vertical FOV around the horizon and (b) the distribution pattern of the beams as illustrated in Figure 2. On the other hand, both Lidars have the same behavior for density and accuracy magnitudes. What can be clearly identified from the histograms in Figure 9 is the inverse relation between the density and accuracy on one side versus coverage from the other side. Accordingly, a compromise of these contradicting properties has been decided by considering the tilting angle range of 30°–50°. This angular orientation range is in general the most suitable for experimenting and evaluating the two Lidars performance for mobile mapping applications as will be presented in Section 4.

3. Experiment Design

The experimental test was applied in a simulated urban environment (Figure 10) where different buildings, light poles, traffic lights, moving cars, and trees are existed. The simulated drive was applied to a mobile mapping system equipped with the mentioned 64-channel Lidars of OS-1-64 and Pandar64 at a height of 2 m above the ground with a constant driving speed of 36 km/h.

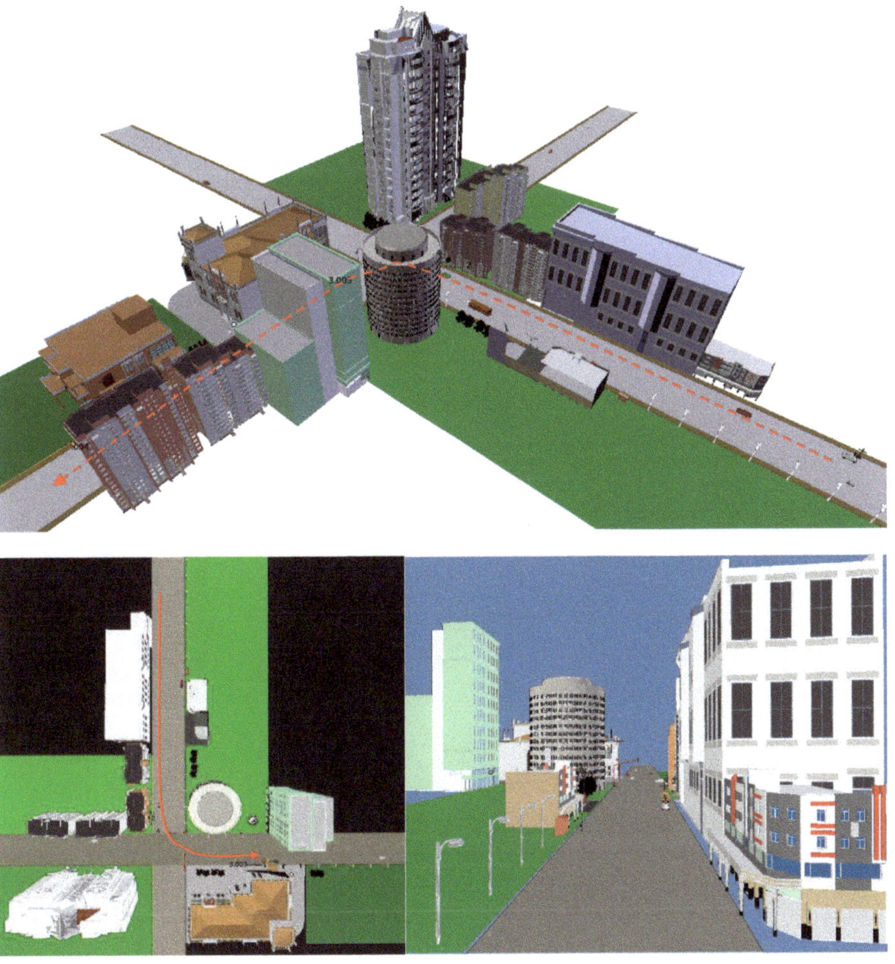

Figure 10. Virtual reality simulation urban model.

Five tilting angles of 30°, 35°, 40°, 45°, and 50° were investigated in the scanning for both Lidars. It should be noted that the simulations omitted any limited FOV and the existence of occlusions that might occur because of the vehicle body or the camera system itself and assumed a perfect occlusion-free mobile mapping system.

The accuracy evaluations of the two Lidars are represented by standard deviations derived in the propagation of error as described in Section 2.2. A standard deviation of ±3 cm in range and ±0.1° in angles were used in the calculations related to the manufacturers' specifications.

Two density checks are applied: (1) overall ground and off ground density check using the CloudCompare tool [24]; (2) density check on two selected slices on the ground (20 m length) and off ground (40 m height) as shown in Figure 11. The density tests are applied for all the aforementioned tilting angles of both Lidars as will be shown in Section 4.

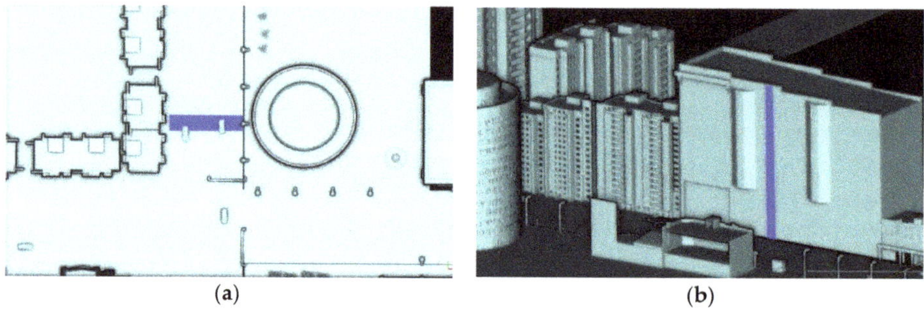

(a) (b)

Figure 11. Selected testing point cloud slices in blue. (**a**) Ground slice. (**b**) Facade slice.

A third test type was applied to check the density achieved on a traffic board sign located at distance of five meters from the trajectory, which increased as the car turned to the left road in the driving scenario (Figure 12). The test checked the effectiveness of cylindrical posts detection at the different tilting angles for the two Lidar types.

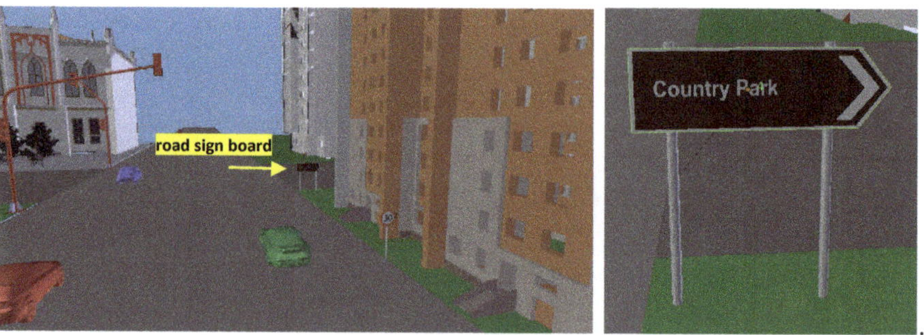

Figure 12. Traffic sign testing.

4. Results

In this section, the performance evaluation of the two Lidar types OS-1-64 and Pandar64 will be shown by a simulated drive with an MMS equipped with one Lidar device each time. The simulations are applied using the same trajectory and a constant vehicle speed of 36 km/h. It should be noted that we did not apply simulations at different driving speeds because of the linear dependency between the produced point clouds data and the mapping vehicle speed.

4.1. Simulation Results of Pandar64

For every tilting angle of 30°, 35°, 40°, 45°, and 50°, the point cloud density was computed on both the ground and off ground. The results are represented as shown in Figure 13, where the density increases on ground features whenever the tilting angle increases.

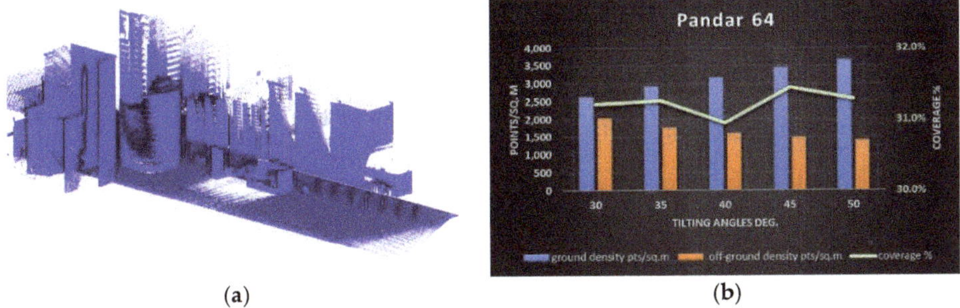

(**a**)   (**b**)

Figure 13. Pandar64 point cloud density on the ground and off ground features and their achieved coverage percentage. (**a**) Point cloud @30°. (**b**) Graphical representation of the coverage and density achieved.

The density check on the selected slices of Figure 11 was also evaluated and represented in Figure 14. It should be noted that the façade slice was 5 m away from the mapping trajectory.

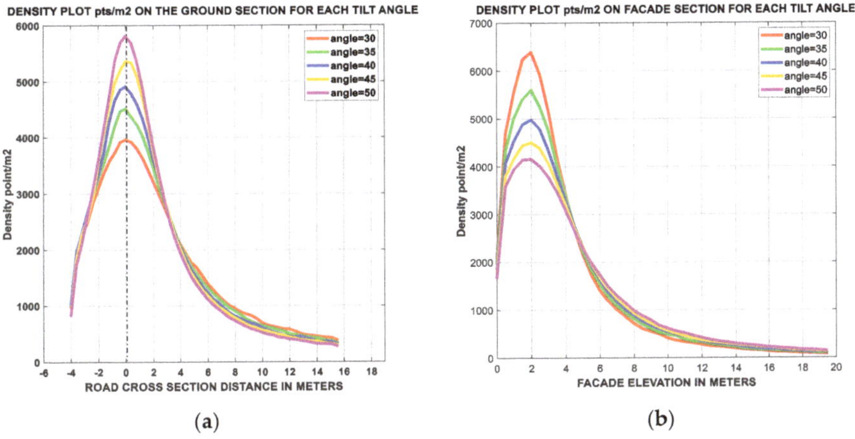

(**a**)   (**b**)

Figure 14. Pandar64 density checks on selected point clouds slices. (**a**) Density computed on the ground section. (**b**) Density computed on the facade section.

It should be noted that the 0 value of the x-axis indicates the scanning trajectory on the ground section in Figure 14a, while the start elevation of the façade section is shown in Figure 14b. For facades, higher densities are achieved up to a height of 4 m, and then drops quickly whenever the elevation increased. While the high scanning density on the ground ($\approx$2000 pts/m^2) is achieved within 8 m road width.

A large produced amount of data is shown in Figures 13 and 14 in terms of the achieved density of the point clouds at every tilting angle. This amount of density, if the achieved accuracy is not counted, is highly suitable for engineering surveys, asset management inventory, roadway condition assessment, powerline clearance, 3D designs, as built surveys, and other general mapping applications, as indicated in [16].

Coverage percentage was evaluated by comparing the resulted point cloud at each tilt angle to the reference model (Figure 15). The coverage was found to be slightly fluctuating as shown in Figure 13, where it starts at a high percentage of 30°, decreases at 40°, and then improves at 45°. However, the difference is within 0.25%.

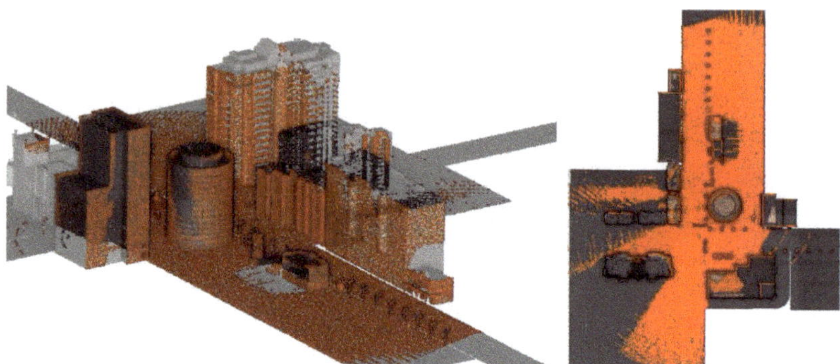

Figure 15. Illustration of point cloud coverage amount with respect to the reference model.

As mentioned earlier, the third parameter for the performance evaluation is the accuracy in the positions of the scanned points. Relative accuracy computed on both the ground and off ground features, and was found to improve whenever the tilting angle increased, as shown in Figure 16a. Furthermore, to illustrate the predicted accuracy at every tilting angle, a cumulative distribution function CDF chart [21] is plotted for the whole point clouds in Figure 16b.

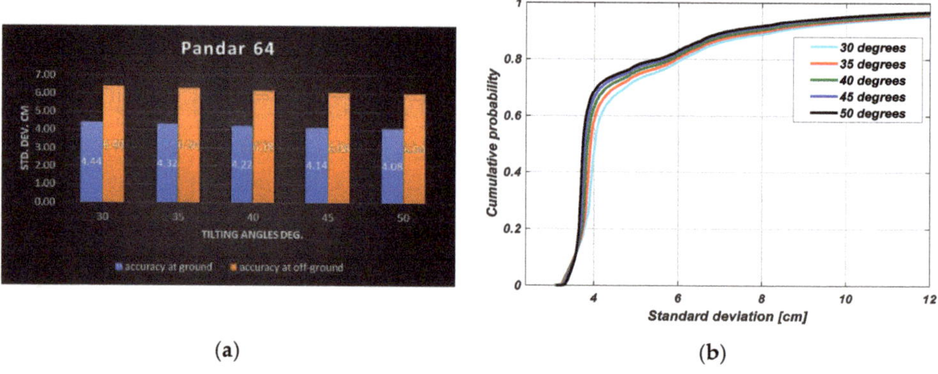

(a)　　　　　　　　　　　　　　(b)

Figure 16. (a) Histogram illustrating the achieved relative accuracy using Pandar64. (b) CDF accuracy chart.

The CDF chart shows the accuracy of the point cloud at each tilting angle using Equation (4). A better accuracy is achieved at larger tilting angles, which is logical since the incident angles of the laser beams are getting closer to right angles at the road surface and facades at the trajectory sides.

The estimated accuracy at 50° and 30° Lidar tilt angles is illustrated in Figure 17a,b, respectively, where every point with ≥5 cm standard deviation in position is colored red. For a better illustration, the point clouds were also tested for both selected slices of Figure 11 on the ground and on the façade. Figure 17c,d show the estimated accuracy at every Lidar tilting angle for both testing slices, where the points that have a standard deviation ≥5 cm are colored red. The summarized average standard deviations at both testing slices are listed in Table 2.

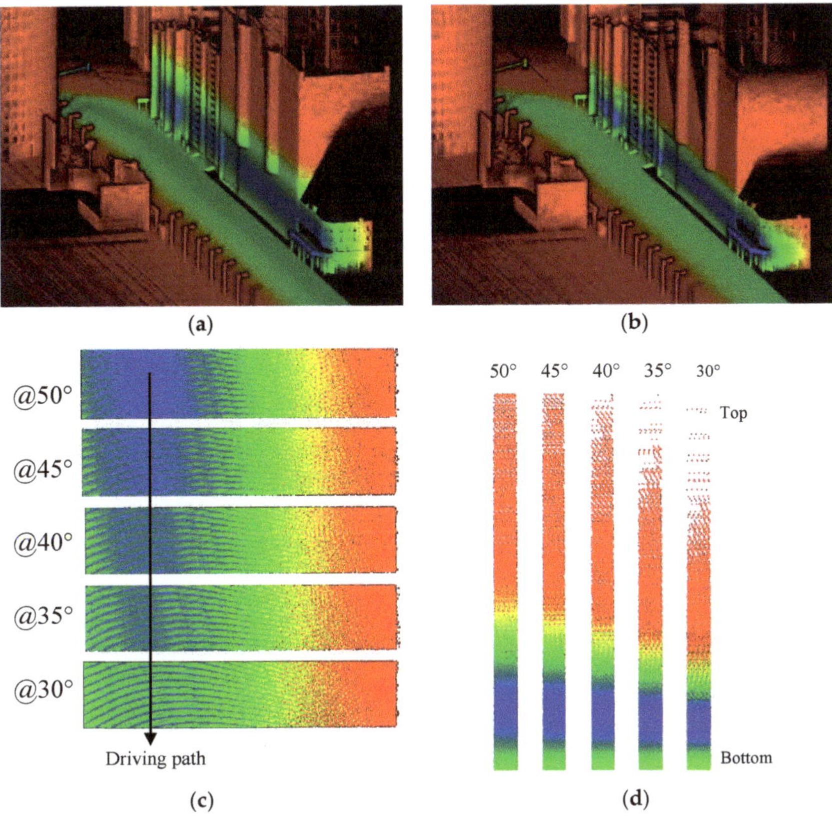

Figure 17. Accuracy labeled point clouds scanned by Pandar64 (red $\sigma \geq 5$ cm) (**a**) Accuracy at 50° Lidar tilt angle. (**b**) Accuracy at 30° Lidar tilt angle. (**c**) Accuracy estimated at the ground slice for every tilting angle. (**d**) Accuracy estimated at the façade slice for every tilting angle.

Table 2. The estimated average standard deviation on the ground and facade slices using Pandar64.

Lidar Tilt Angle	Average Standard Deviation [cm]	
	Façade Slice	**Ground Slice**
50°	3.75 ± 0.95	3.82 ± 0.42
45°	3.76 ± 1.03	3.87 ± 0.46
40°	3.76 ± 1.06	3.93 ± 0.51
35°	3.75 ± 1.08	4.01 ± 0.58
30°	3.76 ± 1.13	4.12 ± 0.68

The plots show a higher relative accuracy at the 50° angle compared to the 30° angle. This difference in accuracy is clearly related to the incidence angle of the scanning beams as mentioned.

Figure 18 illustrates examples for the incidence angle magnitudes achieved on the ground at three different Lidar tilting angles of 15°, 30°, and 50°. The figure illustrates how the increase of the laser beams incidence angles on the ground is related to the increase of the Lidar tilt angle.

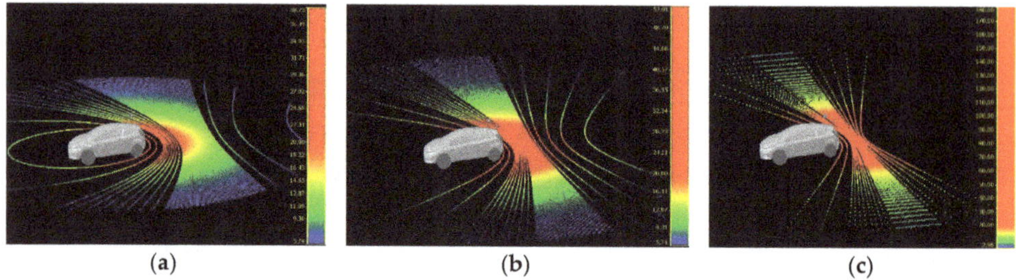

(a)   (b)   (c)

Figure 18. The relation between the Pandar64 Lidar tilt angle and the incidence angle where red points having ≥20° incidence angle on the ground. (**a**) @ 15° tilting angle: the average angle of incidence ≈14° ± 5°. (**b**) @ 30° tilting angle: the average angle of incidence ≈23° ± 9°. (**c**) @ 50° tilting angle: the average angle of incidence ≈44° ± 22°.

The third test as mentioned is to test the scanning goodness of a traffic sign at a road corner (Figure 12). Traffic sign test shows that the average number of scan points is decreasing from an average of 838 pts/m^2 at 30° tilt angle to 507 pts/m^2 at 50° tilting angle (Figure 19a). CloudCompare Ransac shape detection plugin is used to detect the sign cylindrical legs geometric primitives (Figure 19b). Accordingly, the deviation of the two detected cylindrical legs radii is computed with respect to the reference model radius (Figure 19c).

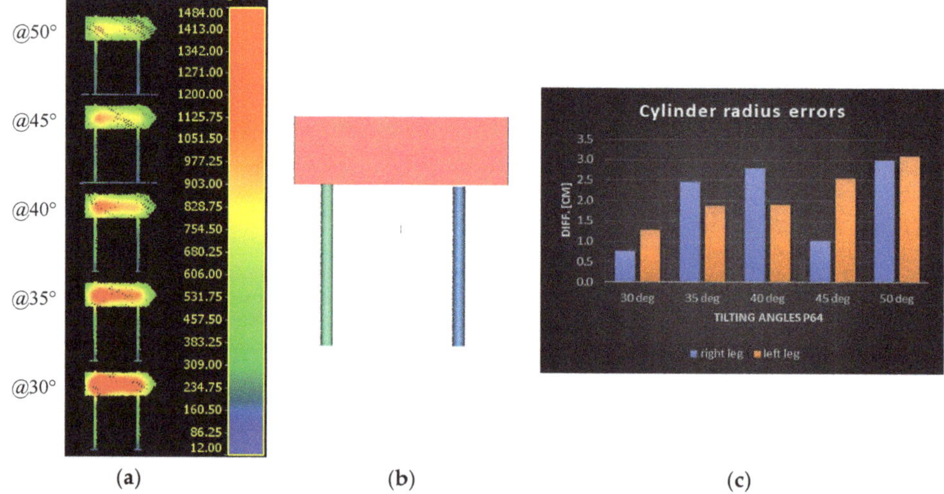

(a)   (b)   (c)

Figure 19. (**a**) Density of points at the road board sign at each tilting angle (red ≥ 1200 pts/m^2). (**b**) An example of the detected primitives of the board sign. (**c**) Histogram of differences in the estimated cylindrical board legs radii.

Assuming good set up parameters were entered for the cylinder fitting algorithm, we can say that, except for the point cloud produced at 30° tilt, the cylinder fitting at the other angles produced extra small cylinders as a false positive detection.

4.2. Simulation Results of OS-1-64

Similar to the tests applied for Pandar64 in Section 4.1, density was computed for the point clouds produced from OS-1-64 for every tilting angle at 30°, 35°, 40°, 45°, and 50°. The point cloud density was computed on both the ground and off ground. The results are shown in Figure 20, where the density increases on ground features whenever the tilting angle increases.

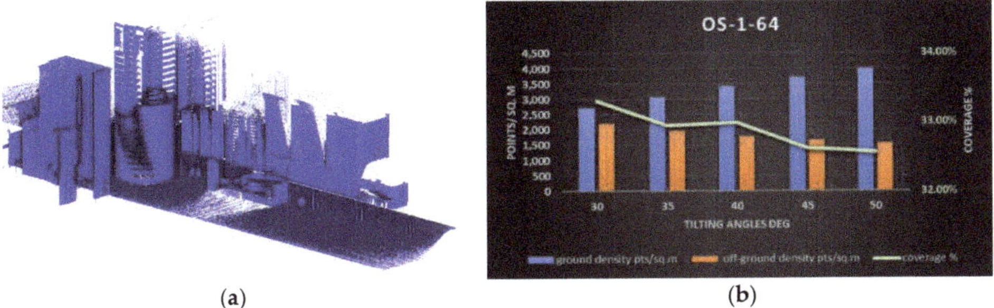

(a)   (b)

Figure 20. Point cloud density of OS-1-64 on the ground and off ground features and their achieved coverage percentage. (**a**) Point cloud @30°. (**b**) Graphical representation of the coverage and density achieved.

The density check on the selected slices of Figure 11 is also evaluated and represented in Figure 21. It should be noted that a 0 value on the x-axis indicates the scanning trajectory on the ground section in Figure 21a, while the start elevation on the façade section is shown in Figure 21b.

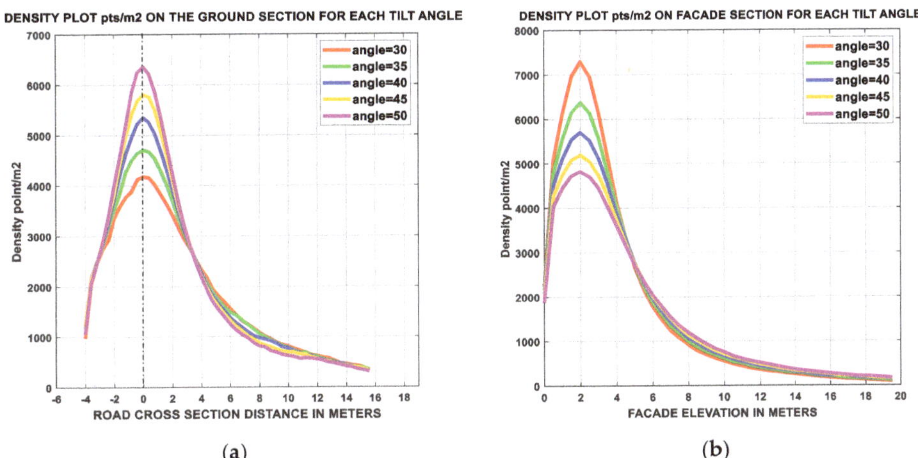

(a)   (b)

Figure 21. OS-1-64 density checks on selected point clouds slices. (**a**) Density computed on the ground section. (**b**) Density computed on the 5 m distant facade section.

Figures 20 and 21 indicate a large produced amount of data in terms of the achieved density of the point clouds at every tilting angle.

Furthermore, coverage percentage is evaluated by comparing the resulted point cloud at each tilt angle to the reference model. The coverage is found to be decreasing as shown in Figure 20 whenever the tiling angle increases.

The performance evaluation in terms of accuracy in the positions of the scanned points was also applied for the point clouds produced by the OS-1-64. Similar to the Pandar64 case, relative accuracy computed on both ground and off ground features was found to be improving whenever the tilting angle increased, as shown in Figure 22a. Furthermore, to illustrate the accuracy predicted at every tilting angle, a CDF chart was plotted for the whole point clouds in Figure 22b.

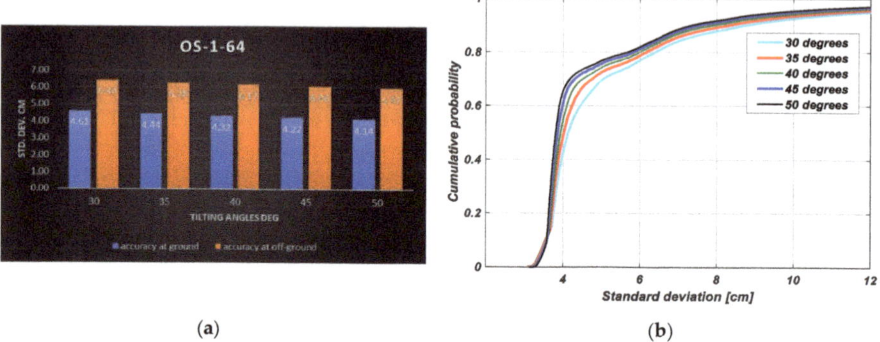

(a)

(b)

Figure 22. (**a**) Histogram illustrating the achieved relative accuracy using OS-1-64. (**b**) CDF accuracy chart.

The CDF chart shows the accuracy of the point cloud at each tilting angle using the error propagation described in Section 2.2. The estimated accuracy at 50° and 30° Lidar tilt angles is illustrated in Figure 23a,b respectively, where every point with ≥5 cm standard deviation in position is colored red. For a better illustration, the point clouds were also tested for both selected slices of Figure 11 on the ground and on the façade. Figure 23c,d show the estimated accuracy at every Lidar tilting angle for both testing slices. The summarized average standard deviations at both testing slices are listed in Table 3.

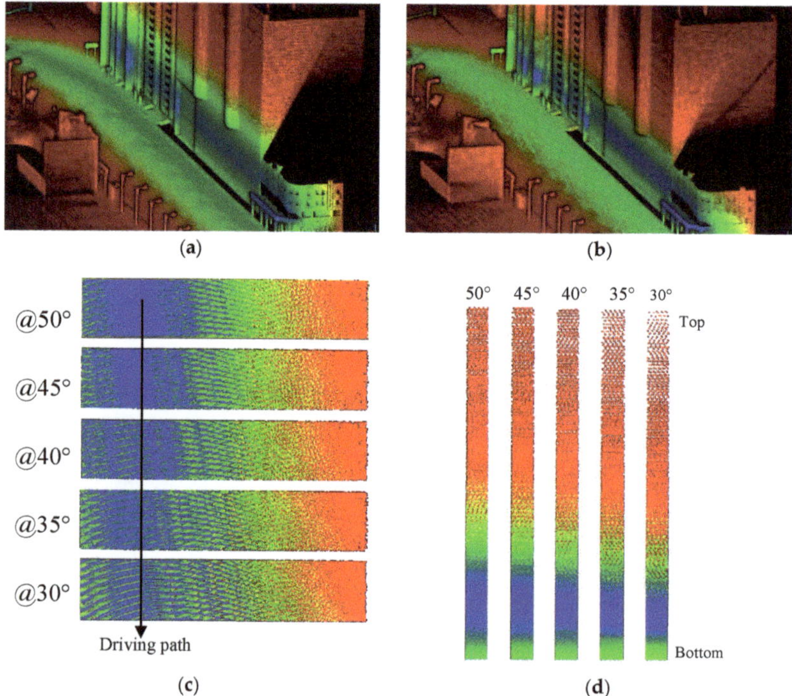

Figure 23. Accuracy labeled point clouds scanned by OS-1-64 (red $\sigma \geq 5$ cm). (**a**) Accuracy at 50° Lidar tilt angle. (**b**) Accuracy at 30° Lidar tilt angle. (**c**) Accuracy estimated at the ground slice for every tilting angle. (**d**) Accuracy estimated at the façade slice for every tilting angle.

Table 3. The estimated average standard deviation on the ground and facade slices using OS-1-64.

	Average Standard Deviation [cm]	
Tilt Angle	**Façade Slice**	**Ground Slice**
50°	3.85 ± 1.62	3.91 ± 0.86
45°	3.87 ± 1.75	3.99 ± 0.97
40°	3.88 ± 1.85	4.08 ± 1.09
35°	3.91 ± 1.94	4.20 ± 1.30
30°	3.95 ± 2.08	4.37 ± 1.53

This difference in accuracy at each Lidar tilt angle is clearly related to the incidence angle of the scanning beams. Figure 24 illustrates examples for the incidence angle magnitudes achieved on the ground at three different Lidar tilting angles of 15°, 30°, and 50°.

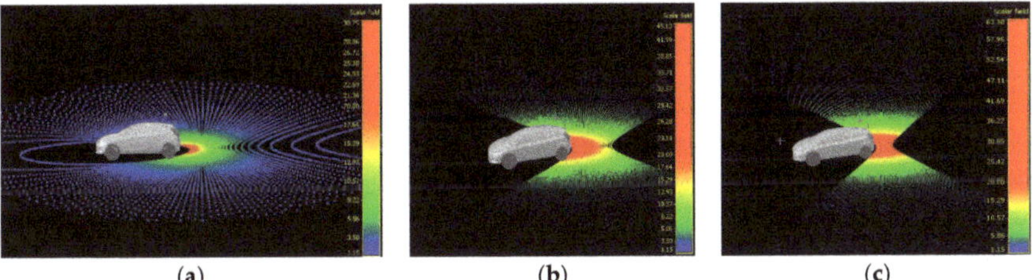

(a) (b) (c)

Figure 24. The relation between the OS-1-64 Lidar tilt angle and the incidence angle where red points having >20° incidence angle on the ground. (**a**) @ 15° tilting angle: the average angle of incidence ≈12° ± 8°. (**b**) @ 30° tilting angle: the average angle of incidence ≈20° ± 11°. (**c**) @ 50° tilting angle: the average angle of incidence ≈30° ± 16°.

As explained, the third test is to evaluate the scanning efficiency on a traffic sign at a road corner (Figure 12). Traffic sign test shows that the average number of scan points is decreasing from an average of 848 pts/m^2 at 30° tilt angle to 539 pts/m^2 at 50° tilting angle (Figure 25a). The deviation of the two detected cylindrical legs radii is computed with respect to the reference model radius and shown in Figure 25b.

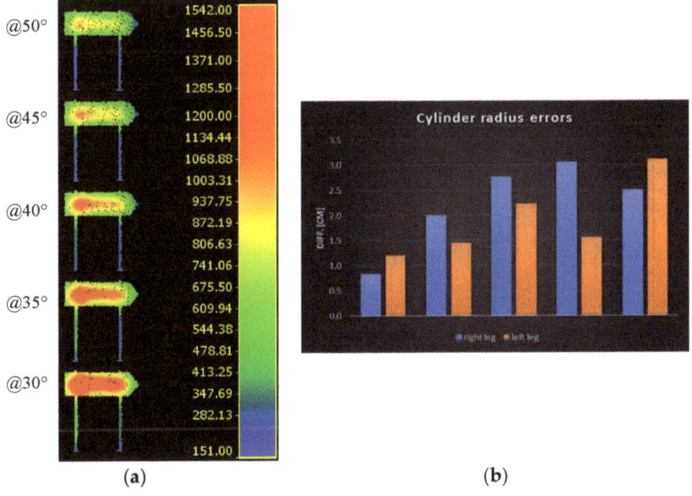

(a) (b)

Figure 25. (**a**) Density of points at the road board sign at each tilting angle (red ≥ 1200 pts/m^2). (**b**) Histogram of differences in the estimated cylindrical board legs radii.

Similar to the observation mentioned using Pandar64, the point cloud produced at 30° tilt angle is more suitable in the sense of less fitting errors and detection.

5. Discussion

The experimental results and analysis of the two Lidar types of Hesai Pandar64 and Ouster OS-1-64 are shown in Sections 4.1 and 4.2, respectively. The results are based on a 3D simulation scanning applied at different angular orientations (30° to 50°) for both types. The produced point clouds are analyzed in terms of density of points pts/m^2, coverage, and accuracy.

Every Lidar type showed its own advantages and the performance was found to be comparable. The performance evaluation is based on the three mentioned parameters of density, coverage, and accuracy, and they are prioritized as follows:

- Density criterion: since most of the mobile mapping applications are aimed for road asset inventory tenders, the higher density magnitude on the ground surface was preferred.
- Coverage criterion: with the variety of feature classes off the ground like traffic signs, trees, vehicles, people, buildings, etc. It was important to have a relatively high coverage off the ground in a mobile mapping project.
- Accuracy criterion: accuracy is preferred to be as high as possible if it does not conflict with the above two criteria or when comply with one of them.

Based on the results achieved as shown in Section 4, we can observe the following:

- For both Lidar types, the tilting angle increase was proportional to the achieved density of points on the ground surface, as shown in Figures 13 and 20, while it was inversely proportional to the achieved density of points on the off-ground features such as the road sign shown in Figures 19 and 25.
- For both Lidar types, the tilting angle at 30° was efficient for running the detection and fitting of a cylindrical object shape of a road sign legs, as shown in Figures 19 and 25.
- The achieved point clouds accuracy of the scanned objects was proportional to the increase of the Lidar tilting angle. However, this was negatively affecting the coverage attained on the lower parts of the building facades facing the trajectory as shown in Figure 17a,b and Figure 23a,b. On the other hand, larger tilting angle negatively affected the coverage attained on the higher parts of the building facades aside the trajectory as shown in Figure 17c,d and Figure 23c,d.
- The impact of the beams distribution on the coverage performance is shown in Figures 17 and 23. The upper parts of the building aside façade slice (Figure 11) is fairly scanned using OS-1-64 at every tilting angle while not fully scanned at lower tilting angles using Pandar64. This is related to the fact that Pandar64 laser beams are concentrated at the horizontal plane and then larger tilting angles are required to cover the facades higher parts.
- Furthermore, the accuracy improvement while increasing the Lidar tilting angle was related to the increase in the incidence angle of the scanning beams when hitting the objects as illustrated in Figures 18 and 24.

Accordingly, it was found that the ideal recommended angle for an MMS using a single Hesai Pandar64 Lidar is 45° (Figure 26), since it will ensure: (1) high coverage of the surrounding features, (2) high density for the ground features, and (3) a relative accuracy of ≤10 cm for 95% of the scanned data.

On the other hand, the ideal recommended angle for an MMS using Ouster OS-1-64 Lidar is 35° (Figure 26), since it will ensure: (1) high coverage of the surrounding features, (2) high density for both ground and off-ground features, and (3) a relative accuracy of ≤10 cm for 95% of the scanned data.

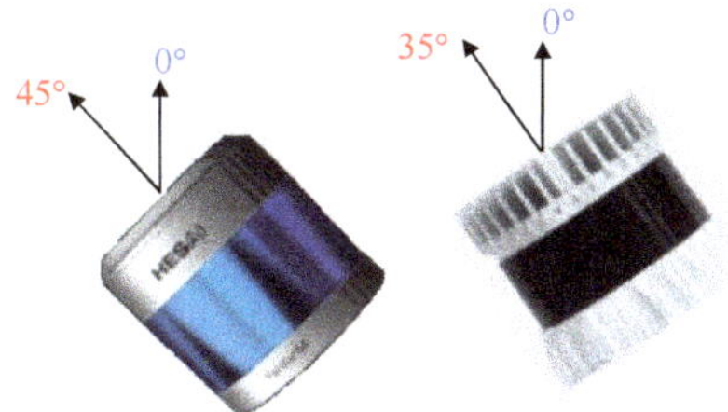

Figure 26. Ideal angular orientation of the selected Lidar devices. (**left**) Pandar64. (**right**) Ouster OS-1-64.

6. Conclusions

In this paper, two state-of-the-art multi-beam spinning Lidar types were investigated, namely, Ouster OS-1-64 and Hesai Pandar64. The investigations assumed a mobile mapping system equipped with a single multi beam Lidar. A 3D simulated urban scene scanning was applied at a constant driving speed of 36 km/h and at different angular orientations (30°–50°) for both Lidar types. The produced point clouds of ground and off ground features were analyzed in terms of density of points pts/m^2, coverage, and accuracy. Based on the extensive analysis applied for both Lidar outputs, it was found that the ideal angular orientation of the Ouster OS-1-64 Lidar type is 35°, while the ideal angular orientation of the Hesai Pandar64 is 45°, as illustrated in Figure 26. These orientation angles are recommended mainly for mobile mapping applications where the major customers demand is for engineering surveys, asset management inventory, roadway condition assessment, power lines clearance, 3D designs, as built surveys, cultural heritage documentation, and other geospatial applications.

It is worth to mention that the applied simulations did not consider a non-uniform vehicle speed, which is expected to have an impact on the results. Furthermore, the study did not account for different mapping scenarios in the simulated 3D model, which would likely strengthen the conclusions made in this paper.

Future work can be applied to investigate other newly released Lidar types including the solid state Lidars. Moreover, dual Lidar installation can also be investigated.

Funding: This research received no external funding.

Conflicts of Interest: The authors declare no conflict of interest.

References

1. Intellias. Intellias Intelligent Software Engineering. Available online: https://www.intellias.com/the-emerging-future-of-autonomus-driving/ (accessed on 10 October 2019).
2. Deepmap. Hd Mapping and Localization for Safe Autonomy. Available online: https://www.deepmap.ai/ (accessed on 30 September 2019).
3. Shanker, R.; Adam, J.; Scott, D.; Katy, H.; Simon, F.; William, G. *Autonomous Cars: Self-Driving the New Auto Industry Paradigm*; Morgan Stanley Research Global: New York, NY, USA, 2013.
4. Waymo. Available online: https://waymo.com/ (accessed on 13 July 2019).
5. Ouster. Available online: https://ouster.com/ (accessed on 20 November 2018).
6. Quanergy. Available online: https://quanergy.com/ (accessed on 20 November 2018).
7. Lidar, Velodyne. Available online: https://velodynelidar.com/ (accessed on 10 October 2019).
8. Hesai. Available online: https://www.hesaitech.com/en/ (accessed on 27 August 2019).
9. Luminar. Available online: https://www.luminartech.com/ (accessed on 13 July 2019).
10. Blickfeld. Available online: https://www.blickfeld.com/ (accessed on 13 July 2019).
11. Riegl. Riegl Vux-1ha. Available online: http://www.riegl.com/products/newriegl-vux-1-series/newriegl-vux-1ha/ (accessed on 2 June 2019).
12. Cahalane, C.; Lewis, P.; McElhinney, C.P.; McCarthy, T. Optimising Mobile Mapping System Laser Scanner Orientation. *ISPRS Int. J. Geo-Inf.* **2015**, *4*, 302–319. [CrossRef]

13. Wang, Y.; Hsi-Yung, F. Effects of Scanning Orientation on Outlier Formation in 3d Laser Scanning of Reflective Surfaces. *Opt. Lasers Eng.* **2016**, *81*, 35–45. [CrossRef]

14. Habib, A.; Bang, K.I.; Kersting, A.P.; Lee, D.C. Error Budget of Lidar Systems and Quality Control of the Derived Data. *Photogramm. Eng. Remote Sens.* **2009**, *75*, 1093–1098. [CrossRef]

15. Habib, A.F.; Kersting, A.P.; Bang, K. A Point-Based Procedure for the Qulity Control of Lidar Data. In Proceedings of the 6th International Symposium on Mobile Mapping Technology, Sao Paulo, Brazil, 21 July 2009.

16. Olsen, M.J. *Nchrp 15-44 Guidelines for the Use of Mobile Lidar in Transportation Applications*; Transportation Research Board: Washington, DC, USA, 2013.

17. Sairam, N.; Nagarajan, S.; Ornitz, S. Development of Mobile Mapping System for 3d Road Asset Inventory. *Sensors* **2016**, *16*, 367. [CrossRef] [PubMed]

18. Hesai. Pandar64-64-Channel Mechanical Lidar-User's Manual. 2019. Available online: https://www.google.com.hk/url?sa=t&rct=j&q=&esrc=s&source=web&cd=3&ved= 2ahUKEwiGhIDc1sjnAhWPHKYKHTx6BJ0QFjACegQIAxAB&url=https%3A%2F%2Fwww.symphotony. com%2Fwp-content%2Fuploads%2FPandar64-64-Channel-Mechanical-LiDAR-3-1.pdf&usg= AOvVaw2nGw8eGQ85euhs9Ks65_dr (accessed on 2 June 2019).

19. Ouster. Mid-Range Lidar Sensor Os1. Available online: https://ouster.com/products/os1-lidar-sensor/ (accessed on 10 October 2019).

20. Blender. Available online: https://www.blender.org/ (accessed on 15 December 2018).

21. Mezian, C.; Vallet, B.; Soheilian, B.; Paparoditis, N. Uncertainty Propagation for Terrestrial Mobile Laser Scanner. *Int. Arch. Photogramm. Remote Sens. Spat. Inf. Sci.* **2016**, *41*, 331–335. [CrossRef]

22. Alsadik, B. *Adjustment Models in 3d Geomatics and Computational Geophysics: With Matlab Examples*; Elsevier Science: Amsterdam, The Netherlands, 2019.

23. Rohrbach, Felix. Point Density and Point Spacing. Available online: https://felix.rohrba.ch/en/2015/point-density-and-point-spacing/ (accessed on 30 September 2019).

24. CloudCompare. Cloudcompare: 3d Point Cloud and Mesh Processing Software. Available online: https: //www.danielgm.net/cc/ (accessed on 2 February 2018).

remote sensing

MDPI

Article

Information-Based Georeferencing of an Unmanned Aerial Vehicle by Dual State Kalman Filter with Implicit Measurement Equations

Rozhin Moftizadeh *, Sören Vogel , Ingo Neumann , Johannes Bureick and Hamza Alkhatib

Geodetic Institute, Leibniz Universität Hannover, Nienburger Str. 1, 30167 Hannover, Germany; vogel@gih.uni-hannover.de (S.V.); neumann@gih.uni-hannover.de (I.N.); johannes@bureick-verm.de (J.B.); alkhatib@gih.uni-hannover.de (H.A.)
* Correspondence: moftizadeh@gih.uni-hannover.de; Tel.: +49-511-762-14736

Abstract: Georeferencing a kinematic Multi-Sensor-System (MSS) within crowded areas, such as inner-cities, is a challenging task that should be conducted in the most reliable way possible. In such areas, the Global Navigation Satellite System (GNSS) data either contain inevitable errors or are not continuously available. Regardless of the environmental conditions, an Inertial Measurement Unit (IMU) is always subject to drifting, and therefore it cannot be fully trusted over time. Consequently, suitable filtering techniques are required that can compensate for such possible deficits and subsequently improve the georeferencing results. Sometimes it is also possible to improve the filter quality by engaging additional complementary information. This information could be taken from the surrounding environment of the MSS, which usually appears in the form of geometrical constraints. Since it is possible to have a high amount of such information in an environment of interest, their consideration could lead to an inefficient filtering procedure. Hence, suitable methodologies are necessary to be extended to the filtering framework to increase the efficiency while preserving the filter quality. In the current paper, we propose a Dual State Iterated Extended Kalman Filter (DSIEKF) that can efficiently georeference a MSS by taking into account additional geometrical information. The proposed methodology is based on implicit measurement equations and nonlinear geometrical constraints, which are applied to a real case scenario to further evaluate its performance.

Keywords: georeferencing; MSS; IEKF; DSIEKF; geometrical constraints; 6-DoF; DTM; 3D city model

Citation: Moftizadeh, R.; Vogel, S.; Neumann, I.; Bureick, J.; Alkhatib, H. Information-Based Georeferencing of an Unmanned Aerial Vehicle by Dual State Kalman Filter with Implicit Measurement Equation. *Remote Sens.* **2021**, *13*, 3205. https://doi.org/10.3390/rs13163205

Academic Editor: Dimitrios D. Alexakis

Received: 10 June 2021
Accepted: 10 August 2021
Published: 12 August 2021

Publisher's Note: MDPI stays neutral with regard to jurisdictional claims in published maps and institutional affiliations.

1. Introduction

Multi-Sensor-Systems (MSS) refer to single platforms that are equipped with various sensors, which are frequently used in engineering to capture various aspects of an environment. Currently, considerable development in industry and technology on one hand, and an increasing demand towards lightweight yet inexpensive MSS on the other hand have enabled light, small-scale, and low-cost sensors to be used for different purposes.

One of the contemporary applications that requires such sensors is Unmanned Aerial Vehicles (UAV), which can be used for in many disciplines. In engineering geodesy, they are frequently used to capture the surrounding environment using laser scanners, cameras, and other sensors that are mounted on them. For further processing of the measurements, taken from different sensors, it is usually required to have the position and orientation of the platform with respect to a superordinate (global) coordinate system.

This whole procedure is also referred to as MSS "georeferencing". The position and orientation each consist of three elements, and therefore the MSS pose is usually referred to as the Six Degrees of Freedom (6-DoF). The most straightforward method of deriving the 6-DoF is to use the Global Navigation Satellite System (GNSS) and Inertial Measurement Unit (IMU). However, in urban areas, the accuracy of GNSS data is greatly affected due to multipath and shadowing effects, and even total signal loss is possible.

In general, low-cost GNSS receivers deliver positions with accuracy at the meter level, unless good GNSS conditions are available and differential techniques are applied, which can increase the accuracy up to the decimeter or even centimeter level [1]. On the other hand, IMUs are always subject to drifting, which makes their data unreliable over time. Detailed information regarding different error sources and biases that the IMU units are subject to and how large these errors could become over time are given in [2–4].

A lightweight and low-cost IMU delivers the roll and pitch angles with an accuracy of 0.1° and the heading angle with an accuracy of 0.8° [1]. According to [5], a minimal accelerometer bias of one least significant bit could lead to a position error of around 200 m in a period of 100 s. Therefore, the development of appropriate algorithms that can best compensate for such possible errors or losses in the GNSS and IMU datasets and, thus, will increase the accuracy of georeferencing is an inevitable request that should be fulfilled.

In the current paper, a Dual State Iterated Extended Kalman Filter (DSIEKF) algorithm is introduced, which deals with MSS georeferencing by using implicit measurement equations and taking nonlinear geometrical constraints of the surrounding environment into account. The algorithm was previously applied to a simulated environment in [6]. However, to further improve its functionality and performance, a real case scenario is tested within the current paper, which consists of a UAV equipped with a 3D laser scanner, a GNSS, and an IMU among other sensors. Subsequently, necessary modifications are applied to the algorithm to better handle the complications, which are usually irresistible when dealing with real environments.

1.1. MSS Georeferencing

For MSS georeferencing, various strategies have been proposed. The decision for a specific strategy should be based on the measurement scenario and the environment in which the MSS moves. For indoor applications, a general overview of the georeferencing strategies is given in [7]. For instance and as investigated by [8], the "Ultra-Wide-Band (UWB)" technology as one of the indoor positioning strategies can be mentioned. In such an approach, the propagation properties of radio waves are used for localization purposes. However, since the focus of the current paper is on outdoor applications, the proposed methodologies are highlighted in the following.

In general, outdoor georeferencing strategies can fit into the three categories of "direct", "indirect", and "data-driven" [9,10]. In "direct" (sensor-driven) georeferencing, GNSS [10] and IMU [11] are usually used to derive the 6-DoF of a MSS with respect to a global coordinate system. In this case, the superordinate coordinate system can also be set up arbitrarily using a total station or a laser tracker, and the MSS can be localized with respect to it [12,13]. Such direct georeferencing techniques could be beneficial for multiple purposes. For example, in a study by [14], UAV images were georeferenced by means of network-based continuously operating reference stations and a differential-based real-time kinematic georeferencing system without using ground control points.

In "indirect (target-driven)" georeferencing, already referenced targets to a superordinate coordinate system are used to derive the 6-DoF of the MSS with respect to the same global coordinate system. In this case, the targets can be of any type, such as flat markers with specific patterns [15] or simple 3D geometries, such as spheres or cylinders [16]. Moreover, the available referenced datasets could serve as possible means to link the MSS pose to a global coordinate system. In this case, the georeferencing is referred to as "data-driven". The referenced datasets, in this case, could be 3D point clouds [17], digital surface models, or 3D city models [7,18,19].

1.2. Georeferencing by Means of Filtering Techniques

As a general strategy, filtering techniques are usually used to encounter possible errors within various datasets and, hence, to derive the unknown parameters in the most accurate way possible. A number of strategies are proposed in the literature, among which the work by [20] could be mentioned, in which a nonlinear $H\infty$ filter with fuzzy adaptive bound and adaptive disturbances attenuation is proposed. The main idea is to overcome data fusion problems, such as modeling errors, by means of a robust nonlinear filter. In addition, the authors in [21] proposed a key frame filtering process to properly combine the laser scanner and IMU data for the purpose of localization in Simultaneous Localization and Mapping (SLAM).

One of the well-known filtering methods is Kalman Filter (KF), which has been the basis of many pose estimation algorithms. In classic KF—which is also referred to as Linear Kalman Filter (LKF)—both the system and measurement equations are linear; however, if either of these equations are nonlinear, a linearization should be applied, which is done by Taylor series expansion with respect to a certain state [22]. Such a realization of KF is referred to as the Extended Kalman Filter (EKF). The authors in [23] used EKF to combine the GNSS and IMU data in the context of sensor fusion of a UAV's local sensors.

In [24], EKF is used for collaboration purposes between several UAVs in order to increase the accuracy of the estimated 6-DoF by combining multiple SLAM algorithms. In [25], the Kullback–Leibler divergence metric is introduced as a measure for the accuracy of the update step approximations of KF variants, which further proves the inaccuracy of EKF. To overcome such an issue, a re-linearization with respect to the most recent estimates could be performed, which, in [26], is shown to be an application of the Gauss–Newton method to approximate a solution. Such a procedure is generally referred to as the Iterated Extended Kalman Filter (IEKF).

In [27], a statistical linear regression is proposed to be performed instead of the first-order Taylor series, which can improve both the EKF and IEKF. So far researches have been mainly focused on IEKF with explicit measurement equations. In such equations—which are also referred to as Gauss–Markov–Models (GMM)—the observations and states are separated from each other. In other words, it is possible to rewrite the measurement equations in such a way as to have all the measurements on one side and to place the unknown parameters on the other side of the equations.

However, sometimes it is also possible to have measurement equations in which such a separation cannot be applied, which indicates implicit measurement equations that are also referred to as Gauss–Helmert–Models (GHM). In [28,29], such IEKF algorithms with GHM were used. Nonetheless, for the first time in the context of MSS georeferencing, the authors in [30] introduced such a combination of IEKF with implicit measurement equations, which was then applied by [31], the authors in [1,32] for localizing different kinds of MSS in various environments.

Sometimes, the system behavior or environmental conditions might vary over time causing the system or observation model(s) to change. In such a case, estimations are limited to not only the system states but also the model(s) parameters, which, due to their dependency, should be done simultaneously. A solution to such a challenge is the Dual State (DS) estimation framework, which is based on grouping the states into two separate vectors and using each vector to estimate the other, iteratively [33–37].

Moreover, sometimes the environment in which the MSS moves through has valuable information that could serve as additional knowledge to increase the accuracy of the estimated results even more. Such prior information, which is imposed on the states, is generally referred to as "constraints" and, in the case of geometrical content, is called "geometrical constraints". Examples of geometrical constraints are the intersection lines between facades of adjacent buildings or perpendicular facades of a building, etc. [30].

Depending on the mathematical formulation of the constraints, they can be of linear or nonlinear type. A good overview of the possible strategies in KF to apply linear and nonlinear constraints is given in [38]. In the case of the latter form, a linearization is usually suggested, which then enables using already developed algorithms for linear geometrical constraints as given, e.g., in [39–42]. A maximum-likelihood-based filtering technique that can be used for both linear and nonlinear constraints without the necessity for linearization in case of the latter constraints form was given in [43]. In [44], linear and nonlinear systems subject to linear and nonlinear constraints along with various proposed solutions were highlighted.

The authors in [45] suggested an algorithm to handle second-order scalar equality constraints. In [46], a smoothly constrained KF was suggested that can be used for any type of nonlinear constraints. The authors in [30,31] integrated equality and inequality constraints into the IEKF framework with implicit measurement equations for MSS georeferencing by means of projection and Probability–Density–Function (PDF) truncation methods, respectively. The derived algorithm by [30] was further used by [1] for georeferencing a UAV in a simulated environment.

The authors in [6] adapted the work of [1] to the DS estimation framework—the result of which is called DSIEKF, which was applied to a simulated environment. Furthermore, in one of the analyzed real scenarios by [32], parts of the environment were taken as geometrical constraints that were imposed on the estimated states, which depicts the impact of such constraints on the final results.

1.3. Contribution

As previously mentioned in [6], the DSIEKF algorithm was introduced for the matter of MSS georeferencing, which was applied to a simulated environment. However, when it comes to real scenarios, complications exist that need to be dealt with. Such a dealing might require slight modifications to the original algorithm making it more promising to be applied to a practical problem. Therefore, the current paper is focused on the application of the DSIEKF algorithm to a real case scenario, which further proves the functionality of the methodology.

The UAV, as well as the measured data are within the scope of a joint research project between the Geodetic Institute (GIH) and Institute of Photogrammetry and Geoinformation (IPI) of Leibniz Universität Hannover (LUH). The considered environment is a courtyard at GIH that is surrounded by multiple infrastructures through which the MSS moves. Several sensors are installed on the UAV platform, among which the 3D laser scanner is the main interest within the current paper. Unlike [6], in this paper, the GNSS and IMU data were only considered for the filter initialization and neglected for the further epochs due to low sensor qualities.

The main idea, as of [1,30], is to link the data of the scanner to the buildings of the environment by means of an available 3D city model in order to properly georeference the UAV. 3D digital city models contain spatial and georeferenced data for different parts of a city (buildings, sites, etc.) [47]. They exist in different levels of detail, which could be used for different purposes. In this paper, the second Level of Detail (LoD-2) 3D city model was used, which is available for many cities in Germany. The same UAV data is also analyzed by [32] in the framework of IEKF for the matter of georeferencing; however, no geometrical constraints are imposed on the final estimations in this case.

1.4. Outline

The paper is organized as follows: In Section 2, the DSIEKF methodology, its use for UAVs, and the general workflow is explained in detail. Section 3 is focused on the real case scenario and its vital aspects that are highlighted in the current paper. In Section 4, the results along with a detailed discussion are presented. Finally, Section 6 is dedicated to the general outcome of the current work and an overview of the potential future research in the same field.

2. DSIEKF for UAV Georeferencing

As previously mentioned, the main goal of this paper is to accurately localize a prototype UAV by adapting a realization of KF that is already developed by [30] (referred to as IEKF algorithm) to the DS estimation framework. The final algorithm is called DSIEKF. The prototype UAV is explained in more detail in Section 3.1.

Such an accurate localization requires multiple aspects to be covered. The main concept is to benefit from the possible geometrical information that the surrounding environment offers. Such additional information could be, e.g., the parallel or perpendicular lines or planes of the environment. A scheme of the whole situation is illustrated in Figure 1 where different MSS are shown to be moving in an inner-city environment with high-rise buildings.

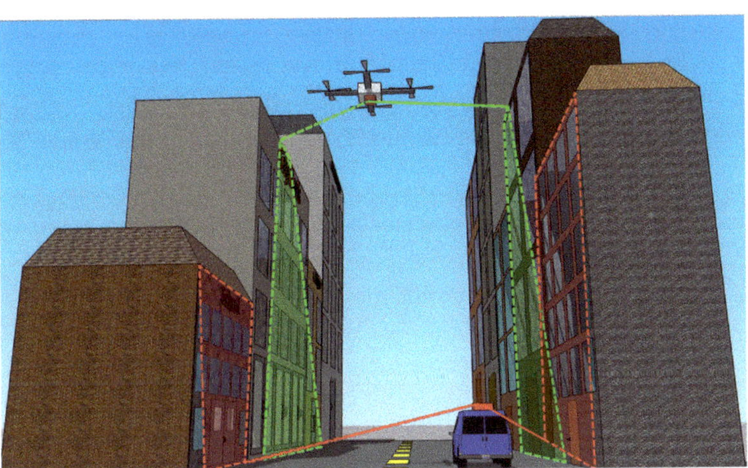

Figure 1. Scheme of different Multi-Sensor-Systems (MSS) capturing an environment. The green and red planes on the buildings are indicators for the scanner measurements.

2.1. General Idea

The main idea is to establish a reliable connection between the sensor measurements and the surrounding environment that the MSS moves through. In the current paper, only the 3D scanner measurements are taken into consideration. Therefore, the first step is to know how the surrounding environment is defined and how this definition could be linked to the scanned data. In the LoD-2 3D city models, the buildings' facades are defined by means of planes with defined parameters and vertices in a global coordinate system.

On the other hand, the ground is defined by a Digital Terrain Model (DTM), which contains a network of cells each having 2D coordinates plus the height information in a global coordinate system. One of the ways to link the measured data to the surrounding environment is to assign each scanned point either to a facade in the 3D city model or to a cell of the DTM. The scanned data are derived in the local coordinate system of the 3D scanner.

Therefore, if these data can be transformed to the same coordinate system in which the planes of the buildings and the heights of the ground are defined, the aforementioned connection could be established by assigning them to either the buildings or the ground. The taken georeferencing procedure in the current paper is in the "data-driven" category (explained in Section 1.1), which can be explained as follows:

Each scanned point is assigned to the closest DTM cell by calculating the 2D Euclidean distances. Afterward, the difference in the transformed height of the assigned scanned point—in the global coordinate system—and that of the DTM cell is derived. Next, the closest building's facade to that point is derived, and then the point is assigned to either the DTM cell or the plane based on a certain defined threshold. A scheme of the assignment to the planes of the buildings is shown in Figure 2. In this figure, the green dots show the assigned scanned data to the building facades, which are shown by a grey plane.

The coordinate system of the far left of this figure is an indicator of the global coordinate system to which the scanned data should be transformed. Furthermore, the red dotted line of the far right side of this figure shows the distance threshold that is used through the assignment process. The idea is that the transformed scanned points with larger distances to the planes than the threshold are not considered through the whole filtering technique. The assignment methodology is taken from [48,49].

As can be seen, multiple points will be assigned to each plane of the 3D city model. Similarly, to each DTM cell, various scanned points are assigned, among which, the smallest point in height is selected as the ground representative at that specific location (Figure 3). The main issue is that the transformation could only be done by means of the MSS pose, which is unknown. Therefore, the main idea is to iteratively apply the transformation until the assignments are satisfactory. Such satisfaction lies in having appropriate transformation parameters, which, in turn, means having the correct MSS pose.

When it comes to real environments, such a "correctness" is challenging to be judged due to unknown "true" values; however, various criteria from different aspects could serve as means to decide upon the trustworthiness of the estimated pose values. In the current paper, the final results from the DSIEKF algorithm are compared with those derived from the IEKF algorithm that is also applied to the same environment.

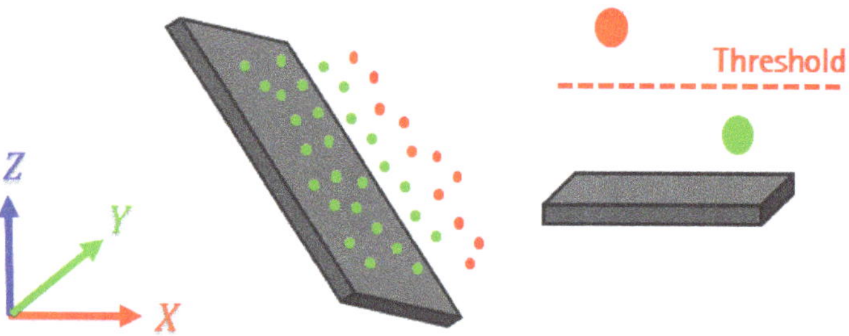

Figure 2. Scheme of the assigned scanned data to the building facades.

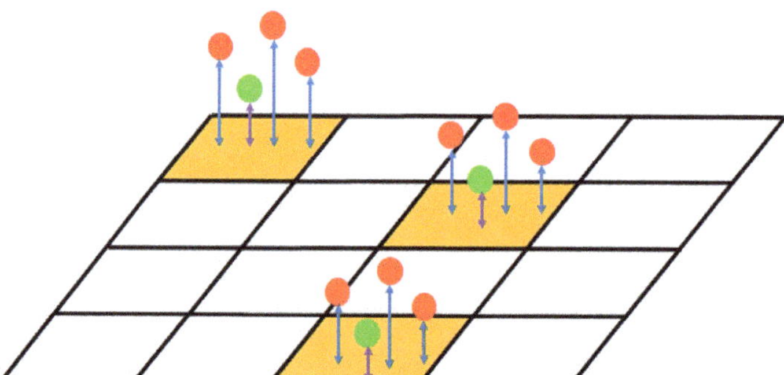

Figure 3. Scheme of the assigned points to the DTM cells. In each cell, the minimum assigned point in height (the green dots) is selected as the ground representative at that location.

2.2. Transformation to Global Coordinate System

To assign the scanned data to the planes of the buildings or the ground, they should be transformed to the global coordinate system by means of the transformation parameters. These parameters are, indeed, the 6-DoF of the MSS at each epoch in time. The transformation could be done as follows:

$$\mathbf{P}_{glo} = \mathbf{t} + \mathbf{R} \cdot \mathbf{P}_{loc} \tag{1}$$

where $\mathbf{P}_{glo}$ is the transformed laser scanner point cloud from the local to the global coordinate system. $\mathbf{t}$ and $\mathbf{R}$ are the translation parameters vector and rotation matrix, respectively.

$\mathbf{P}_{loc}$ is the laser scanner point cloud in its local coordinate system. $\mathbf{R}$ is derived according to (2).

$$\mathbf{R} = \mathbf{R}_\omega \cdot \mathbf{R}_\phi \cdot \mathbf{R}_\kappa \tag{2}$$

where $\mathbf{R}_\omega$, $\mathbf{R}_\phi$, and $\mathbf{R}_\kappa$ are rotations around the X, Y, and Z axes of the global coordinate system, respectively. These rotation matrices can be derived from (3).

$$\mathbf{R}_\omega = \begin{bmatrix} 1 & 0 & 0 \\ 0 & cos\omega & -sin\omega \\ 0 & sin\omega & cos\omega \end{bmatrix}, \ \mathbf{R}_\phi = \begin{bmatrix} cos\phi & 0 & sin\phi \\ 0 & 1 & 0 \\ -sin\phi & 0 & cos\phi \end{bmatrix}, \ \mathbf{R}_\kappa = \begin{bmatrix} cos\kappa & -sin\kappa & 0 \\ sin\kappa & cos\kappa & 0 \\ 0 & 0 & 1 \end{bmatrix} \tag{3}$$

2.3. DSIEKF (Dual State Iterated Extended Kalman Filter)

An environment can have a great deal of geometrical information that would make the filtering approach inefficient if they were to be considered. This is especially true when it comes to IEKF where iterations are applied in the update step to overcome the linearization errors. Therefore, in DSIEKF, the main idea is to divide the state vector into two separate parts that should be estimated, interactively [6]. The reasoning behind such a division in the current paper is the fact that one vector can be fixed in size during all the epochs; whereas, the other vector can become larger over time due to the possible increasing information flow of the surrounding environment.

However, since such prior knowledge is taken from a reliable source—like the LoD-2 3D city models—the convergence of the second vector occurs faster than the first one. Consequently, the estimation time is decreased due to fixing the second vector after several iterations, and therefore only one vector needs to be estimated, which is considerably smaller in size than the case in which all the states are included in one vector (the IEKF methodology). A scheme of DSIEKF is presented in Figure 4, which shows the general idea of this algorithm and its application to the real scenario.

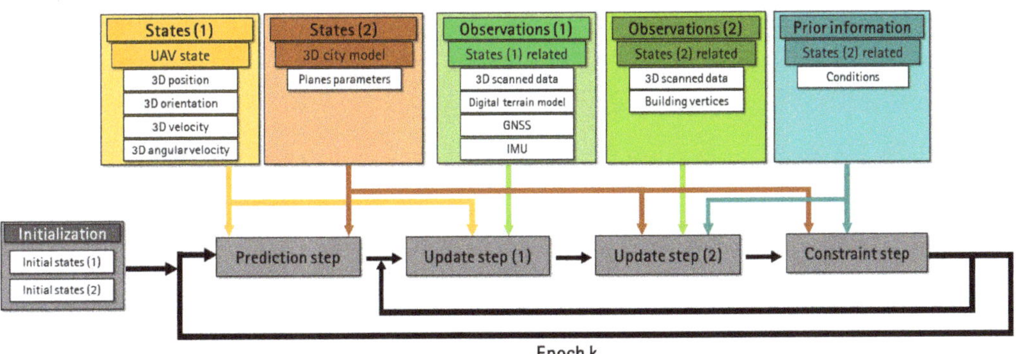

Figure 4. Simplified workflow of the DSIEKF algorithm.

2.3.1. State Vector

As mentioned previously, there are two separate state vectors that should be estimated in the developed DSIEKF algorithm (see (4)). One that is related to the MSS and consists of the 6-DoF as well as the translational and angular velocities. The second state vector is related to the additional information of the environment, which, in the context of the current paper, consists of the plane parameters. The plane parameters are extracted from the LoD-2 3D city model and are taken as unknown parameters due to errors, such as generalization, which are irresistible while developing city models.

However, since the given information in such models are reliable enough, the convergence of these parameters occurs faster than the first state vector, which is a beneficial aspect of the DSIEKF algorithm, as also explained in Section 2.3. The two state vectors are:

$$
\mathbf{x}^{(1)}_{state,k} = \left[\underbrace{t_{x,k}, t_{y,k}, t_{z,k},}_{\text{translations}} \underbrace{\omega_k, \phi_k, \kappa_k,}_{\text{orientations}} \overbrace{V_{x,k}, V_{y,k}, V_{z,k}}^{\text{translational velocities}}, \overbrace{\Omega_{\omega,k}, \Omega_{\phi,k}, \Omega_{\kappa,k}}^{\text{angular velocities}} \right]^T
$$

$$
= [\mathbf{T}_k, \mathbf{O}_k, \mathbf{V}_k, \mathbf{\Omega}_k]^T \tag{4}
$$

$$
\mathbf{x}^{(2)}_{plane,k} = \left[\underbrace{\mathbf{n}_{1,k}; d_{1,k};}_{\text{plane 1}} \underbrace{\mathbf{n}_{2,k}; d_{2,k};}_{\text{plane 2}} \cdots ; \underbrace{\mathbf{n}_{E,k}; d_{E,k}}_{\text{plane E}} \right]
$$

where $\mathbf{T}$, $\mathbf{O}$, $\mathbf{V}$, and $\mathbf{\Omega}$ are translations, orientations, translational velocities, and angular velocities, respectively. $\mathbf{n}$ and d for each plane are representatives of its normal vector and distance to origin, respectively.

2.3.2. Observation Vector

Similar to the state vector that is separated into two parts, the observations are also divided into two categories each corresponding to one set of the unknown parameters. Correspondence in this case means to have observations that can be linked to the unknown parameters via mathematical formulations. The mathematical formulations, which are referred to as observation models, will then be used in the update step of the filter to modify the predicted states.

As given in (4), the first state vector consists of the MSS pose as well as its velocities. Two sets of observations could be linked to this state vector. One is the measured scanned points by means of the laser scanner and the second are the GNSS and IMU data. The scanned points are related to the first state vector through the transformation equation (see (1)), which should be used while assigning the points to the planes of the 3D city model as well as the DTM cells. These observations are measured in the local coordinate system of the scanner and can be written as follows:

$$
\mathbf{l}^{loc}_{scan,k} = \left[\mathbf{l}^{loc}_{scan,1,k}; \mathbf{l}^{loc}_{scan,2,k}; \cdots ; \mathbf{l}^{loc}_{scan,e,k}; \cdots ; \mathbf{l}^{loc}_{scan,E,k} \right]
$$

$$
\mathbf{l}^{loc}_{scan,e,k} = \left[\mathbf{P}^{loc}_{e,1,k}; \mathbf{P}^{loc}_{e,2,k}; \cdots ; \mathbf{P}^{loc}_{e,i,k}; \cdots ; \mathbf{P}^{loc}_{e,N_e,k} \right] \tag{5}
$$

$$
\mathbf{P}^{loc}_{e,i,k} = \left[x^{loc}_{e,i,k}, y^{loc}_{e,i,k}, z^{loc}_{e,i,k} \right]^T
$$

where $\mathbf{l}^{loc}_{scan,k}$ are all the scanned points assigned to the planes in epoch k, which could be divided into smaller groups each belonging to a specific plane. $\mathbf{l}^{loc}_{scan,e,k}$ shows the scanned points in epoch k that belong to the e-th plane. $\mathbf{P}^{loc}_{e,i,k}$ is the i-th 3D point—with x, y, and z coordinates in the local coordinate system of the scanner—that lies on the e-th plane and N_e is the total number of assigned 3D points to plane e, which are measured in epoch k, respectively.

Similarly, the scanned points assigned to the DTM cells are as follows:

$$
\mathbf{l}^{loc}_{DTM,k} = \left[\mathbf{l}^{loc}_{DTM,1,k}; \mathbf{l}^{loc}_{DTM,2,k}; \cdots ; \mathbf{l}^{loc}_{DTM,c,k}; \cdots ; \mathbf{l}^{loc}_{DTM,C,k} \right]
$$

$$
\mathbf{l}^{loc}_{DTM,c,k} = \mathbf{P}^{loc}_{c,k} = \left[x^{loc}_{c,k}, y^{loc}_{c,k}, z^{loc}_{c,k} \right]^T \tag{6}
$$

where $\mathbf{l}^{loc}_{DTM,k}$ are all the scanned points assigned to the DTM cells in epoch k, which could be divided into smaller groups each belonging to a specific cell. Each group contains one 3D point $\mathbf{l}^{loc}_{DTM,c,k}$, which shows the scanned point in epoch k—with x, y, and z coordinates in the local coordinate system of the scanner—that belongs to the c-th cell. C shows the total number of DTM cells in epoch k to which a scanned point is assigned.

On the other hand, if the GNSS and IMU data are available in each epoch, they can also be related to the first state vector. As the first state vector contains the 6-DoF, which are directly the GNSS and IMU measurements. Consequently, the first observation vector corresponding to the first state vector is as follows:

$$
\mathbf{l}_k^{(1)} = \left[\mathbf{1}_{scan,k}^{loc}; \mathbf{1}_{DTM,k}^{loc}; \mathbf{1}_{pose,k}^{glo} \right]
$$

$$
\mathbf{1}_{pose,k}^{glo} = \left[\underbrace{t_{x,k}^{GNSS}, t_{y,k}^{GNSS}, t_{z,k}^{GNSS}}_{\text{GNSS data}}, \underbrace{\omega_k^{IMU}, \phi_k^{IMU}, \kappa_k^{IMU}}_{\text{IMU data}} \right]^T \tag{7}
$$

However, as previously mentioned, due to technical issues, GNSS and IMU data ($\mathbf{1}_{pose,k}^{glo}$) were not reliable enough for the prototype UAV, and hence they are not considered in the current paper.

With a similar justification as for the first state vector, the corresponding observations of the second state vector should also be selected. In the current paper, these measurements are not only the scanned data of the laser scanner but also the corner points of the 3D city model. These two sets of measurements could be related to the second state vector via the plane equation.

Otherwise stated, both the transformed scanned data of the laser scanner as well as the corner points of the facades in the 3D city model should satisfy the mathematical equation of their corresponding plane if, and only if, the plane parameters are estimated correctly. Equation (8) shows the corner points measurements with a similar definition as given for Equation (5). The only difference is that the corner points are already in the global coordinate system, and hence the transformation is not applied to them.

$$
\begin{aligned}
\mathbf{1}_{corner,k}^{glo} &= \left[\mathbf{1}_{corner,1,k}^{glo}; \mathbf{1}_{corner,2,k}^{glo}; \cdots; \mathbf{1}_{corner,e,k}^{glo}; \cdots; \mathbf{1}_{corner,E,k}^{glo} \right] \\
&= \left[\mathbf{P}_{e,1,k}^{glo}; \mathbf{P}_{e,2,k}^{glo}; \cdots; \mathbf{P}_{e,i,k}^{glo}; \cdots; \mathbf{P}_{e,N_c,k}^{glo} \right] \\
\mathbf{P}_{e,i,k}^{glo} &= \left[X_{e,i,k}^{glo}, Y_{e,i,k}^{glo}, Z_{e,i,k}^{glo} \right]^T
\end{aligned}
\tag{8}
$$

Consequently, the second observation vector related to the second state vector can be written as follows:

$$
\mathbf{l}_k^{(2)} = \left[\mathbf{1}_{scan,k}^{loc}; \mathbf{1}_{corner,k}^{glo} \right]
\tag{9}
$$

2.3.3. System Equations

System equations are used in the prediction step of the DSIEKF to model the dynamic behavior of the MSS. Using such equations, it is possible to derive the state of the MSS in the current epoch based on its states in the previous epoch. In [50], a detailed dynamic model for UAVs is given that takes into account significant effects and disturbances. In the current paper, however, a simple linear model is used. The reason for this is the high number of observations as well as the integration of the additional information in the update step, which we think could compensate the simply used dynamic model. In the proposed methodology of DSIEKF, two system equations are used each for one of the state vectors, which can be written as (10):

$$
\begin{aligned}
\mathbf{x}_k^{(1)} &= \mathbf{f}^{(1)} \left(\mathbf{x}_{k-1}^{(1)}, \mathbf{u}_{k-1}^{(1)}, \omega_{k-1}^{(1)} \right) \\
\mathbf{x}_k^{(2)} &= \mathbf{f}^{(2)} \left(\tilde{\mathbf{x}}_k^{(2)}, \mathbf{u}_k^{(2)}, \omega_k^{(2)} \right)
\end{aligned}
\tag{10}
$$

where $\mathbf{x}_k^{(1)}$, $\mathbf{f}^{(1)}$, $\mathbf{x}_{k-1}^{(1)}$, $\mathbf{u}_{k-1}^{(1)}$, and $\omega_{k-1}^{(1)}$ all belong to the first state vector and indicate the states in epoch k, the corresponding (nonlinear) system equation, the states in epoch $k - 1$, the control unit, and the system noise. A similar definition applies to the elements of the second Equation in (10), except that, for the second state vector, the already available information in the current epoch $(\tilde{\mathbf{x}}_k^{(2)})$ are to be modified. In the current paper, linear system equations are claimed to be sufficient within the prediction step. Therefore, such equations for both of the state vectors can be written as follows:

$$
\begin{aligned}
\mathbf{x}_k^{(1)} &= \mathbf{F}_x^{(1)} \cdot \mathbf{x}_{k-1}^{(1)} + \omega_{k-1}^{(1)} \\
\mathbf{x}_k^{(2)} &= \mathbf{F}_x^{(2)} \cdot \tilde{\mathbf{x}}_k^{(2)} + \omega_k^{(2)} \\
\mathbf{F}_x^{(1)} &= \begin{bmatrix} \mathbf{I}_{[6\times6]} & diag([\Delta\tau,\Delta\tau,\Delta\tau,\Delta\tau,\Delta\tau,\Delta\tau]) \\ \mathbf{0}_{[6\times6]} & \mathbf{I}_{[6\times6]} \end{bmatrix} \\
\mathbf{F}_x^{(2)} &= \mathbf{I}_{[4\cdot E_k \times 4\cdot E_k]}
\end{aligned}
\tag{11}
$$

where $\mathbf{F}_x^{(1)}$ and $\mathbf{F}_x^{(2)}$ are the "transition matrices", $\mathbf{I}$ is an identity matrix, $\Delta\tau$ is the time period between two consecutive epochs, and E_k is the total number of planes that are detected in epoch k.

The Variance Covariance Matrix (VCM) of the system noise consists of two parts: a part related to the MSS states and a part related to the planes of the 3D city model. To build the predicted VCM of the MSS states, the concept of "Continuous White Noise Acceleration Model" of [51] is used within the current paper. In this model, the main idea is the contribution of the MSS acceleration to the VCM of the system noise. The authors in [51] claimed that, in systems with constant velocity, the expected acceleration over all the epochs is zero; however, the velocity undergoes slight changes, and therefore the VCM of the system noise is affected by the resulting random acceleration. Inspired by such a claim, the VCM of the system noise within the context of this paper is formulated as follows:

$$
\begin{aligned}
\Sigma_{\omega\omega}^{(1)} &= \Sigma_{\omega\omega,state^{(1)}} = \mathbf{Q} \cdot \tilde{\mathbf{q}} \\
\mathbf{Q} &= \begin{bmatrix} \frac{(\Delta\tau)^3}{3} \cdot \mathbf{I}_{6\times6} & \mathbf{0}_{6\times6} \\ \mathbf{0}_{6\times6} & (\Delta\tau) \cdot \mathbf{I}_{6\times6} \end{bmatrix} \\
\tilde{\mathbf{q}} &= diag(\mathbf{q}_T, \mathbf{q}_O, \mathbf{q}_V, \mathbf{q}_\Omega) \\
\mathbf{q}_T &= [a, a, a] \; , \quad \mathbf{q}_O = [b, b, b] \; , \quad \mathbf{q}_V = [c, c, c] \; , \quad \mathbf{q}_\Omega = [d, d, d]
\end{aligned}
\tag{12}
$$

where $\Sigma_{\omega\omega}^{(1)}$ is the VCM of the system noise corresponding to the first state vector, $\mathbf{Q}$ is derived based on the particular solution of the superposition law that is applied to the system noise (more detail provided in [51]), and $\tilde{\mathbf{q}}$ are the continuous-time process noise intensities that appear as parameters to be designed, which are assumed to be constant over time for all the states parameters. Numerical values considered for the application within the current paper are given in Table 1.

Table 1. Applied accuracy and design parameters ($\tilde{q}$) of the system noise values.

Initial State Accuracy	$\sigma_{T,0} = 0.5$ m $\sigma_{O,0} = 0.2°$ $\sigma_{V,0} = 1$ m/s $\sigma_{\Omega,0} = 1°/\text{s}$ $\sigma_{n,0} = 0.001$ $\sigma_{d,0} = 0.03$ m $\sigma_{CP,0} = 0.03$ m
System Noise	$\Delta\tau = $ approx. 0.1 s between each two epochs $a = 270$ m^2/s^3 $b = 0.08$ rad^2/s^3 $c = 0.02$ m^2/s^3 $d = 7.62 \times 10^{-4}$ rad^2/s^3 $\sigma_{n,\omega} = 0$ $\sigma_{d,\omega} = 0$
Measurement Noise	$\sigma_{LS} = 0.02$ m $\sigma_{CP} = 0.03$ m $\sigma_{DTM} = 0.2$ m

2.3.4. Measurement Equations

As already mentioned in Section 2.3.2, observation models link the measurements to the unknown parameters. These models are then used in the update step of the filter to modify the predictions of the states. As also explained in Section 1.2, observation models are of two types referred to as GMM and GHM. They can be formulated as follows (the first equation corresponding to GMM and the second equation corresponding to GHM):

$$\mathbf{l}_k + \boldsymbol{\nu}_k = \mathbf{H}_{x,k}\, \mathbf{x}_k$$
$$\mathbf{h}(\mathbf{l}_k + \boldsymbol{\nu}_k, \mathbf{x}_k) = \mathbf{0} \tag{13}$$

where $\mathbf{l}$, $\mathbf{x}$, $\mathbf{H}_x$, $\boldsymbol{\nu}$, $\mathbf{h}$ are the measurements vector, state vector, design matrix, measurements noise with VCM $\boldsymbol{\Sigma}_{\nu\nu}$, and the nonlinear observation model in epoch k, respectively.

For the first state vector, if GNSS and IMU data are available, an observation model of GMM type could be used, which can be written as follows:

$$\mathbf{l}^{glo}_{pose,k} + \boldsymbol{\nu}_k = \mathbf{x}^{(1)}_{state,pose,k}$$

$$\mathbf{x}^{(1)}_{state,pose,k} = \left[\underbrace{t_{x,k}, t_{y,k}, t_{z,k}}_{\text{translations}}, \underbrace{\omega_k, \phi_k, \kappa_k}_{\text{orientations}} \right]^T \tag{14}$$

where $\mathbf{l}^{glo}_{pose,k}$, $\boldsymbol{\nu}_k$, and $\mathbf{x}^{(1)}_{state,pose,k}$ are the GNSS and IMU data as measurements, observations noise, and the unknown 6-DoF, respectively. These 6-DoF which are part of the first state vector are the pose parameters.

However, in the current paper, due to unavailable GNSS and IMU data, such GMM observation models are not considered. Nonetheless, the prototype UAV is also equipped with a 3D laser scanner, which constantly captures the surrounding environment. Furthermore, corner points of the buildings within the 3D city model are considered as additional observations to ensure higher redundancy and, hence, more accurate estimations.

On the other hand, as explained in Section 2.3.1, one of the state vectors to be estimated contains geometrical parameters of the surrounding environment. In order to derive this state vector, a suitable relationship should be established that links it to the 3D scanned data from the laser scanner and corner points of the 3D city model. Such a relationship could be the Hesse form of a plane, which has a general type as the second Equation in (13) and, hence, is a GHM:

$$\mathbf{n}_{xk} \cdot \mathbf{X}_k^{glo} + \mathbf{n}_{yk} \cdot \mathbf{Y}_k^{glo} + \mathbf{n}_{zk} \cdot \mathbf{Z}_k^{glo} - \mathbf{d}_k = \mathbf{0} \tag{15}$$

where $\mathbf{n}_x$, $\mathbf{n}_y$, and $\mathbf{n}_z$ are the planes' normal vector parameters. $\mathbf{d}$ is the planes' distance parameter vector. $\mathbf{X}_k^{glo}$, $\mathbf{Y}_k^{glo}$, and $\mathbf{Z}_k^{glo}$ are the global point coordinates. In case of the scanned points by means of a 3D laser scanner, a transformation should be applied, which is to bring the points from the local coordinate system to the global one. Such a transformation is done according to (1). In case of corner points, no transformation is required, since these points are already in the same coordinate system in which the plane parameters are defined.

As explained in Section 2.1, the ground is defined by relating the scanned points to the DTM of the environment. Such a relation is derived based on the following equation of GHM type, which compares the transformed height of the scanned point in the global coordinate system to that of the DTM cell:

$$\mathbf{Z}^{glo,DTM} - \mathbf{Z}_{DTM} = 0 \tag{16}$$

where $\mathbf{Z}^{glo,DTM}$ and $\mathbf{Z}_{DTM}$ are the transformed height of the scanned points to the global coordinate system and height information of the DTM, respectively. Unlike the plane parameters that are taken as unknown parameters through the filter, the height information is taken as deterministic values, and therefore they are not modified.

2.3.5. Nonlinear Equality Constraints

Aside from the additional geometrical information of the environment, sometimes it is also possible to engage certain restrictions that lead to more accurate estimations. Such restrictions, which are also taken from the surroundings, appear as mathematical formulations that could be integrated into the filter. In the current paper, the unity of the planes' normal vectors are taken as the geometrical constraints that have to be fulfilled by the final estimated plane parameters.

There are a number of methods for applying geometrical constraints on the filtered states [38]. In the current paper, the projection approach is used, mainly since it is the chosen strategy in [1,30,31] as well. The considered geometrical constraint could be formulated as follows:

$$\sqrt{n_x^2 + n_y^2 + n_z^2} = 1 \tag{17}$$

where n_x, n_y, and n_z are the normal vector parameters of each plane. This geometrical constraint is of nonlinear type and, therefore, a Taylor series expansion is used for linearization within the projection method in [1,6]. The linearized form of a nonlinear constraint (as (18)) by means of the first degree Taylor polynomial is as given in (19).

$$g(x_k) = b \tag{18}$$

where g is a nonlinear mathematical function that has an exact value of b in a true state x_k.

$$g(x_k) = g(\hat{x}_k^-) + g'(\hat{x}_k^-)(x_k - \hat{x}_k^-) \tag{19}$$

where $\hat{x}_k^-$ and g' are the predicted values of x_k and the derivative of g with respect to $\hat{x}_k^-$, respectively.

2.3.6. Initialization

Initializing KF realizations is always a challenging task, as inappropriate initial values lead to filter divergence and, hence, unreliable outputs. Therefore, it is preferable to choose such values, reasonably and based on proper judgments of the overall scenario. For the current paper, the initial translation and orientation parameters are taken from the GNSS and IMU data at the beginning of the experiment. The initial velocities are taken as zero values, and the initial plane parameters are extracted directly from the LoD-2 3D city model. Similar to the parameters of the state, their VCM should also be reasonably initiated.

In (20)–(27), the order and arrangement of the initial state parameters as well as their VCM is given. Numerical values considered for the application within the current paper are given in Table 1.

$$\mathbf{x}^{(1)}_{state,0} = \left[\mathbf{T}^{GNSS}_0, \mathbf{O}^{IMU}_0, \mathbf{V}_0, \mathbf{\Omega}_0 \right]^T \tag{20}$$

$$\mathbf{T}^{GNSS}_0 = \left[t^{GNSS}_{x,0}, t^{GNSS}_{y,0}, t^{GNSS}_{z,0} \right]^T \ , \ \mathbf{O}^{IMU}_0 = \left[\omega^{IMU}_0, \phi^{IMU}_0, \kappa^{IMU}_0 \right]^T \tag{21}$$

$$\mathbf{V}_0 = [0,0,0]^T \ , \ \mathbf{\Omega}_0 = [0,0,0]^T \tag{22}$$

$$\mathbf{x}^{(2)}_{plane,0} = [\mathbf{n}_{1,0}; d_{1,0}; \mathbf{n}_{2,0}; d_{2,0}; \ldots; \mathbf{n}_{E,0}; d_{E,0}] \tag{23}$$

$$\mathbf{\Sigma}^{(1)}_{xx,0} = diag(\sigma^2_{T,0}, \sigma^2_{O,0}, \sigma^2_{V,0}, \sigma^2_{\Omega,0}) \tag{24}$$

$$\sigma^2_{T,0} = diag(\sigma^2_{T,0}, \sigma^2_{T,0}, \sigma^2_{T,0}) \ , \ \sigma^2_{O,0} = diag(\sigma^2_{O,0}, \sigma^2_{O,0}, \sigma^2_{O,0}) \tag{25}$$

$$\sigma^2_{V,0} = diag(\sigma^2_{V,0}, \sigma^2_{V,0}, \sigma^2_{V,0}) \ , \ \sigma^2_{\Omega,0} = diag(\sigma^2_{\Omega,0}, \sigma^2_{\Omega,0}, \sigma^2_{\Omega,0}) \tag{26}$$

$$\mathbf{\Sigma}^{(2)}_{xx,0} = diag(\sigma^2_{n,0}, \sigma^2_{d,0}, \sigma^2_{n,0}, \sigma^2_{d,0}, \ldots, \sigma^2_{n,0}, \sigma^2_{d,0}) \tag{27}$$

2.4. Workflow

The workflow of the developed DSIEKF that is applied to the real scenario is given in Algorithm 1. For better comprehension of the overall procedure, a sketch of the algorithm is also provided in Figure 5. The given algorithm is based on the one given by [6] but modified and made consistent with the real environment. The modification lies mainly in the VCM of the system noise and the integration of the DTM model into the filter as explained in Sections 2.1 and 2.3.3, respectively.

Furthermore, the information-based georeferencing methodology—referred to as IEKF in the current paper—that was developed by [30] and applied by [1] on a simulated UAV environment is also considered. The main reason for doing so is to compare the two methodologies. When it comes to real environments, no true values exist, and therefore comparing the outputs of two algorithms on a common scenario could be one way to examine the final results. In lines 1 to 4 of Algorithm 1, the corresponding system and observation models of the two state vectors are defined.

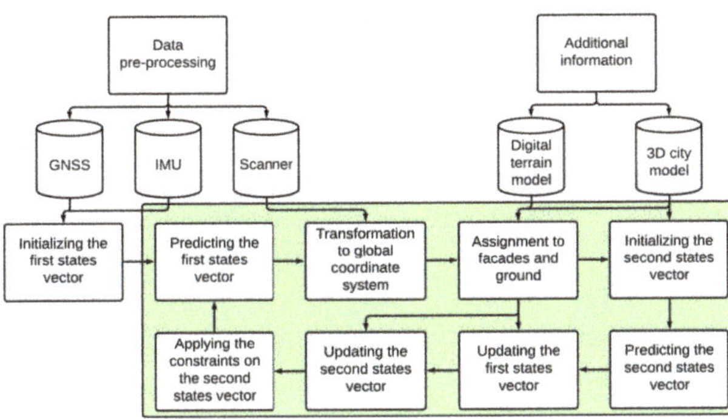

Figure 5. Sketch of the DSIEKF workflow in the context of the current paper. Steps within the green area are applied in each epoch. the GNSS and IMU are only used for the initialization step in the current paper.

Algorithm 1: Dual State Iterated Extended Kalman Filter (DSIEKF) with nonlinear equality constraints.

1   System model 1: $\mathbf{x}_k^{(1)} = \mathbf{f}^{(1)}(\mathbf{x}_{k-1}^{(1)}, \mathbf{u}_{k-1}^{(1)}, \omega_{k-1}^{(1)}), \quad \omega_{k-1}^{(1)} \sim N\left(0, \Sigma_{\omega\omega}^{(1)}\right)$

2   System model 2: $\mathbf{x}_k^{(2)} = \mathbf{f}^{(2)}(\tilde{\mathbf{x}}_k^{(2)}, \mathbf{u}_k^{(2)}, \omega_k^{(2)}), \quad \omega_k^{(2)} \sim N\left(0, \Sigma_{\omega\omega}^{(2)}\right)$

3   Observation model 1: $\mathbf{h}^{(1)}(\mathbf{1}_k^{(1)} + v_k^{(1)}, \mathbf{x}_k^{(1)}, \mathbf{x}_k^{(2)}) = 0, \quad v_k^{(1)} \sim N\left(0, \Sigma_{vv}^{(1)}\right)$

4   Observation model 2: $\mathbf{h}^{(2)}(\mathbf{1}_k^{(2)} + v_k^{(2)}, \mathbf{x}_k^{(1)}, \mathbf{x}_k^{(2)}) = 0, \quad v_k^{(2)} \sim N\left(0, \Sigma_{vv}^{(2)}\right)$

5   Initialization state vector 1: $\hat{\mathbf{x}}_0^{(1+)} = \mathbf{x}_0^{(1)}, \quad \Sigma_{\hat{x}\hat{x},0}^{(1+)} = \Sigma_{xx,0}^{(1)}, \quad k = 1$

6   Initialization of a matrix for storing the filtered second state vector: $\mathbf{C} = [\,]$

7   **while** $k < \mathrm{K}$ **do**

8   *Extraction of the Captured Planes Data from the 3D City Model*

9     $\hat{\mathbf{x}}_0^{(2+)} = \mathbf{x}_0^{(2)}, \quad \Sigma_{\hat{x}\hat{x},0}^{(2+)} = \Sigma_{xx,0}^{(2)}$

10   Only those planes that are <u>not</u> in $\mathbf{C}$ will be filtered.

11   *Prediction Step*

12    $\mathbf{F}_{x,k}^{(1)} = \partial\mathbf{f}^{(1)}/\partial\mathbf{x}^{(1)}\big|_{\hat{\mathbf{x}}_{k-1}^{(1+)}, \mathbf{u}_{k-1}^{(1)}, \omega_{k-1}^{(1)}}, \quad \mathbf{F}_{\omega,k}^{(1)} = \partial\mathbf{f}^{(1)}/\partial\omega^{(1)}\big|_{\hat{\mathbf{x}}_{k-1}^{(1+)}, \mathbf{u}_{k-1}^{(1)}, \omega_{k-1}^{(1)}}$

13    $\hat{\mathbf{x}}_k^{(1-)} = \mathbf{F}_{x,k}^{(1)}\hat{\mathbf{x}}_{k-1}^{(1+)} + \omega_{k-1}^{(1)}, \quad \Sigma_{xx,k}^{(1-)} = \mathbf{F}_{x,k}^{(1)}\Sigma_{\hat{x}\hat{x},k-1}^{(1)}\mathbf{F}_{x,k}^{(1)^T} + \mathbf{F}_{\omega,k}^{(1)}\Sigma_{\omega\omega}^{(1)}\mathbf{F}_{\omega,k}^{(1)^T}$

14    $\mathbf{F}_{x,k}^{(2)} = \partial\mathbf{f}^{(2)}/\partial\mathbf{x}^{(2)}\big|_{\hat{\mathbf{x}}_k^{(2+)}, \mathbf{u}_k^{(2)}, \omega_k^{(2)}}, \quad \mathbf{F}_{\omega,k}^{(2)} = \partial\mathbf{f}^{(2)}/\partial\omega^{(2)}\big|_{\hat{\mathbf{x}}_k^{(2+)}, \mathbf{u}_k^{(2)}, \omega_k^{(2)}}$

15    $\hat{\mathbf{x}}_k^{(2-)} = \mathbf{F}_{x,k}^{(2)}\hat{\mathbf{x}}_k^{(2+)} + \omega_k^{(2)}, \quad \Sigma_{xx,k}^{(2-)} = (\lambda^{-1} - 1)\Sigma_{\hat{x}\hat{x},k-1}^{(2+)}, \quad \lambda \in (0,1]$

16   *Update Step*

17    $\breve{\mathbf{l}}_{k,0}^{(1)} = \mathbf{1}_k^{(1)}, \quad \check{\mathbf{x}}_{k,0}^{(1)} = \hat{\mathbf{x}}_k^{(1-)}, \quad \breve{\mathbf{l}}_{k,0}^{(2)} = \mathbf{1}_k^{(2)}, \quad \check{\mathbf{x}}_{k,0}^{(2)} = \hat{\mathbf{x}}_k^{(2-)}, \quad \Delta\mathbf{x} = \infty, \quad n = 0$

18   **for** $m = 0 \ldots M - 1$ **do**

19      $\mathbf{H}_{x,k,m}^{(1)} = \partial\mathbf{h}^{(1)}/\partial\mathbf{x}^{(1)}\big|_{\breve{\mathbf{l}}_{k,m}^{(1)}, \check{\mathbf{x}}_{k,m}^{(1)}, \check{\mathbf{x}}_{k,m}^{(2)}}, \quad \mathbf{H}_{l,k,m}^{(1)} = \partial\mathbf{h}^{(1)}/\partial\mathbf{l}^{(1)}\big|_{\breve{\mathbf{l}}_{k,m}^{(1)}, \check{\mathbf{x}}_{k,m}^{(1)}, \check{\mathbf{x}}_{k,m}^{(2)}}$

20      $\mathbf{O}_{k,m}^{(1)} = \mathbf{H}_{x,k,m}^{(1)}\Sigma_{xx,k}^{(1-)}\mathbf{H}_{x,k,m}^{(1)^T}, \quad \mathbf{S}_{k,m}^{(1)} = \mathbf{H}_{l,k,m}^{(1)}\Sigma_{vv}^{(1)}\mathbf{H}_{l,k,m}^{(1)^T}$

21      $\mathbf{K}_{k,m}^{(1)} = \Sigma_{xx,k}^{(1-)}\mathbf{H}_{x,k,m}^{(1)^T}(\mathbf{O}_{k,m}^{(1)} + \mathbf{S}_{k,m}^{(1)})^{-1}, \quad \mathbf{r}_{k,m}^{(1)} = \mathbf{H}_{l,k,m}^{(1)}(\mathbf{1}_k^{(1)} - \breve{\mathbf{l}}_{k,m}^{(1)}) + \mathbf{H}_{x,k,m}^{(1)}(\hat{\mathbf{x}}_k^{(1-)} - \check{\mathbf{x}}_{k,m}^{(1)})$

22      $\check{\mathbf{x}}_{k,m+1}^{(1)} = \hat{\mathbf{x}}_k^{(1-)} - \mathbf{K}_{k,m}^{(1)}(\mathbf{h}^{(1)}(\breve{\mathbf{l}}_{k,m}^{(1)}, \check{\mathbf{x}}_{k,m}^{(1)}, \check{\mathbf{x}}_{k,m}^{(2)}) + \mathbf{r}_{k,m}^{(1)})$

23      $\mathbf{G}_{k,m}^{(1)} = \Sigma_{vv}^{(1)}\mathbf{H}_{l,k,m}^{(1)^T}(\mathbf{O}_{k,m}^{(1)} + \mathbf{S}_{k,m}^{(1)})^{-1}$

24      $\breve{\mathbf{l}}_{k,m+1}^{(1)} = \mathbf{1}_k^{(1)} - \mathbf{G}_{k,m}^{(1)}(\mathbf{h}^{(1)}(\breve{\mathbf{l}}_{k,m}^{(1)}, \check{\mathbf{x}}_{k,m}^{(1)}, \check{\mathbf{x}}_{k,m}^{(2)}) + \mathbf{r}_{k,m}^{(1)})$

25   **while** $\max(\Delta\mathbf{x}) > c_{stop}$ **do**

26        $\breve{\mathbf{l}}_{k,n}^{(2)} = \breve{\mathbf{l}}_{k,m}^{(2)}, \quad \check{\mathbf{x}}_{k,n}^{(2)} = \check{\mathbf{x}}_{k,m}^{(2)}$

27   A similar computation to lines 19 to 24 should be done for the second state vector.

28        $\mathbf{L}_{k,n}^{(2)} = \mathbf{I}_{u_2 \times u_2} - \mathbf{K}_{k,n}^{(2)}\mathbf{H}_{x,k,n}^{(2)}$

29        $\Sigma_{\hat{x}\hat{x},k,n+1}^{(2+)} = \mathbf{L}_{k,n}^{(2)}\Sigma_{xx,k}^{(2-)}\mathbf{L}_{k,n}^{(2)^T} + \mathbf{K}_{k,n}^{(2)}\mathbf{S}_{k,n}^{(2)}\mathbf{K}_{k,n}^{(2)^T}$

30   *Constraint Step*

31        $\mathbf{D} = g'(\hat{\mathbf{x}}_k^{(2-)}), \quad \mathbf{d} = \mathbf{b} - g(\hat{\mathbf{x}}_k^{(2-)}) + g'(\hat{\mathbf{x}}_k^{(2-)})\check{\mathbf{x}}_{k,n}^{(2)}, \quad \mathbf{W} = \underset{j \times j}{\mathbf{I}}$

32        $\check{\mathbf{x}}_{k,n+1}^{(2)} = \check{\mathbf{x}}_{k,n+1}^{(2)} - \mathbf{W}^{-1}\mathbf{D}^T(\mathbf{D}\mathbf{W}^{-1}\mathbf{D}^T)^{-1}(\mathbf{D}\check{\mathbf{x}}_{k,n+1}^{(2)} - \mathbf{d})$

33        $\Sigma_{\hat{x}\hat{x},k,n+1}^{(2+)} = \Sigma_{\hat{x}\hat{x},k,n+1}^{(2+)} - \Sigma_{\hat{x}\hat{x},k,n+1}^{(2+)}\mathbf{D}^T(\mathbf{D}\Sigma_{\hat{x}\hat{x},k,n+1}^{(2+)}\mathbf{D}^T)^{-1}\mathbf{D}\Sigma_{\hat{x}\hat{x},k,n+1}^{(2+)}$

34        $n = n + 1$

35      $\check{\mathbf{x}}_{k,m+1}^{(2)} = \check{\mathbf{x}}_{k,n+1}^{(2)}, \quad \breve{\mathbf{l}}_{k,m+1}^{(2)} = \breve{\mathbf{l}}_{k,n+1}^{(2)}, \quad \Delta\mathbf{x} = \check{\mathbf{x}}_{k,n+1}^{(2)} - \check{\mathbf{x}}_{k,n}^{(2)}$

36    $\hat{\mathbf{x}}_k^{(1+)} = \check{\mathbf{x}}_{k,M-1}^{(1)}, \quad \hat{\mathbf{l}}_k^{(1+)} = \breve{\mathbf{l}}_{k,M-1}^{(1)}$

37    $\mathbf{S}_{k,M-1}^{(1)} = \mathbf{h}_{l,k,M-1}^{(1)}\Sigma_{vv}^{(1)}\mathbf{H}_{l,k,M-1}^{(1)^T}, \quad \mathbf{l}_k^{(1)} = \mathbf{I}_{u_1 \times u_1} - \mathbf{K}_{k,M-1}^{(1)}\mathbf{h}_{x,k,M-1}^{(1)}$

38    $\Sigma_{\hat{x}\hat{x},k}^{(1+)} = \mathbf{l}_k^{(1)}\Sigma_{xx,k}^{(1-)}\mathbf{L}_k^{(1)^T} + \mathbf{K}_{k,M-1}^{(1)}\mathbf{S}_{k,M-1}^{(1)}\mathbf{K}_{k,M-1}^{(1)^T}$

39    $\hat{\mathbf{x}}_k^{(2+)} = \check{\mathbf{x}}_{k,n+1}^{(2)}, \quad \hat{\mathbf{l}}_k^{(2+)} = \breve{\mathbf{l}}_{k,n+1}^{(2)}$

40    $\Sigma_{\hat{x}\hat{x},k}^{(2+)} = \Sigma_{\hat{x}\hat{x},k,n+1}^{(2+)}$

41    $\mathbf{C} = [\mathbf{C}; (\text{idx}, \hat{\mathbf{x}}_k^{(2+)}, \Sigma_{\hat{x}\hat{x},k}^{(2+)})]$, "idx" are indices of the captured planes in the current epoch.

42    $k = k + 1$

In line 5, the first state vector and its corresponding VCM are initialized. In line 6, a matrix is initialized, which will be filled in as given in line 41. The main aim is to store the modified planes in this matrix so that they are not modified anymore if they are captured in the next epochs (as also indicated in line 10). In each epoch, the scanned data are assigned to the surrounding building models, and thus the planes information could directly be extracted from the 3D city model.

This extracted information along with the considered accuracy values are taken as the initialized values of the second state vector and its corresponding VCM (as given in line 9). Lines 12 to 15 of the algorithm deal with predicting both of the state vectors and their corresponding VCM. For predicting the VCM of the second state vector (line 15), a user-defined factor λ is used, which is referred to as the "forgetting factor" and ranges from 0 to 1 [33]. The main idea of using such a factor is to have an exponentially decaying weighting on the past data [33].

For the current paper, this factor was chosen to be 0.5. Lines 17 to 40 show the update steps in which both of the predicted state vectors are modified by taking into their corresponding observations. The first update step (lines 19 to 24) deals with modifying the first state vector and its corresponding observations within an outer loop indicated by m. Afterward, the second state vector and its corresponding observations should be modified, which is indicated in line 27 and happens within an inner loop indicated by n.

Therefore, similar computations as of lines 19 to 24 should be repeated but this time for the second state vector. In this update step, anywhere that the first state vector and its corresponding observations are needed, the modified ones from lines 22 and 24 should be used. Similarly, for updating the first state vector, anywhere that the second state vector and its corresponding observations appear, the most recent modified ones should be used.

Since the geometrical constraints should be applied on the second state vector, they are applied in the inner loop while updating the second state vector (lines 31 to 33). The inner loop is only entered if $\max(\Delta\mathbf{x})$ exceeds a certain threshold indicated by c_{stop}. $\Delta\mathbf{x}$ is the difference between the updated second state vector in two consecutive iterations. c_{stop} is chosen to be 10^{-4} in the current paper.

3. Application

In geodesy, capturing the surrounding environment for further analytical purposes is a common task that is sought in different ways based on the demand. In certain cases, static measurements are sufficient to partly survey the environment; whereas, in other cases, kinematic measurements are preferred in order to capture different parts of it. Various possibilities exist when it comes to kinematic measurements among which using trolleys or vehicles equipped with multiple sensors could be mentioned. However, sometimes the environment is complicated enough to not allow using the aforementioned kinematic means.

In those cases, the UAVs are proper choices for mapping purposes, which can usually be steered in desired heights and along favorable paths. They usually have a compact and light-weight design, which enables their flexible movements. As mentioned previously, to investigate different aspects of UAV-based problems, the GIH and IPI at LUH developed a prototype Unmanned Aerial System (UAS) where a UAV is equipped with GNSS, IMU, a 3D laser scanner, and camera units within the course of a joint project. In Figure 6, a scheme of such a UAV is presented followed by further details in the succeeding sections.

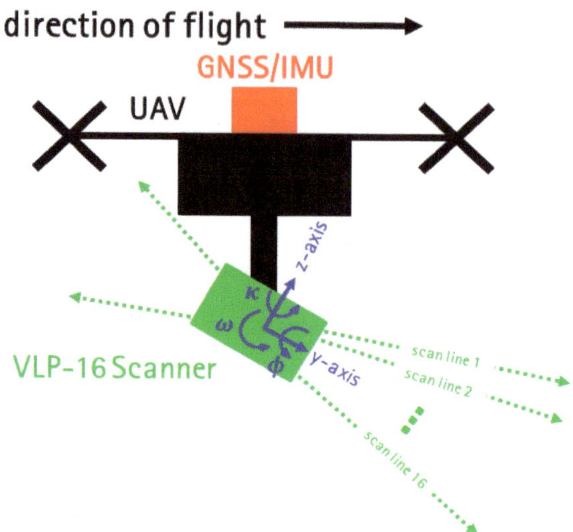

Figure 6. Modified sketch of the UAV platform setup from [1]. The *x*-axis of scanner's local coordinate system is in the reader's line of sight.

3.1. Overview of the Real Environment

The mentioned prototype UAV is equipped with GNSS, IMU, a 3D laser scanner, and camera units mounted on a gimbal for better stabilization and damping. The 3D laser scanner is a Velodyne Puck (VLP-16) with 16 individual scan lines, a 30° × 360° field of view, and a 3 cm range accuracy [52]. The rotation frequency was selected to be 10 Hz, which corresponds to 30,000 3D points per 360° rotation. In order to have a reliable synchronization and data combination with the other mounted sensors, the GNSS time stamps are applied to the scanned data via additional GNSS units. Moreover, although GNSS and IMU data are available through the whole measurement process, they are neglected in this paper. Camera units are used in the joint project; however, they are neglected here, since integrating camera images in the filter is out of the scope of the current paper. An overview of the whole system can be seen in Figure 7.

Figure 7. Overview of the prototype kinematic laser scanner-based UAV [53].

Within the project, an inner courtyard was selected as the environment in which the UAV carried out the measurements at a height of almost 2 m. A map-based representation of the area along with its corresponding images is shown in Figure 8. Such a selected spacious environment has perfect conditions for evaluating the performance of both the DSIEKF and IEKF algorithms.

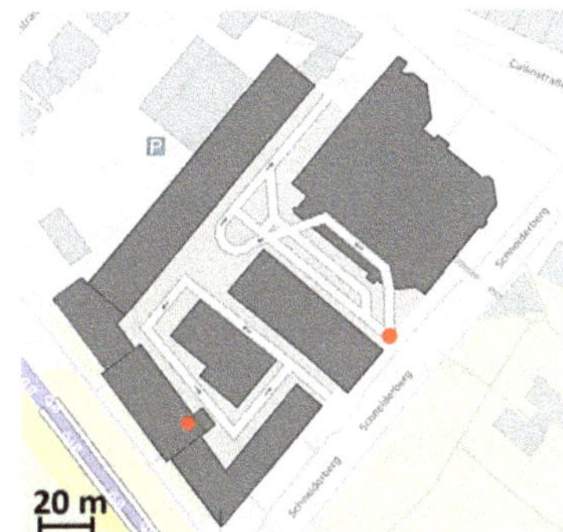

Figure 8. An overview of the measured area. The map-based representation is the middle figure, and on its left and right side are the images corresponding to the left and right red circles, respectively [32].

3.2. Additional Complementary Information

As previously indicated, the IEKF and DSIEKF algorithms are based on available and reliable information that could be extracted from the environment in which the MSS moves through. In the current paper, such information is selected to be the LoD-2 3D city model of Hanover along with the DTM due to their availability for the environment of interest. These complementary information are extracted from [54,55].

In Figure 9, an overview of such additional information for the current project is presented. The DTM in the current paper refers to the German height reference network (Deutsches Haupthöhennetz—DHHN2016) which has a grid width of 1 m with the accuracy of height information less than 0.3 m [55]. Due to the generalization process, the LoD-2 3D city model has deviations up to the centimeter or even decimeter level compared to the real environment, which is the main reason for taking such additional information as unknown parameters to be estimated in the filtering process.

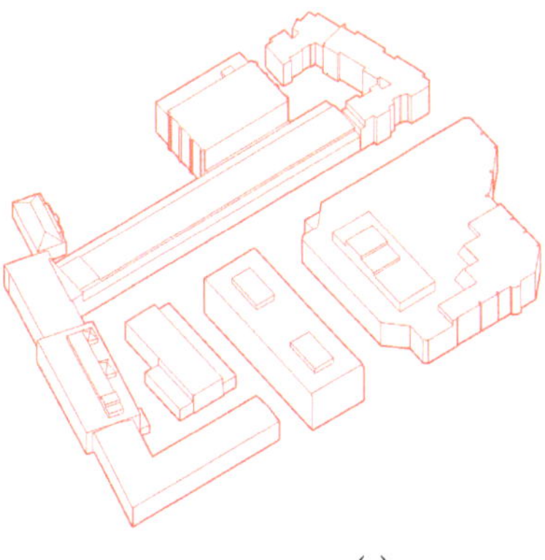

(a)

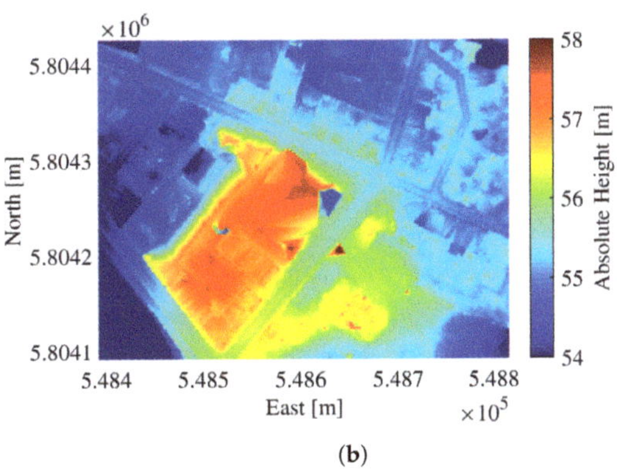

(b)

Figure 9. Object information of the measured environment: (**a**) LoD-2 3D city model, and (**b**) DTM of the area [32].

3.3. Analysis Material and Requisites

The considered values for the initial state accuracy, system, and measurement noise are selected as given in Table 1. In this table, the subscripts "LS" and "CP" refer to the laser scanner and corner points data and the definition of the other subscripts is consistent with those given in the mathematical formulations of this paper. The considered initial state accuracy of translations ($\sigma_{T,0}$), orientations ($\sigma_{O,0}$), and translational and angular velocities ($\sigma_{V,0}$ and $\sigma_{\Omega,0}$) are selected based on the installed sensors on the UAV platform.

For example, the selected value of $\sigma_{T,0} = 0.5$ m is based on the accuracy of the used GNSS unit. Since no uncertainty information of either the plane parameters or the corner points is given in the LoD-2 3D city model, they were selected based on the accuracy of this digitized model, which is usually at the centimeter or even decimeter level. Therefore, the values of $\sigma_{n,0}$, $\sigma_{d,0}$, and $\sigma_{CP,0}$ are user-defined. As previously mentioned in Section 2.3.3, the accuracy of system noise is based on several parameters that should be designed.

Such a design should be based on the system dynamics as well as the affecting elements of its surrounding environment, such as wind. However, modeling these influencing factors is complicated and not in the scope of the current paper. Therefore, to select the values of *a*, *b*, *c*, and *d*, a sensitivity analysis was performed in order to see the effect of different values on the overall estimations and, hence, select those values that lead to more optimized solutions.

Furthermore, in the current paper, the accuracy of the measurement noise for the corner points and DTM heights was also selected based on the accuracy of the LoD-2 and DTM models. Additionally, the initial state vectors were selected accurately enough for the algorithms to not encounter divergence problems. Such a realization is based on a comparison between the first initial values of the GNSS and IMU data and the geographic coordinates of the captured measurement area.

Furthermore, processing all the measurements is, on one hand, computationally inefficient and, on the other hand, unnecessary due to the similar information content of nearby observations. Therefore, down-sampling is an irresistible step that is usually taken into account while analyzing real scenarios. In the current paper, it is applied by defining spatial boxes with a grid size of 1 m around the laser scanner measurements and averaging the neighbored observations.

For that, the already built-in MATLAB library (pcdownsample) was used. Performing this step led to a considerable decrease in the number of scanned data. As an example, in the first epoch, the scanned measurements were reduced in size from around 22,000 points to about 1700 points. However, by applying this down-sampling library, the original observations were changed to artificial ones due to averaging. Nonetheless, since the average is based on the real measurements, it is claimed that they can be treated as original observations to test the performance of the algorithms on a real environment.

In the following, the final results alongside the related discussions are presented. The algorithms are implemented in MATLAB R2019b programming language which are then performed on a windows-based computer with Intel(R) Xeon(R) CPU E5-1650 v2 @ 3.50 GHz 3.50 GHz processor and 64 GB RAM.

4. Results

In the following, the results of the DSIEKF along with those derived from the IEKF algorithm are depicted. As previously mentioned, the main difference between these two methodologies is the way the unknown parameters are treated and estimated within the whole filtering process. Therefore, having both of the algorithms applied to the same real scenario provides a good opportunity to not only compare their performance but also to validate the results of each methodology with respect to the other.

In general, "validation" can be done when the true values of the unknowns are available; however, in real environments, such values are never known. In principle, deriving the reference solutions in practical applications is feasible, but it requires great effort, which might not be always efficient. The authors in [32] derived a linear reference trajectory for a practical application by means of a laser tracker.

However, due to limited range measurements of a laser tracker, such a reference solution is only carried out for a small part of the trajectory. Therefore, other strategies should be taken to evaluate the final results. In the current paper, the procedure for doing so is to compare the estimations of both the algorithms with each other.

In Figure 10, the derived trajectories by means of the two algorithms are given from the top view. the derived trajectories have no considerable divergence from the real covered path by the UAV for they have all remained in the same environment where the UAV moved along.

In Figures 11 and 12, the absolute differences between the estimated translation and orientation parameters over all the epochs are shown, respectively. In these figures, Δt_x, Δt_y, and Δt_z correspond to the absolute differences between the estimated translation parameters along X, Y, and Z axes. Similarly, $\Delta \omega$, $\Delta \phi$, and $\Delta \kappa$ represent the absolute differences between the estimated orientations along the three main axes.

According to Figure 11, the difference between the estimations of t_x and t_y are not as large as t_z. The reason for that is having multiple building facades in the surroundings that provide additional information for the filters. These building models are also modified within the filters, and therefore any errors from sources, such as generalization, are well compensated for through the estimation process. Consequently, both of the filters deliver similar t_x and t_y estimations.

In Figure 12, the κ estimations from the two filters are similar. The reason is due to the small movements of the UAV around the Z axis and the suitable initialization in both filters. In Table 2, the maximum of these values over 1902 epochs is given. the two algorithms greatly disagree with each other in t_z estimations. The reason is claimed to be the DTM height information that is taken as deterministic values in both of the algorithms.

As explained in Section 3.2, the accuracy of height information is less than 0.3 m. Unlike the plane parameters that are taken as unknowns to be modified by the filters, this height information is considered constant, which subsequently leads to less qualitative results compared to the other transformation parameters. Such a claim is also approved by taking the mean standard deviation of the estimated parameters (in 1902 epochs) that are shown in Table 3.

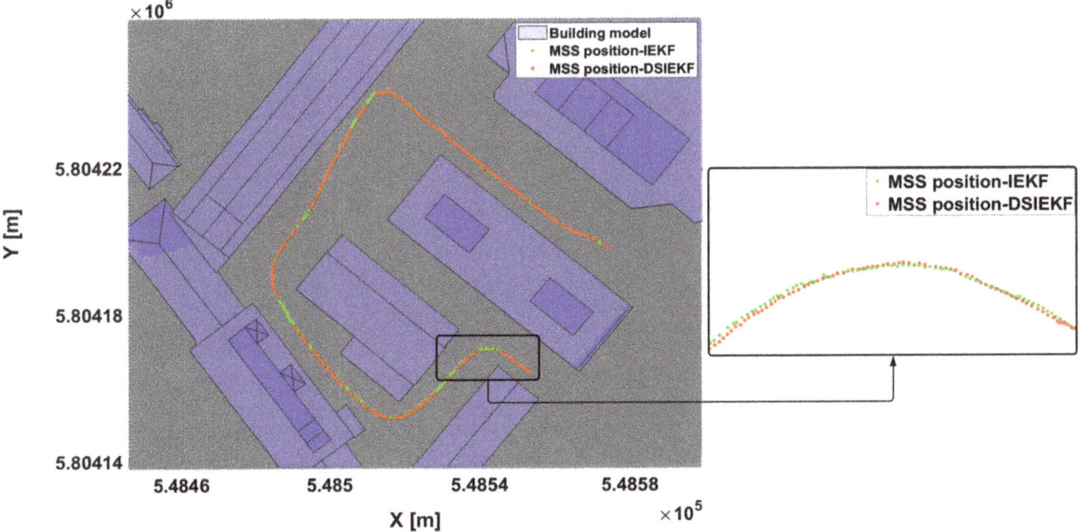

Figure 10. The estimated trajectories in 2D (top view).

Table 2. The maximum difference between the estimated pose parameters over 1902 epochs by means of IEKF and DSIEKF.

Δ_{t_x} [m]	Δ_{t_y} [m]	Δ_{t_z} [m]	Δ_ω [°]	Δ_ϕ [°]	Δ_κ [°]
0.4	0.6	2	0.06	0.08	0.02

Table 3. Measures of different variables over 1902 epochs for each algorithm. T: Duration of the update step, σ: Mean standard deviation of the state parameters.

Algorithm	T [min]	σ_{t_x} [mm]	σ_{t_y} [mm]	σ_{t_z} [mm]	σ_ω [°]	σ_ϕ [°]	σ_κ [°]
IEKF	28	6	6.2	9.4	0.03	0.03	0.004
DSIEKF	15	1.5	1.6	8.8	0.02	0.02	0.003

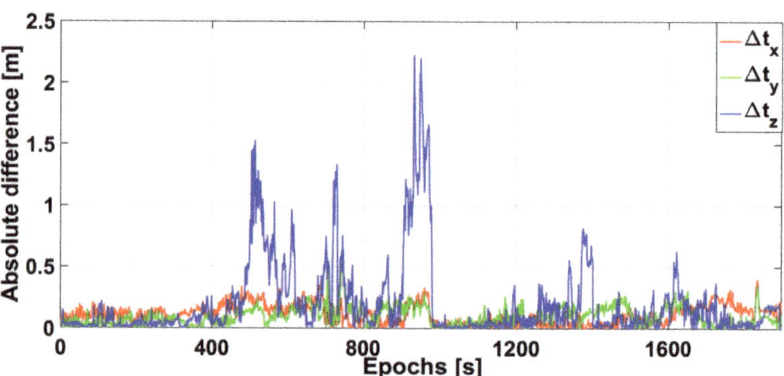

Figure 11. Absolute differences between the estimated translation parameters by means of IEKF and DSIEKF.

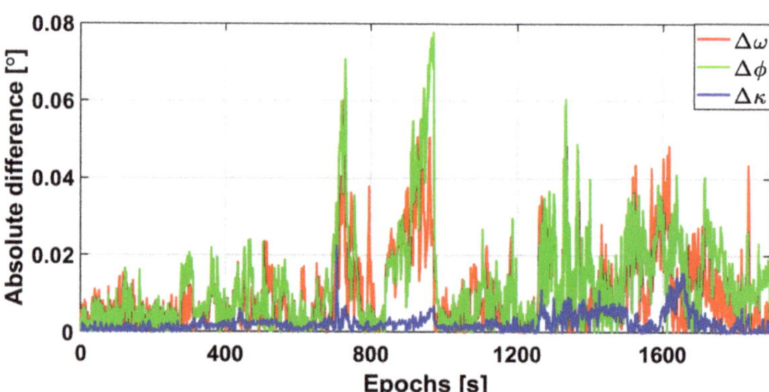

Figure 12. Absolute differences between the estimated orientation parameters by means of IEKF and DSIEKF.

The standard deviation of the estimated translation parameter along the Z axis was larger than the other ones in both of the algorithms. Moreover, by comparing the standard deviation of the other estimated pose parameters, it is seen that, in general, not only did the DSIEKF take a shorter time to run but it also delivered better estimations.

Figures 13 and 14 show the maximum absolute differences between the estimated planes' normal vectors and distances to the origin and their given values in the 3D city model, respectively. In Figure 13 and for each epoch, the maximum value was derived for the absolute differences between the three components of all the normal vectors. Similarly, in Figure 14 and for each epoch, the maximum value was derived for the absolute differences between the distance parameters.

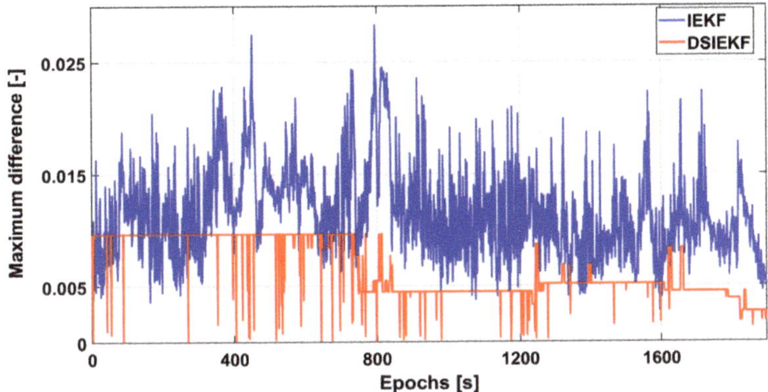

Figure 13. The maximum difference between the estimated normal vectors of the planes by means of the algorithms and the available information from the 3D city model.

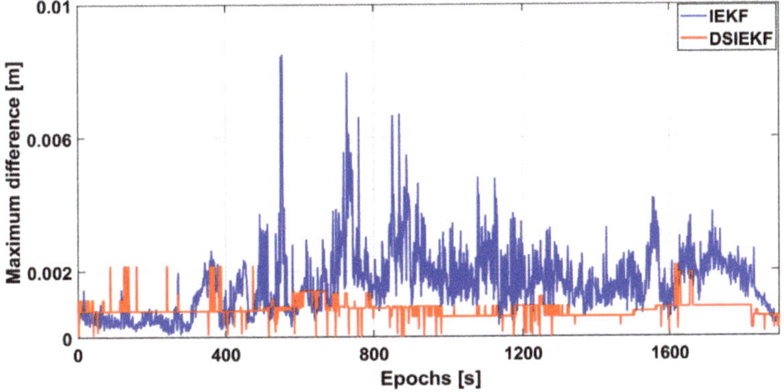

Figure 14. The maximum difference between the estimated distance parameters of the planes by means of the algorithms and the available information from the 3D city model.

DSIEKF delivers estimations that are closer to the available values in the 3D city model than those derived by IEKF. Having such results approves the aforementioned justification of the error propagation effect even more. However, as mentioned in Section 3.2, the LoD-2 3D city model itself is prone to errors due to generalization, and therefore having closer estimated plane parameters to such a model should not be interpreted as superiority of DSIEKF over IEKF in sense of accuracy.

Nonetheless, no error-free true information of the surrounding environment exists, and, in the context of these two filters, the LoD-2 3D city model is taken as a reliable source. Therefore, based on such a consideration, it could be claimed that DSIEKF delivers more reliable estimations than IEKF, although the difference between the two filters is not significant.

Finally, the duration of the update time for both of the algorithms is shown in Figure 15. The reason for only showing the update time instead of the overall time is to show how the main difference between the two algorithms—which is the update step—influences the computation time. In Table 3, it is given that, for 1902 epochs, IEKF takes 28 min to finish the update steps, while DSIEKF takes only 15 min. Therefore, the DSIEKF algorithm is about 45% faster than the IEKF.

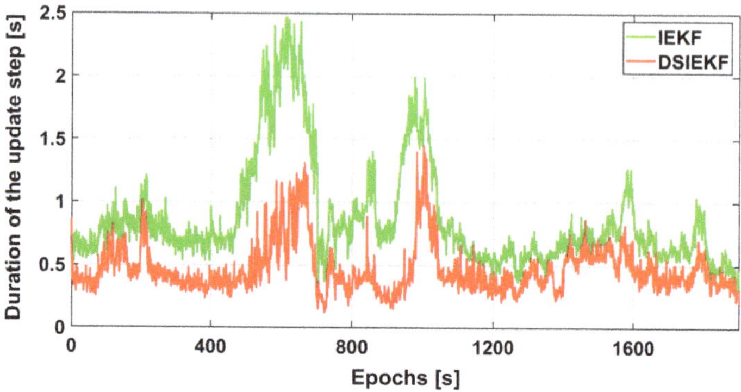

Figure 15. Duration of the update time.

The reason for having faster computations in DSIEKF is, first, the significantly smaller matrices in size to be inverted than in the IEKF algorithm. Such matrices can be found, e.g., in line 20 of Algorithm 1, which, due to their size, are inverted faster in the following lines of the same algorithm than in IEKF. Secondly, after several iterations, lines 25 to 34 of Algorithm 1 are neglected, which leads to a considerable decrease in the overall estimation time of the update step.

5. Discussion

As shown in Section 4, outcomes of the DSIEKF for georeferencing a real case UAV are compared to those of the IEKF. The main reasons for doing so are, first, the similar principle that forms the basis of these two algorithms. Secondly, as also stated before, no ground truth solutions are available for this case study in order to validate the DSIEKF performance. The final estimated states along the trajectory by the two algorithms are shown to have a maximum difference of 60 cm, with an average standard deviation of around 2 and 6 mm by DSIEKF and IEKF, respectively.

The height estimations are shown to have a maximum difference of around 2 m by the two methodologies with an average standard deviation of around 9 mm for both the algorithms. Having such a significant difference between the height estimations is claimed to be due to the inaccurate height information that are taken as deterministic values into account. The orientation estimations from the two algorithms are shown to be in the same range. A maximum difference of around 0.1 degree was the largest difference between the estimations of the two algorithms, which was derived for the orientation around the Y axis.

In general, the DSIEKF estimations were derived with smaller standard deviations, which could be an indicator for the better performance of this methodology compared to the IEKF. It is claimed that such results are due to the state division strategy that is taken in DSIEKF, which avoids propagating the measurements uncertainties in the iterations of the update step, repeatedly. in other words, in the IEKF algorithm, where all unknown parameters in a vector are considered, the number of iterations required to reach an optimized solution is unnecessarily increased to fulfill a specified threshold that should be fulfilled by all estimated parameters.

Having all the unknown parameters in one vector to be estimated would mean to have more measurements that are iteratively used, which, consequently, produces a higher error propagation effect. Conversely, in DSIEKF, the second state vector is estimated in a lower number of iterations, which, in tur,n means considering fewer observations for deriving the pose parameters and, thus, less propagated uncertainties. The update time needed to carry out the estimations for the whole scenario by DSIEKF is nearly half of time taken by the IEKF algorithm, which is another important aspect of the DSIEKF methodology.

6. Conclusions and Outlook

The focus of this paper is to show the functionality of the IEKF and DSIEKF algorithms in a real environment. As mentioned, the IEKF algorithm was first introduced by [30], and was later used by [1] to georeference a UAV in a simulated environment. The authors in [6] adapted the IEKF algorithm to the DS estimation framework and developed a new algorithm, called DSIEKF, which was also applied to a simulated environment for georeferencing a UAV.

The real environment of the current work consists of a prototype UAV equipped with a GNSS, an IMU, a 3D laser scanner, and camera units that move in an inner courtyard. Due to inappropriate GNSS and IMU data, they were only used for initialization and disregarded further on in the analysis steps. Moreover, due to the existence of multiple buildings in the environment, their information was considered by means of available LoD-2 3D city model within the algorithms. The existing DTM was used as the object information for defining the ground, which was not defined within the 3D city models.

The results of the analysis showed a maximum deviation of around 2 m in the translation parameters and 0.1 degrees in the orientation parameters between the two algorithms. Since, on the one hand, no true trajectory exists and, on the other hand, all the algorithms gave rational 6-DoF estimates, no priority could be given to them, and they can only be compared relative to each other.

However, it could be claimed that, at least for this measurement scenario, the DSIEKF was more efficient as it took about half the required update time in the IEKF algorithm. Additionally, it was shown that the estimated building planes by DSIEKF were closer to their values given in the 3D city model. However, due to existing generalization errors within LoD-2 3D city models, the DSIEKF algorithm could not be prioritized in general.

In the current paper, the measurements were down-sampled by averaging between the neighboring measurements. Therefore, the observations that are used within the algorithms were not the real measurements. Consequently, further research should consider down-sampling methods that, while preserving the homogeneous distribution of the observations, lead to the least possible information loss. For that, we are planning to integrate methods, such as the Optimum Single MLS Dataset introduced by [56], in our developed methodology.

In this approach, the idea is to preserve those 3D points that best represent the structure of a scanned object by using the Visvalingam–Whyatt [57] and Douglas–Peuker [58] techniques. Moreover, making the algorithms robust against measurement outliers as well as wrong object information are other aspects that are sought in future work. Furthermore, the behavior of the MSS over time might become too complicated to be described by simple models, or the system and observation noise might have distributions other than the Gaussian.

Therefore, suitable strategies are needed to deal with such challenges, among which, the particle filter framework could be mentioned. Consequently, for our future research, we are aiming at developing an efficient particle filter algorithm that can handle implicit and explicit measurement equations and, hence, can deal with the aforementioned problems. Furthermore, when it comes to real environments, no true trajectory exists, and therefore validating algorithms becomes challenging and complicated.

Hence, finding proper ways to evaluate the performance of the developed methodologies is an irresistible task that should be fulfilled. Performing measurements by means of a laser tracker, integrating camera images into the filters, or comparison of the final transformed scanned points to a reference point cloud are examples of suitable solutions for validation that are within the scope of future research.

Finally, with multiple MSSs in an environment, it is beneficial to establish a network between them in order to exchange useful information that could lead to more accurate georeferencing results. Therefore, we intend to also consider the concept of dynamic nodes within our developed algorithm(s) in future works.

Author Contributions: Conceptualization, R.M., S.V., I.N., J.B. and H.A.; methodology, R.M., S.V., J.B. and H.A.; software, R.M.; formal analysis, R.M., S.V. and J.B.; investigation, R.M., S.V., I.N. and H.A.; writing—original draft preparation, R.M., S.V., I.N., J.B. and H.A.; writing—review and editing, R.M., S.V., I.N., J.B. and H.A.; visualization, R.M.; supervision, H.A. and I.N.; All authors have read and agreed to the published version of the manuscript.

Funding: This research was funded by the Deutsche Forschungsgemeinschaft (DFG, German Research Foundation) and as part of the Research Training Group i.c.sens [RTG 2159].

Institutional Review Board Statement: Not applicable.

Informed Consent Statement: Not applicable.

Data Availability Statement: Not applicable.

Acknowledgments: We would like to gratefully thank Dmitri Diener for providing the measurement data.

Conflicts of Interest: The authors declare no conflict of interest.

References

1. Bureick, J.; Vogel, S.; Neumann, I.; Unger, J.; Alkhatib, H. Georeferencing of an unmanned aerial system by means of an iterated extended Kalman filter Using a 3D city model. *J. Photogramm. Remote Sens. Geoinf. Sci.* **2019**, *87*, 229–247. [CrossRef]
2. Bonnor, N. Principles of GNSS, inertial, and multisensor integrated navigation systems—Second Edition Paul D. Groves Artech House. *J. Navig.* **2014**, *67*, 191–192. [CrossRef]
3. Titterton, D.; Weston, J.L.; Weston, J. *Strapdown Inertial Navigation Technology*, 2nd ed.; The American Institute of Aeronautics and Astronautics; IET: London, UK, 2004.
4. Woodman, O.J. *An Introduction to Inertial Navigation*; Technical Report UCAM-CL-TR-696; University of Cambridge, Computer Laboratory: Cambridge, UK, 2007.
5. Pirník, R.; Hruboš, M.; Nemec, D.; Mravec, T.; Božek, P. Integration of inertial sensor data into control of the mobile platform. In *Advances in Intelligent Systems and Computing, Proceedings of the Federated Conference on Software Development and Object Technologies, Žilina, Slovakia, 19–20 November 2015*; Springer International Publishing: Cham, Switzerland, 2015; pp. 271–282.
6. Moftizadeh, R.; Bureick, J.; Vogel, S.; Neumann, I.; Alkhatib, H. Information-based georeferencing by dual state iterated extended Kalman filter with implicit measurement equations and nonlinear geometrical constraints. In Proceedings of the IEEE 23rd International Conference on Information Fusion (FUSION), Rustenburg, South Africa, 6–9 July 2020.
7. Vogel, S.; Alkhatib, H.; Neumann, I. Accurate indoor georeferencing with kinematic multi sensor systems. In Proceedings of the IEEE International Conference on Indoor Positioning and Indoor Navigation (IPIN), Alcala de Henares, Spain, 4–7 October 2016; pp. 1–8.
8. Yang, B.; Yang, E.; Yu, L. Vision and UWB-based anchor self-localisation system for UAV in GPS-denied environment. *J. Phys. Conf. Ser.* **2021**, *1922*, 012001. [CrossRef]
9. Schuhmacher, S.; Böhm, J. Georeferencing of Terrestrial Laser Scanner Data for Applications in Architectural Modeling. In *Proceedings of the ISPRS 3D-ARCH: Virtual Reconstruction and Visualization of Complex Architectures*; El-Hakim, S., Remondinoand, F., Gonzo, L., Eds.; Mestre: Venice, Italy, 2005; Volume XXXVI.
10. Paffenholz, J.-A. *Direct Geo-Referencing of 3D Point Clouds with 3D Positioning Sensors*; Deutsche Geodätische Kommission (DGK): München, Germany, 2012.
11. Talaya, J.; Alamus, R.; Bosch, E.; Serra, A.; Kornus, W.; Baron, A. Integration of a terrestrial laser scanner with GPS/IMU orientation sensors. *Proc. XXth ISPRS Congr.* **2004**, *35*, 1049–1055.
12. Dennig, D.; Bureick, J.; Link, J.; Diener, D.; Hesse, C.; Neumann, I. Comprehensive and highly accurate measurements of crane runways, profiles and fastenings. *Sensors* **2017**, *17*, 1118. [CrossRef]
13. Hartmann, J.; Trusheim, P.; Alkhatib, H.; Paffenholz, J.A.; Diener, D.; Neumann, I. High accurate pointwise (geo-) referencing of a k-TLS based multi sensor system. *ISPRS Ann. Photogramm. Remote Sens. Spat. Inf. Sci.* **2018**, *4*, 81–88. [CrossRef]
14. Zeybek, M. Accuracy assessment of direct georeferencing UAV images with onboard global navigation satellite system and comparison of CORS/RTK surveying methods. *Meas. Sci. Technol.* **2021**, *32*, 065402. [CrossRef]
15. Abmayr, T.; Härtl, F.; Hirzinger, G.; Burschka, D.; Fröhlich, C. A correlation based target finder for terrestrial laser scanning. *JAG* **2008**, *2*, 131–137. [CrossRef]
16. Elkhrachy, I.; Niemeier, W. Stochastic Assessment of Terrestrial Laser Scanner. In Proceedings of the ASPRS Annual Conference, Reno, NV, USA, 1–5 May 2006.
17. Soloviev, A.; Bates, D.; Van Graas, F. Tight coupling of laser scanner and inertial measurements for a fully autonomous relative navigation solution. *Navigation* **2007**, *54*, 189–205. [CrossRef]
18. Li-Chee-Ming, J.; Armenakis, C. Determination of UAS Trajectory in a Known Environment from FPV Video. *ISPRS Int. Arch. Photogramm. Remote Sens. Spat. Inf. Sci.* **2013**, *XL-1/W2*, 247–252. [CrossRef]

19. Dehbi, Y.; Lucks, L.; Behmann, J.; Klingbeil, L.; Plümer, L. Improving GPS Trajectories Using 3D City Models and Kinematic Point Clouds. In Proceedings of the ISPRS Annals of the Photogrammetry, Remote Sensing and Spatial Information Sciences, Volume IV-4/W9, 2019, 4th International Conference on Smart Data and Smart Cities, Kuala Lumpur, Malaysia, 1–3 October 2019.
20. Outamazirt, F.; Djaidja, D.; Boudjimar, C.; Louadj, K.; Nemra, A.; Li, F.; Yan, L. Multi-sensor fusion approach based on nonlinear $H\infty$ filter with interval type 2 fuzzy adaptive parameters tuning for unmanned vehicle localization. *Proc. Inst. Mech. Eng. Part I J. Syst. Control Eng.* **2021**, *235*, 881–897. [CrossRef]
21. Zhang, Y. Localization scheme based on key frame selection and a reliable plane using lidar and IMU. *Appl. Opt.* **2021**, *60*, 5430–5438. [CrossRef] [PubMed]
22. Denham, W.F.; Pines, S. Sequential estimation when measurement function nonlinearity is comparable to measurement error. *AIAA J.* **1966**, *4*, 1071–1076. [CrossRef]
23. Tailanián, M.; Paternain, S.; Rosa, R.; Canetti, R. Design and implementation of sensor data fusion for an autonomous quadrotor. In Proceedings of the IEEE International Instrumentation and Measurement Technology Conference (I2MTC) Proceedings, Montevideo, Uruguay, 12–15 May 2014; pp. 1431–1436.
24. Forster, C.; Lynen, S.; Kneip, L.; Scaramuzza, D. Collaborative monocular SLAM with multiple micro aerial vehicles. In Proceedings of the IEEE/RSJ International Conference on Intelligent Robots and Systems, Tokyo, Japan, 3–7 November 2013; pp. 3962–3970.
25. Morelande, M.R.; García-Fernández, Á.F. Analysis of Kalman filter approximations for nonlinear measurements. *IEEE Trans. Signal Process.* **2013**, *61*, 5477–5484. [CrossRef]
26. Bell, B.; Cathey, F. The iterated Kalman filter update as a Gauss-Newton method. *IEEE Trans. Autom. Control* **1993**, *38*, 294–297. [CrossRef]
27. García-Fernández, Á.F.; Svensson, L.; Morelande, M.R.; Särkkä, S. Posterior linearization filter: Principles and implementation using Sigma points. *IEEE Trans. Signal Process.* **2015**, *63*, 5561–5573. [CrossRef]
28. Dang, T. An iterative parameter estimation method for observation models with nonlinear constraints. *Metrol. Meas. Syst.* **2008**, *15*, 421–432.
29. Steffen, R. A robust iterative Kalman filter based on implicit measurement equations Robuster Iterativer Kalman-Filter mit Implizierten Beobachtungsgleichungen. *Photogramm. Fernerkund. Geoinf.* **2013**, *2013*, 323–332. [CrossRef]
30. Vogel, S.; Alkhatib, H.; Neumann, I. Iterated extended Kalman filter with implicit measurement equation and nonlinear constraints for information-based georeferencing. In Proceedings of the IEEE 21st International Conference on Information Fusion (FUSION), Cambridge, UK, 10–13 July 2018; pp. 1209–1216.
31. Vogel, S.; Alkhatib, H.; Bureick, J.; Moftizadeh, R.; Neumann, I. Georeferencing of laser scanner-based kinematic multi-sensor systems in the context of iterated extended Kalman filters using geometrical constraints. *Sensors* **2019**, *19*, 2280. [CrossRef]
32. Vogel, S. *Kalman Filtering with State Constraints Applied to Multi-Sensor Systems and Georeferencing*; Deutsche Geodätische Kommission (DGK): München, Germany, 2020.
33. Haykin, S. *Kalman Filtering and Neural Networks*; John Wiley & Sons: Hoboken, NJ, USA, 2004.
34. Wan, E.A.; Nelson, A.T. Dual Kalman filtering methods for nonlinear prediction, smoothing and estimation. In *Advances in Neural Information Processing Systems*; Mozer, M.C., Jordan, M.I., Petsche, F., Eds.; The MIT Press: Cambridge, MA, USA, 1997; pp. 793–799.
35. Popovici, A.; Zaal, P.; Pool, D.M. Dual extended Kalman filter for the identification of time-varying human manual control behavior. In Proceedings of the AIAA Modeling and Simulation Technologies Conference, Denver, CO, USA, 5–9 June 2017; p. 3666.
36. Khodadadi, H.; Jazayeri-Rad, H. Applying a dual extended Kalman filter for the nonlinear state and parameter estimations of a continuous stirred tank reactor. *Comput. Chem. Eng.* **2011**, *35*, 2426–2436. [CrossRef]
37. Wenzel, T.A.; Burnham, K.J.; Blundell, M.V.; Williams, R.A. Dual extended Kalman filter for vehicle state and parameter estimation. *Veh. Syst. Dyn.* **2006**, *44*, 153–171. [CrossRef]
38. Simon, D. Kalman filtering with state constraints: A survey of linear and nonlinear algorithms. *IET Control Theory A* **2010**, *4*, 1303–1318. [CrossRef]
39. Simon, D.; Simon, D.L. Kalman filtering with inequality constraints for turbofan engine health estimation. *IEE Proc. Control Theory Appl.* **2006**, *153*, 371–378. [CrossRef]
40. Gupta, N.; Hauser, R. Kalman filtering with equality and inequality state constraints. *arXiv* **2007**, arXiv:0709.2791.
41. Sircoulomb, V.; Hoblos, G.; Chafouk, H.; Ragot, J. State estimation under nonlinear state inequality constraints. A tracking application. In Proceedings of the IEEE 16th Mediterranean Conference on Control and Automation, Ajaccio, France, 25–27 June 2008; pp. 166–1674.
42. Simon, D.; Simon, D.L. Constrained Kalman filtering via density function truncation for turbofan engine health estimation. *Int. J. Syst. Sci.* **2010**, *41*, 159–171. [CrossRef]
43. Wang, L.S.; Chiang, Y.T.; Chang, F.R. Filtering method for nonlinear systems with constraints. *IEE Proc. Control Theory Appl.* **2002**, *149*, 525–531. [CrossRef]
44. Teixeira, B.O.; Chandrasekar, J.; Tôrres, L.A.; Aguirre, L.A.; Bernstein, D.S. State estimation for linear and nonlinear equality-constrained systems. *Int. J. Control* **2009**, *82*, 918–936. [CrossRef]
45. Yang, C.; Blasch, E. Kalman filtering with nonlinear state constraints. *IEEE Trans. Aerosp. Electron. Syst.* **2009**, *45*, 70–84. [CrossRef]

46. De Geeter, J.; Van Brussel, H.; De Schutter, J.; Decréton, M. A smoothly constrained Kalman filter. *IEEE Trans. Pattern Anal. Mach. Intell.* **1997**, *19*, 1171–1177. [CrossRef]
47. Döllner, J.; Baumann, K.; Buchholz, H. *Virtual 3D City Models as Foundation of Complex Urban Information Spaces*; CORP: Vienna, Austria, 2006.
48. Unger, J.; Rottensteiner, F.; Heipke, C. Integration of a generalised building model into the pose estimation of uas images. *ISPRS Arch.* **2016**, *41*, 1057–1064.
49. Unger, J.; Rottensteiner, F.; Heipke, C. Assigning tie points to a generalised building model for UAS image orientation. *ISPRS Arch.* **2017**, *42*, 385–392. [CrossRef]
50. Ibrahim, I.N.; Pavol, B.; Al Akkad, M.A.; Karam, A. Navigation control and stability investigation of a hexacopter equipped with an aerial manipulater. In Proceedings of the 2017 21st International Conference on Process Control, Strbske Pleso, Slovakia, 6–9 June 2017; pp. 204–209.
51. Bar-Shalom, Y.; Li, X.R.; Kirubarajan, T. *Estimation with Applications to Tracking and Navigation: Theory Algorithms and Software*; John Wiley & Sons: Hoboken, NJ, USA, 2004.
52. *Puck Real-Time 3D LiDAR Sensor*; Velodyne LiDAR: Morgan Hill, CA, USA, 2017. Available online: http://velodynelidar.com (accessed on 12 August 2021).
53. Bureick, J.; Vogel, S.; Neumann, I.; Diener, D.; Alkhatib, H. *Geo-Referenzierung von Unmanned Aerial Systems über Laserscannermessungen und 3D-Gebäudemodelle*; Terrestrisches Laserscanning 2019 (TLS 2019), Schriftenreihe des DVW; Wißner-Verlag: Augsburg, Germany, 2019; Volume 96, pp. 63–74.
54. Landeshauptstadt Hannover. Produktinformation: Digitales 3D-Stadtmodell. FB Planen und Stadtentwicklung. *Bereich Geoinf.* 2017. Available online: https://www.hannover.de/content/download/641518/15209724/file/Produktblatt_3DStadtmodell.pdf (accessed on 12 November 2020).
55. Landeshauptstadt Hannover. Produktinformation: Digitales Geländemodell (DGM1). FB Planen und Stadtentwicklung. *Bereich Geoinf.* 2017. Available online: https://www.hannover.de/content/download/641402/15208189/file/Produktblatt_DGM.pdf (accessed on 12 November 2020).
56. Błaszczak-Bąk, W.; Koppanyi, Z.; Toth, C. Reduction method for mobile laser scanning data. *ISPRS Int. J. Geo-Inf.* **2018**, *7*, 285. [CrossRef]
57. Visvalingam, M.; Whyatt, J.D. Line generalization by repeated elimination of points. *Cartogr. J.* **2017**, *30*, 144–155.
58. Douglas, D.H.; Peucker, T.K. Algorithms for the reduction of the number of points required to represent a digitized line or its caricature. *Can. Cartogr.* **1973**, *10*, 112–122. [CrossRef]

Article

A V-SLAM Guided and Portable System for Photogrammetric Applications

Alessandro Torresani [1,2,*], Fabio Menna [1], Roberto Battisti [1] and Fabio Remondino [1]

1   3D Optical Metrology (3DOM) Unit, Bruno Kessler Foundation (FBK), Via Sommarive, 18, 38123 Trento, Italy; fmenna@fbk.eu (F.M.); rbattisti@fbk.eu (R.B.); remondino@fbk.eu (F.R.)
2   Department of Information Engineering and Computer Science (DISI), Universitá degli Studi di Trento, Via Sommarive, 9, 38123 Trento, Italy
*   Correspondence: atorresani@fbk.eu or alessandro.torresani@unitn.it

Abstract: Mobile and handheld mapping systems are becoming widely used nowadays as fast and cost-effective data acquisition systems for 3D reconstruction purposes. While most of the research and commercial systems are based on active sensors, solutions employing only cameras and photogrammetry are attracting more and more interest due to their significantly minor costs, size and power consumption. In this work we propose an ARM-based, low-cost and lightweight stereo vision mobile mapping system based on a Visual Simultaneous Localization And Mapping (V-SLAM) algorithm. The prototype system, named GuPho (Guided Photogrammetric System), also integrates an in-house guidance system which enables optimized image acquisitions, robust management of the cameras and feedback on positioning and acquisition speed. The presented results show the effectiveness of the developed prototype in mapping large scenarios, enabling motion blur prevention, robust camera exposure control and achieving accurate 3D results.

Keywords: V-SLAM; real-time; guidance; embedded-systems; 3D surveying; exposure control; photogrammetry

Citation: Torresani, A.; Menna, F.; Battisti, R.; Remondino, F. A V-SLAM Guided and Portable System for Photogrammetric Applications. *Remote Sens.* **2021**, *13*, 2351. https://doi.org/10.3390/rs13122351

Academic Editors: Ville Lehtola, François Goulette and Andreas Nüchter

Received: 24 April 2021
Accepted: 11 June 2021
Published: 16 June 2021

Publisher's Note: MDPI stays neutral with regard to jurisdictional claims in published maps and institutional affiliations.

1. Introduction

Nowadays there is a strong demand for fast, cost-effective and reliable data acquisition systems for the 3D reconstruction of real-world scenarios, from land to underwater structures, natural or heritage sites [1]. Mobile mapping systems (MMSs) [1–3] probably represent today the most effective answer to this need. Many MMSs have been proposed since the late 80s differing in size, platform (vehicle-based, aircraft-based, drone-mounted), portability (backpack, handheld, trolley [4]) and employed sensors (laser scanners, cameras, GNSS, IMU, rotary encoders or a combination of them).

While most of the available MMSs for the 3D mapping purposes are based on laser scanning [1], systems exclusively based on photogrammetry [5–12] have always attracted great interest for their significant minor cost, size and power consumptions. Nevertheless, image-based systems are highly influenced, in addition to the object texture and surface complexity, by the imaging geometry [13–16], image sharpness [17,18] as well as image brightness/contrast/noise [19–21].

During the past years, great efforts have been made to increase the effectiveness and the correctness of the image acquisitions in the field. In GNSS-enabled environments it is common to take advantage of real-time kinematic positioning (RTK) to acquire the imagery along pre-defined flight/trajectory plans [22–24] to ensure a complete and robust image coverage of the object. However, enabling guided acquisitions in GNSS-denied scenarios, such as indoor, underwater or in urban canyons, is still an open and difficult problem. Existing localization technologies (such as, for example, motion capture systems, Ultra-Wide Band (UWB) [25] in indoor, acoustic systems in underwater environments) are expensive and/or generally require a time-consuming system set up and calibration. Moreover, these environments are often characterized by challenging illumination conditions that require both robust auto exposure algorithms [26,27] and careful acquisitions to avoid motion blur.

We present an ARM-based, low-cost portable prototype of a MMS based on stereo vision that exploits Visual Simultaneous Localization and Mapping (V-SLAM) to keep track of its position and attitude during the survey and enable a guided and optimized acquisition of the images. V-SLAM algorithms [28–31] were born in the robotic community and have enormous potential: they act as a standalone local positioning system and they also enable a real-time 3D awareness of the surveyed scene. In our system (Figure 1), the cameras send high frame rate (5 Hz) synchronized stereo pairs to a microcomputer, where they are down-scaled on the fly and fed into the V-SLAM algorithm that updates both the position of the MMS and the coarse 3D reconstruction of the scene. These quantities are then used by our guidance system not only to provide distance and motion blur feedback, but also to decide whether to keep and save the corresponding high-resolution image, according to the desired image overlap of the acquisition. Moreover, the known depth and image location of the stereo matched tie-points are used to reduce the image domain considered by the auto exposure algorithm, giving more importance to objects/areas lying at the target acquisition distance. Finally, the system can also be set to drive the acquisition of a third higher resolution camera triggered by the microcomputer when the device has reached a desired position. To the best of our knowledge, our system is the first example of a portable camera-based MMS that integrates a real-time guidance system designed to cover both geometric and radiometric image aspects without assuming coded targets or external reference systems. The most similar work to our prototype is Ortiz-Coder at al. [9], where the integrated V-SLAM algorithm has been used to provide both a real-time 3D feedback of the acquired scenery and select the dataset images. This work, however, does not include a proper guidance system that covers both geometric and radiometric image issues (i.e., distance and motion blur feedback, image overlap control or guided auto exposure), it has a limited portability as it requires a consumer laptop to run the V-SLAM algorithm, and does not provide metric results due to the monocular setup of the system. Table 1 reports a concise panorama of the most representative and/or recent portable MMS in both commercial and research domains. The table reports, in addition to the format of MMS and the sensor configuration, also the system architecture of the main unit and, in case of camera-based devices, whether the system includes a real-time guidance system.

Figure 1. The proposed handheld, low-cost, low-power prototype (named GuPho) based on stereo cameras and V-SLAM method for 3D mapping purposes. It includes a real-time guidance system aimed at enabling optimized and less error prone image acquisitions. It uses industrial global shutter video cameras, a Raspberry Pi 4 and a power bank for an overall weight less than 1.5 kg. When needed, a LED panel can also be mounted in order to illuminate dark environments as shown in the two images on the left.

The experiments, carried out in controlled conditions and in a real-case outdoor and challenging scenarios, have confirmed the effectiveness of the proposed system in enabling optimized and less error prone image acquisitions. We hope that these results could inspire the development of more advanced vision MMSs capable of deeply understanding the acquisition process and of actively [32,33] helping the involved operators to better accomplish their task.

Table 1. Some representative and/or recent portable MMSs in the research (R) and commercial (C) domains (DOM.). They are categorized according to the format (FORM.) as handheld (H), smartphone (S), backpack (B), trolley (T), sensing devices, system architecture (SYST. ARCH.) and presence, in photogrammetry-based systems, of an onboard real-time guidance system (R.G.S.).

MMS	FORM.	SENSING DEVICE	SYST. ARCH.	R.G.S.	DOM.
Holdener et al. [7]	H	Cameras (multi-view)	ARM	No	R
Ortiz-Coder et al. [9]	H	Camera	X86/X64	Partial	R
Nocerino et al. [12]	S	Camera	ARM	No	R
Karam et al. [34]	B	Laser	X86/X64	-	R
Lauterbach et al. [35]	B	Laser	X86/X64	-	R
Kalisperakis et al. [36]	T	Laser-cameras (multi-view)	ARM	-	R
Vexcel Ultracam Panther	B	Laser-cameras (multi-view)	X86/X64	-	C
GeoSLAM Horizon	H	Laser-camera	X86/X64	-	C
Leica Pegasus:backpack	B	Laser-cameras (multi-view)	X86/X64	-	C
Kaarta Stencil 2	H	Laser-camera	X86/X64	-	C
Ours	H	Cameras (stereo)	ARM	Yes	R

1.1. Brief Introduction to Visual SLAM

V-SLAM algorithms have witnessed a growing popularity in recent years and nowadays are being used in different cutting-edge fields ranging from autonomous navigation (from ubiquitous recreational UAVs to the more sophisticated robots used for scientific research and exploration on Mars), autonomous driving and virtual and augmented reality, to cite a few. The popularity of V-SLAM is mainly motivated by its capabilities to enable both position and context awareness using solely efficient, lightweight and low-cost sensors: cameras. V-SLAM has been highly democratized in the last decade and many open-source implementations are currently available [37–44]. V-SLAM methods can be roughly seen [28] as the real-time counterparts of Structure from Motion (SfM) [45,46] algorithms. Both indeed estimate the camera position and attitude, called localization in V-SLAM and the 3D structure of the imaged scene, called mapping in V-SLAM, solely from image observations. They differ however in the input format and processing type: SfM is generally a batch process, and it is designed to work on sparse and time unordered bunches of images. V-SLAM instead processes high frame rate, and time ordered sets of images in a real-time and sequential manner, making use of the estimated motion and dead reckoning technique to improve tracking. Real-time is generally achieved through a combination of lower image resolutions, background optimizations with fewer parameters, local image subsets [38] and/or faster binary feature detectors and descriptors [47–49]. On the highest-level, V-SLAM algorithms are usually categorized in direct [39,40] and indirect [37,38,42–44,49] methods, and in filter-based [37] or graph-based methods [38,42–44,49]. We refer the reader to other surveys [28–30] to have a better and deep review about V-SLAM.

1.2. Paper Contributions

The main contributions of the article are:

- Realization of a flexible, low-cost and highly portable V-SLAM-based stereo vision prototype—named GuPho (Guided Photogrammetric System)—that integrates an image acquisition system with a real-time guidance system (distance from the target object, speed, overlap, automatic exposure). To the best of our knowledge, this is the first example of a portable system for photogrammetric applications with a real-time guidance control that does not use coded targets or any other external positioning system (e.g., GNSS, motion capture).

- Modular system design based on power efficient ARM architecture: data computation and data visualization are split among two different systems (Raspberry and smartphone) and they can exploit the full computational resources available on each device.
- An algorithm for assisted and optimized image acquisition following photogrammetric principles, with real-time positioning and speed feedback (Sections 2.3.1 and 2.3.3).
- A method for automatic camera exposure control guided by the locations and depths of the stereo matched tie-points, which is robust in challenging lighting situations and prioritizes the surveyed object over background image regions (Section 2.3.4).
- Experiments and validation on large-scale scenarios with long trajectory (more than 300 m) and accuracy analyses in 3D space by means of laser scanning ground truth data (Section 3).

2. Proposed Prototype

2.1. Hardware

A sketch of the hardware components is shown in Figure 2. The main computational unit of the system is a Raspberry Pi 4B. It is connected via USB3 to two synchronized global shutter 1 Mpixel (1280 × 1024) cameras placed in stereo configuration. They are produced by Daheng Imaging (model MER-131-210U3C) and mount 4 mm F2.0 lenses (model LCM-5MP-04MM-F2.0-1.8-ND1). The choice of this hardware was made for the high flexibility in terms of possible imaging configurations (great variety of available lenses), software and hardware synchronization, low power consumption and overall cost of the device (about 1000 Eur). The stereo baseline and angle can be adapted as needed making the system flexible to different use scenarios. The system output is displayed to the operator on a smartphone or tablet connected to the Raspberry with an Ethernet cable. This choice relieves the Raspberry from managing a GUI and leaves more resources to the V-SLAM algorithm. The very low demanding power of the ARM architectures and cameras allows the system to be powered by a simple 10,400 mAh 5 V 3.0 A power bank, which guarantees approximately two hours of continuous working activity.

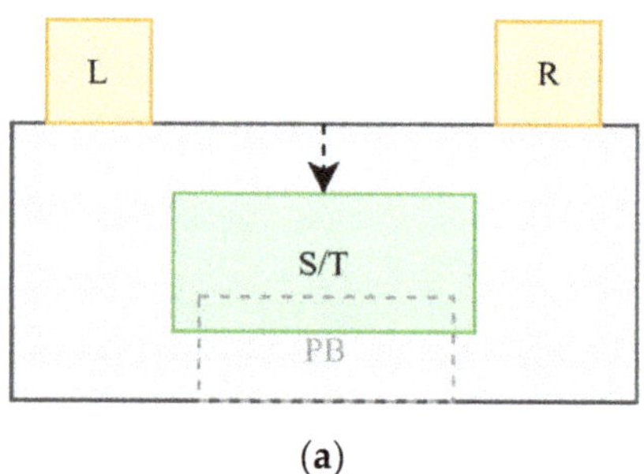

(a)

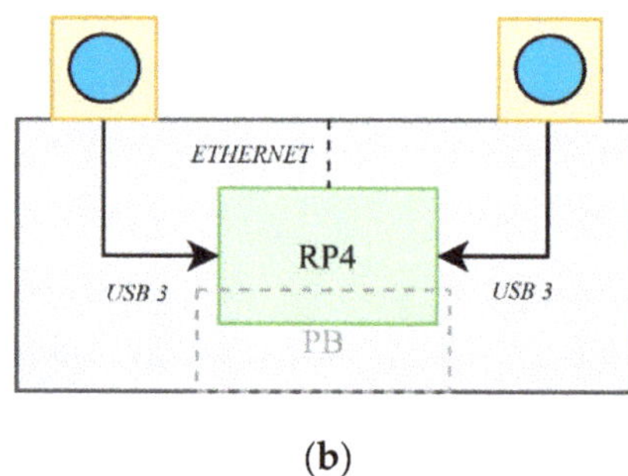

(b)

Figure 2. A sketch of the prototype. (**a**) Rear view: the operator can monitor the system output on a smartphone/tablet (S/T). (**b**) Front view: the Raspberry Pi 4B (RP4) is connected with USB3 to the cameras (L, R) and with Ethernet to the smartphone/tablet (S/T). The power bank (PB) is placed inside the body frame.

2.2. Software

The proposed prototype builds upon OPEN-V-SLAM [43], which is derived from ORB-SLAM2 [42]. OPEN-V-SLAM is an indirect, graph-based V-SLAM method that uses Oriented FAST and Rotated BRIEF (ORB) [49] for data association, g2o [50] for local and global graph optimizations and DBow2 [51] for re-localization and loop detection. In addition to achieving state of the art performances in terms of accuracy, and supporting different camera models (perspective, fisheye, equirectangular) and camera configurations (monocular, stereo, RGB-D), the code of this algorithm is well modularized, making the process of integrating the guidance system much easier. Moreover, it exploits Protobuf, Socket.io, and Tree.js frameworks to easily supports the visualization of the system output inside a common web browser. This feature allows to split computation and visualization on different devices. With the aforementioned hardware (Section 2.1) and software configurations, the system can perform can work at 5 Hz using stereo pair images at half linear resolution (640×512 pixel).

2.3. Guidance System

A high-level overview of the system is shown in Figure 3. It is composed of three different modules (dashed boxes): (i) the camera module that manages the imaging sensors, (ii) the OPEN-V-SLAM module that embeds the localization/mapping operations and our guidance system and (iii) the viewer module that shows to the operator the system controls and output.

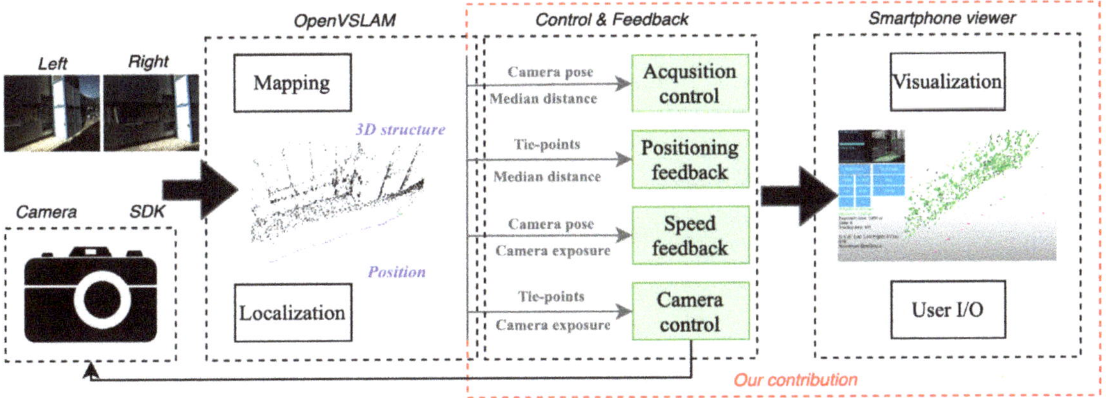

Figure 3. High level overview of the proposed system, highlighting our contribution. The images acquired by the synchronized stereo cameras are used by the V-SLAM algorithm to estimate in real-time the local position of the device and a sparse 3D structure of the surveyed scene. Camera positions and 3D structure are used in real-time to optimize and guide the acquisition.

The realized guidance system is divided in four units:

1. Acquisition control.
2. Positioning feedback.
3. Speed feedback.
4. Camera control.

The first two units are aimed at helping the operator to acquire an ideal imaging configuration according to photogrammetric principles, thus respecting a planned acquisition distance and a specific amount of overlap between consecutive images. The last two are instead targeted at avoiding motion blur and obtaining correctly exposed images.

Let us introduce some notation before presenting more in detail the four modules of the guidance system. Let t denotes the time index of a stereo pair. At a given time t, let L_t, R_t be the left and right images and T_t^L and T_t^R their estimated camera poses. Finally, let d_t be the estimated depth (distance to the surveyed object along the optical axis of the camera). In our approach, d_t is computed as the median depth of the stereo matched tie-points between L_t and R_t. Thanks to the system calibration (interior and relative orientation of the cameras), all the estimated variables have a metric scale.

2.3.1. Acquisition Control

In current vision-based MMSs, it is common to acquire and store images at high frame rate (1 Hz, 2 Hz, 5 Hz, 30 Hz) and delegate at post acquisition methods [52] the task of selecting the images to use in the 3D reconstruction. This approach is extremely memory and power inefficient, especially when the images are saved in raw format.

Our prototype exploits OPEN-V-SLAM to optimize the acquisition of the images and keep their overlap constant. Let (L_s, R_s) be the last selected and stored image pair. Given a new image pair (L_t, R_t) and known their camera poses (T_t^L, T_t^R), the pair is considered part of the acquisition, and stored, if the baseline between the camera center of L_t (resp. R_t) and the camera center of L_s (resp. R_s) is bigger than a target baseline b_t. b_t is adapted in real-time according to the estimated distance to the object d_t. More formally, b_t is defined as:

$$
b_t = \begin{cases}
w\left(\dfrac{d_t}{f}\right)(1 - O_x), & \text{if movement along camera X axis} \\[2ex]
h\left(\dfrac{d_t}{f}\right)(1 - O_y), & \text{if movement along camera Y axis} \\[2ex]
k, & \text{if movement along camera Z axis}
\end{cases}
$$

where w and h are, respectively, the width and height of the camera sensor, f the focal length of the camera, k a constant value depending on the planned ground sample distance (GSD-Section 2.3.2), and O_x, O_y the target image overlaps, respectively, along the X and Y axes in decimals. The different cases ensure that the same image overlap is enforced when the movement occurs along the shortest or the longest dimension of the image sensor (Figure 4). The movement direction is detected in real-time from the largest direction cosine between the camera axes and the displacement vector between the camera centers of L_t (resp. R_t) and L_{t-1} (resp. R_{t-1}). For simplicity, the movement along the camera Z axis (forward, backward) is managed differently using a preset constant baseline k, settable by the user for the specific requirements of the acquisition. This is, in any case, an uncommon situation in standard photogrammetric acquisitions, and mostly typical of acquisitions carried out in narrow tunnels/spaces with fisheye lenses [53]. During the acquisition, the user can visualize the 3D camera poses of the selected images along with the 3D sparse reconstruction of the scenery.

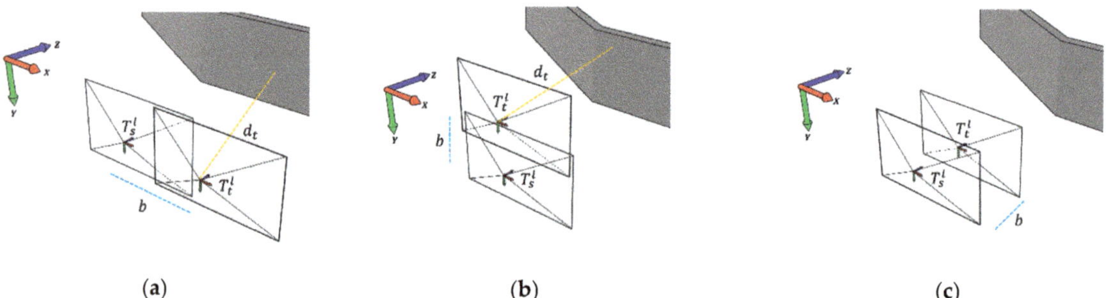

Figure 4. Visualization, relative to the left camera of the stereo system, of the different cases considered by the acquisition control: movement along the camera (**a**) X axis and (**b**) Y axis. (**c**) The movement along the Z axis, less common in photogrammetric acquisitions, is handled with a preset constant baseline k.

2.3.2. Positioning Feedback

The ground sample distance (GSD) [54] is the leading acquisition parameter when planning a photogrammetric survey. It theoretically determines the size of the pixel in object space and, consequently, the distance between the camera and the object, measured along the optical axis. Its value is typically set based on specific application requirements.

Our system allows users to specify a target GSD range (GSD_{min}, GSD_{max}) and helps them to satisfy it throughout the entire image acquisition process. Usually, the GSD of an image is not uniformly distributed and depends on the 3D structure of the imaged scene. Areas lying closer to the camera will have a smaller GSD than areas lying farther away. To properly manage these situations, the system computes the GSD in multiple 2D image locations exploiting the depth of the stereo matched tie-points. The latter are then visualized over the image in the viewer, and colored based on their GSD value. Red tie-points are those having a GSD bigger than GSD_{max}, green those having a GSD in the target range, and blue those having a GSD smaller than GSD_{min}. This gives the user an easy-to-understand feedback about his/her actual positioning even on a limited sized display. Moreover, when the pair (L_s, R_s) has been deemed as part of the image acquisition by the acquisition control (Section 2.3.1), the system automatically updates the average GSD of the 3D tie-points, or landmarks in the V-SLAM terminology, matched/seen at the time s by the stereo pair (L_s, R_s). In this way it is possible to finely triangulate the GSD of the acquired images over the sparse 3D point cloud of OPEN-V-SLAM. The latter can be visualized in real-time in the viewer and, exploiting the same color scheme as above, it can help the user to have a global awareness of the GSD of the acquired images, thus detecting areas where the target GSD was not achieved.

2.3.3. Speed Feedback

Motion blur can significantly worsen the quality of image acquisitions performed in motion. It occurs when the camera, or the scene objects, move significantly during the exposure phase. The exposure time, i.e., the time interval during which the camera sensor is exposed to the light, is usually adjusted, either manually or automatically, during the acquisition. This is accomplished to compensate for different lighting conditions and avoid under/over-exposed images. Consequently, also the acquisition speed should be adapted, especially when the scene is not well illuminated, and the exposure time is relatively long.

Rather than detecting motion-blur with image analysis techniques [55–57], our system tries to prevent it by monitoring the speed of the acquisition device and by warning the user with specific messages on the screen (e.g., using traffic light conventionally colored messages, such as "slow down" or "speed increase possible"). This approach avoids costly image analysis computations and exploits the computations already accomplished by the V-SLAM algorithm. The speed at time t, v_t, is computed dividing the space travelled by the cameras, the Euclidean norm of the last two frames of the left camera centers L_t and L_{t-1}, by the elapsed time interval. For this computation, a main assumption and simplification here is accomplished considering the imaging sensor parallel to the main object's surface, which is typically the case in photogrammetric surveys. A speed warning is raised when:

$$\delta_t \geq \max(GSD_{min}, GSD_{minD})$$

where δ_t is the space travelled by the cameras at speed v_t during the current exposure time, and GSD_{minD} is the GSD of the closest stereo-matched tie-point. A filtering strategy that uses the 5-th percentile of stereo matched distances is used to avoid computing the GSD on outliers. The logic behind this control is that the camera should move, during the exposure time, less than the target GSD. This ensures that the smallest details remain sharp in the images. The max function is used to avoid false warnings when the closest element in the scene is farther than the minimum GSD.

2.3.4. Camera Control for Exposure Correction

The literature of automatic exposure (AE) algorithms is quite rich, and each solution has been usually conceived to fit a particular scenario. For example, in photogrammetric acquisitions, the surveyed object should have the highest priority, even at the cost of worsening the image content of foreground and background image regions. However, distinguishing in the images the surveyed object from the background requires in general a semantic image understanding such as for example using convolutional neural networks (CNNs) [58]. Deploying CNNs on embedded systems is currently an open problem [59] for the limited resource capabilities of these systems. As a workaround, our device assumes that the stereo matched tie-points with a depth value within the target acquisition range (Section 2.3.2) are distributed, in the image, over the surveyed object. This allows us to smartly select the image regions considered by the AE algorithm, masking away the foreground and background areas that will not contribute to the computation of the exposure time. The image regions are built considering 5×5 pixel windows around the locations of the stereo-matched tie-points whose estimated distance lie in the target acquisition range. The optimal exposure time is then computed on the masked image using a simple histogram-based algorithm very similar to [60]. The exposure time is incremented (resp. decremented) by a fixed value Γ when the mean sample value of the brightness histogram is below (resp. above) the middle gray.

3. Experiments

The following sections present indoor and outdoor experiments realized to evaluate the camera synchronization and calibration, guidance system (positioning, speed and exposure control) and accuracy performances of the developed handheld stereo system. All the tests were performed live on the onboard ARM architecture of the system. The stereo images are downscaled to 640×512 pixel (half linear resolution) and rectified before being fed to OPEN-V-SLAM.

3.1. Camera Syncronization and Calibration

In the current implementation of the system, the acquisition of the stereo image pair is synchronized with software triggers, although hardware triggers are possible. We have measured a maximum synchronization error between the left and the right image of 1 ms. This has been tested recording for several minutes the display of a simple millisecond counter system using, on both cameras, a shutter speed of 0.5 ms (Figure 5a).

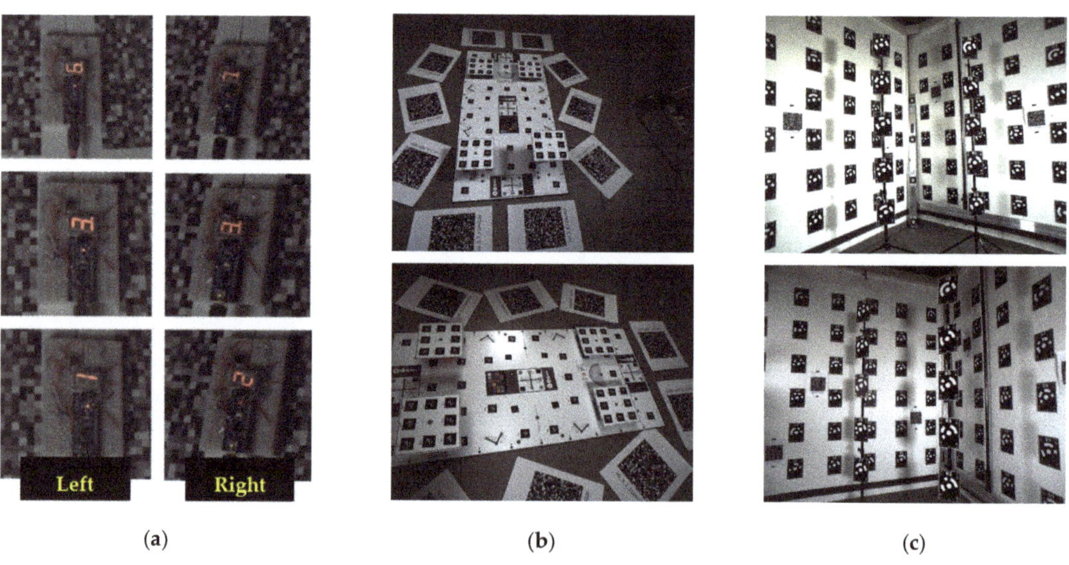

Figure 5. Some images of the millisecond counter: the maximum synchronization error amounts to 1 ms (**a**). Some calibration images showing the small (**b**) and big (**c**) test fields available in the FBK-3DOM lab.

OPEN-V-SLAM, similar to many other V-SLAM implementations, does not perform, for efficiency reasons, self-calibration during the bundle adjustment iterations. Therefore, both the interior parameters of the cameras, and, in stereo setups, the relative orientation and baseline of the cameras, must be estimated beforehand. The calibration has been accomplished by means of bundle adjustment with self-calibration, exploiting test of photogrammetric targets with known 3D coordinates and accuracy. The calibration procedure was carried out twice: one with camera focused at 1 m for the positioning and speed feedback test (Section 3.2), and one with the cameras focused at the hyperfocal distance (1.2 m with circle of confusion of 0.0048 mm and f2.8) for the exposure (Section 3.3) and accuracy (Section 3.4) tests. In the former case we used a small 500×1000 mm test field composed of several resolution wedges, Siemens stars, color checkboards and circular targets placed at different heights (Figure 5b). In the latter case we used a bigger test field (Figure 5c) with 82 circular targets. In both cases, the test fields were imaged from many positions (respectively, 116 and 62 stereo pairs) and orientations (portrait, landscape) to ensure optimal intersection geometry and reduce parameter correlations. In both scenarios, some randomly generated patterns were added to help the orientation of the cameras within the bundle adjustment approach. Table 2 reports the estimated exterior camera parameters with their standard deviations.

Table 2. Calibration results. Exterior parameters of the right camera with respect to the left one.

First Calibration with Focus 1m—Small Test Field							
B *(mm)*	σ_B *(mm)*	ω *(deg)*	σ_ω *(deg)*	φ *(deg)*	σ_φ *(deg)*	κ *(deg)*	σ_κ *(deg)*
245.3167	0.1099	−0.0935	0.0394	−22.1123	0.0121	−0.0239	0.0341
Second Calibration with Focus at Hyperfocal—Large Test Field							
B *(mm)*	σ_B *(mm)*	ω *(deg)*	σ_ω *(deg)*	φ *(deg)*	σ_φ *(deg)*	κ *(deg)*	σ_κ *(deg)*
244.9946	0.2388	−0.0941	0.0136	−22.4721	0.0079	−0.01199	0.0055

Finally, to enable a more accurate feedback on the GSD (Section 3.2), the actual modulation transfer function (MTF) of the lens was measured using an ad-hoc test chart [61]. The chart includes photogrammetric targets that allow the relative pose of the camera to be determined with respect to the chart plane and, consequently, a better estimation of the actual GSD as well as other optical characteristics such as the depth of field. The test chart uses slant-edges according to the ISO 12233 standard. Moreover, it includes resolution wedges along the diagonals with metric scale that allow a direct visual estimation of the limiting resolution of the lens. In our experiment we compared the expected nominal GSD at the measured distance from the chart against the worst of radially and tangentially resolved patterns along the diagonals of the chart as shown in Figure 6. We estimated a ratio of about 2 considering both left and right cameras that was then used in the implemented system for a more accurate positioning feedback that is based on actual spatial resolution values attainable by our low-cost MMS.

3.2. Positioning and Speed Feedback Test

In this experiment the system positioning (Section 2.3.2) and speed (Section 2.3.3) feedback are evaluated in a controlled scenario. The experiment is structured as follow: (i) set a target GSD range of 1–2 mm; (ii) use the small test field (Figure 5b) as target object and start the image acquisition outside of the target GSD range; (iii) move closer to the object until the system says that we are in the target acquisition range; (iv) start different strips over the test field at constant distance alternating strips predicted in the speed range with strips predicted out of the speed range.

Figure 7a shows the system interface at the beginning of the test. On the top left corner there is the live stream of the left camera with the GSD-colored tie-points. Below, under the control buttons, are reported the speed of the device, the median distance to the object, the exposure and gain values of the cameras and the number of 3D tie-points matched by the image. On the right of the interface there is the 3D viewer in which are displayed the GSD-colored point cloud, the current system position (green pyramid) and the 3D positions of the acquired images (pink pyramids). The distance feedback is given at multiple levels: in the tie-point of the live image, in the median distance panel and in the 3D point cloud. All these indicators are red since we are out of the target range. Figure 7b confirms the correct feedback of the system. The thickest lines of the Siemens stars are 2.5 mm thick but they are not clearly visible. Figure 7c depicts the interface upon reaching, according to the system, the target acquisition range. All the indicators (tie-points, current distance and 3D point cloud) are green. Figure 7d confirms again the system prediction and details with 2 mm resolution can be clearly seen.

Figure 7e,f show a close view of the resolution chart in the middle of the test field during the horizontal strips over the object. The former images are taken from strips predicted in speed range, while the latter ones from strips predicted out of range. From these images it is very clear to see how, in the second case, motion blur occurred along the movement direction and evidently reduced the visible details. Despite being the image taken from a distance where theoretically it should be possible to distinguish details at 2 mm, motion blur has made even details at 2.5 hardly visible. On the other hand, when we respected the speed suggested by the system, the images remained sharp satisfying the target acquisition GSD of 2 mm.

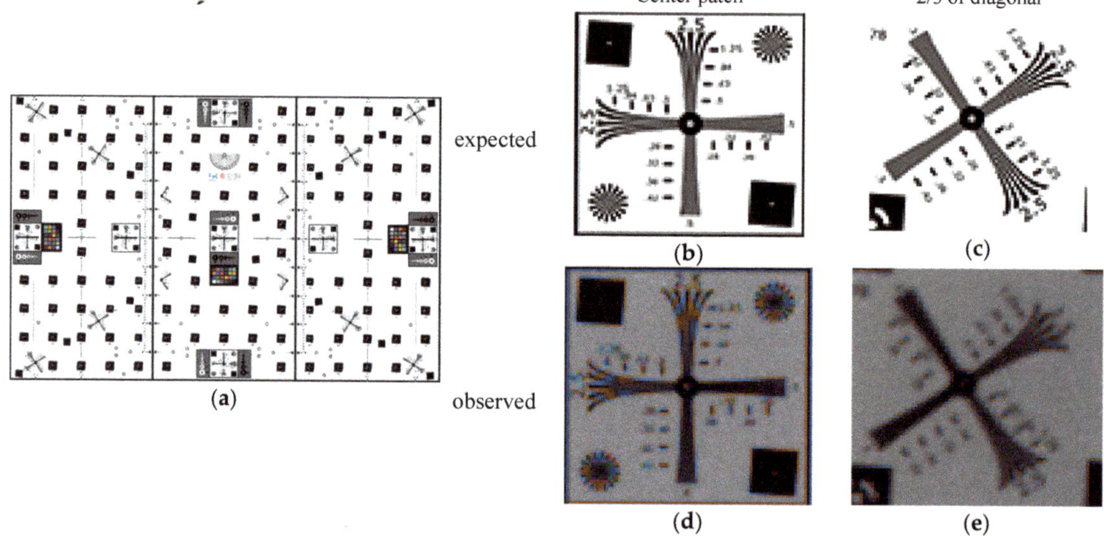

Figure 6. (a) Resolution chart [61] used to experimentally estimate the modulation transfer function of the used lenses. Examples of the expected resolution patches (using the designed test chart at the same resolution of 1280 pixel in width as for the camera used), respectively, at the center (b) and at 2/3 of the diagonal (c) against the imaged ones from one of the two cameras (d,e).

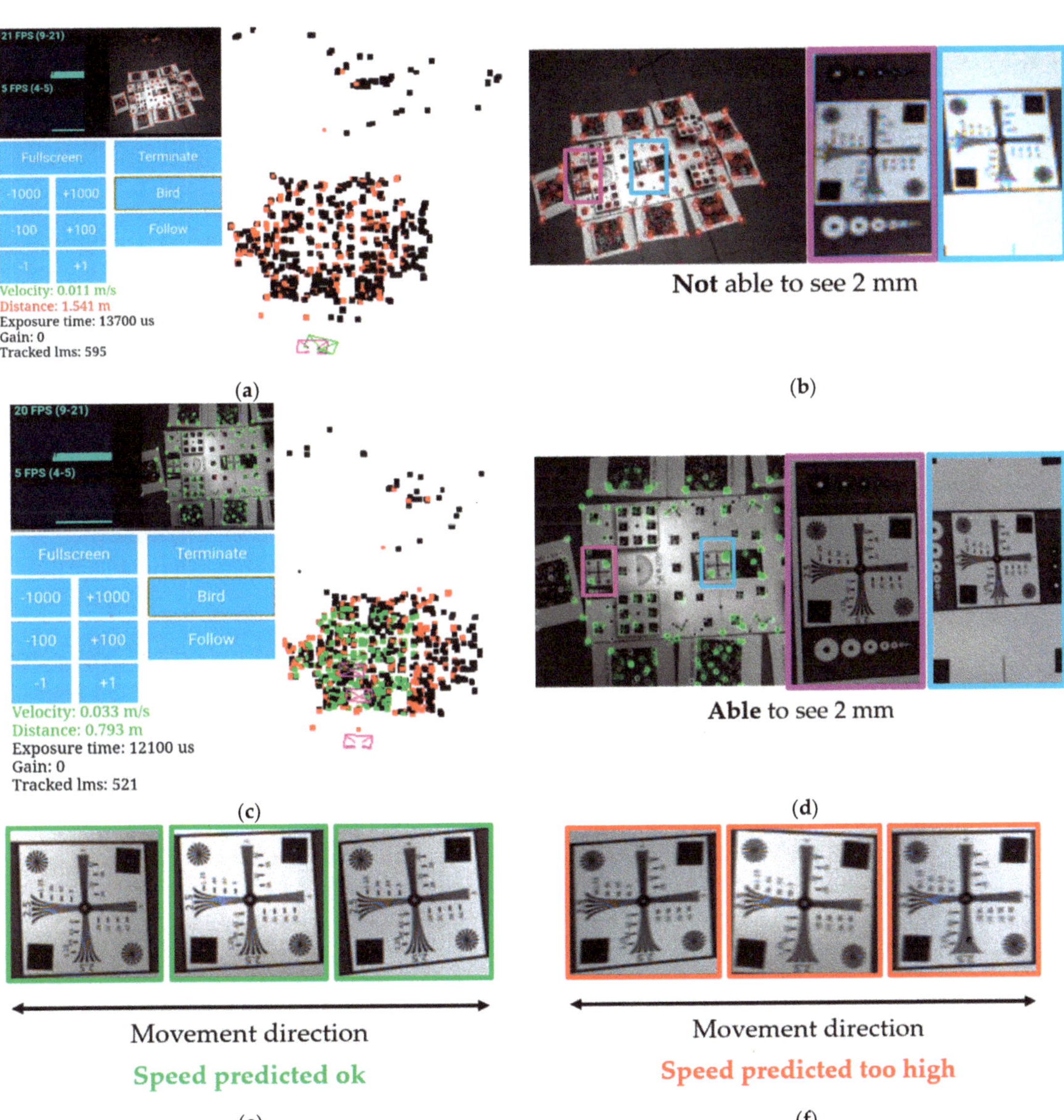

(a)

(b)

(c)

(d)

(e)

(f)

Figure 7. System interface and the positioning and speed feedback test (**a–f**). See Section 3.2 for a detailed explanation.

3.3. Camera Control Test

This section reports how a camera automatic exposure (AE-Section 2.3.4) can be boosted by the V-SLAM algorithm in challenging lighting conditions. A typical situation is when the target object is imaged against a brighter or darker background. Our prototype was used to acquire two consecutive datasets of a building in situations of sky backlight. The two datasets, acquired with the same movements and trajectories, differ a few minutes between each other, so the illumination practically remained the same. The first dataset is processed using the camera auto exposure algorithm, whereas the second sequence uses the proposed masking procedure. A visual comparison of the acquired images in the first

(Figure 8a) and second dataset (Figure 8b) clearly highlights the advantages brought by the proposed masking procedure to support the auto exposure in giving more importance to objects having a specific distance. Having set, for the acquired datasets, a target GSD range between 0.002 and 0.02 m, the masked images have been built in the neighborhoods of the tie-points with depth values between 0.83 and 8.3 m. Figure 8c plots the exposure time of the cameras in the two tests, where it is possible to notice, after the tenth frame, the different behavior of the AE algorithm. Figure 8d shows one of the test images with the extracted tie-points: it can be observed how they are properly distributed on the surveyed object. On the other hand, when the exposure was adjusted on the full image, it was significantly influenced by the sky brightness and underexposed the building.

AE algorithm underexposed the surveyed object, negatively affected by the sky brightness

(**a**)

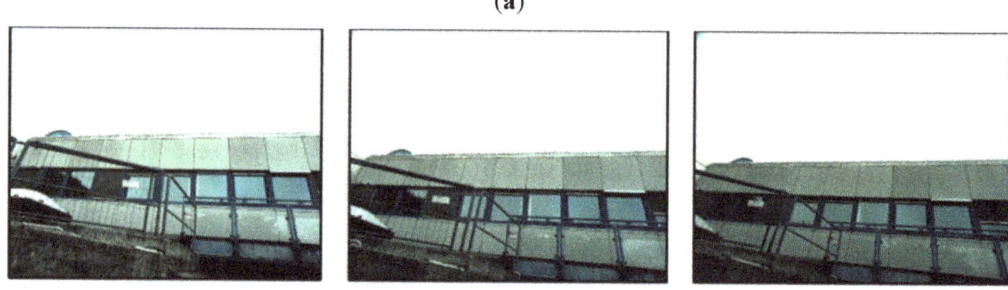

AE algorithm correctly exposed the surveyed object, ignoring the sky overexposure

(**b**)

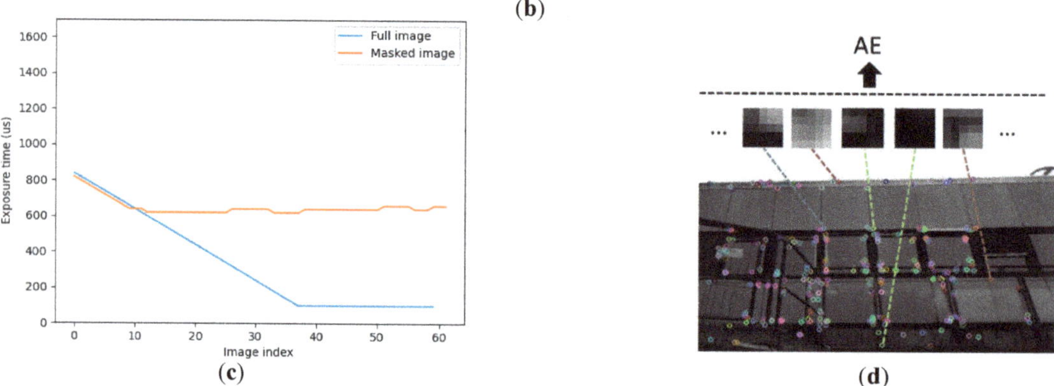

(**c**)　　　　(**d**)

Figure 8. Some of the images acquired when the auto exposure (AE) algorithm uses the whole image (**a**) and the proposed masked image (**b**). (**c**) Plot of the exposure time in the two tests. (**d**) Visualization of the tie-points locations used by the AE algorithm in one of the images.

3.4. Accuracy Test

A large-scale surveying scenario is used to evaluate the image selection procedure and the potential accuracy of the proposed V-SLAM-based device. The surveyed object is a FBK's building, which spans approximately 40×60 m (Figure 9a). The acquisition device was handheld by an operator who surveyed the object by keeping the camera sensors mostly parallel to the building facades. To collect 3D ground truth data, the building was scanned with a Leica HDS7000 (angular accuracy of 125 μrad, range noise 0.4 mm RMS at 10 m) from 21 stations along its perimeter at an approximate distance of 10 m. All the scans were manually cleaned from the vegetation and co-registered (Figure 9b) with a global iterative closest point (ICP) algorithm implemented in MeshLab [62]. The final median RMS of the residuals from the alignment transformation was about 4 mm.

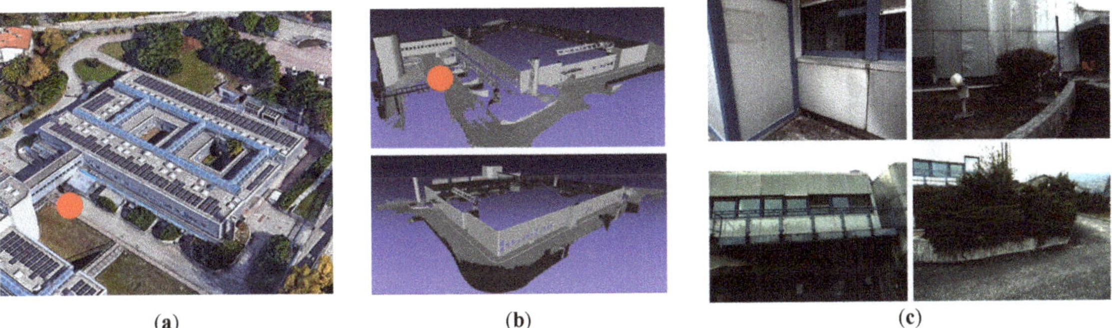

Figure 9. The FBK's building used for the device's accuracy evaluation: (**a**) aerial view (Google Earth) and (**b**) laser scanning mesh model used as ground truth. The red dot is the starting/ending surveying position. (**c**) Some of the acquired images of the building: challenging glasses, reflections, repetitive patterns, poor texture surfaces and vegetation are clearly visible.

The image acquisition was performed by setting the device to acquire images in raw format with 80% overlap and a GSD range between 3 and 30 mm. The acquisition started and ended from the same positions and it lasted roughly 18 min, following device feedback on distance-to-object and speed. The overall trajectory was estimated, after a successful loop closure, about 315 m long. Along the trajectory, the acquisition control selected and saved 271 image pairs (white dots in Figure 10a). For a comparison, the common acquisition strategy of taking images at 1Hz would have acquired approximately 1080 images. The derived V-SLAM-based point cloud is color-coded with the average GSD of the selected images. The acquisition frequency was adapted to the distance to the structure, hence the irregular distance among the image pairs. Some areas of the building, such as 1 and 2 in Figure 10a, were not acquired at the target GSD due to physical limitations to reach closer positions while walking. On the other hand, a van parked in the middle of the street (area 3 in Figure 10a) forced to change the planned trajectory. This unexpected event was automatically managed by the developed algorithm that adapted the baseline in realtime to consider the shorter distance from the building.

The acquired stereo pairs were then processed offline to obtain the dense point cloud of the structure, following two approaches:

1. Directly using the camera poses estimated by our system, without further refining.
2. Re-estimating the camera poses with photogrammetric software, fixing the stereo baseline between corresponding pairs to impose the scale.

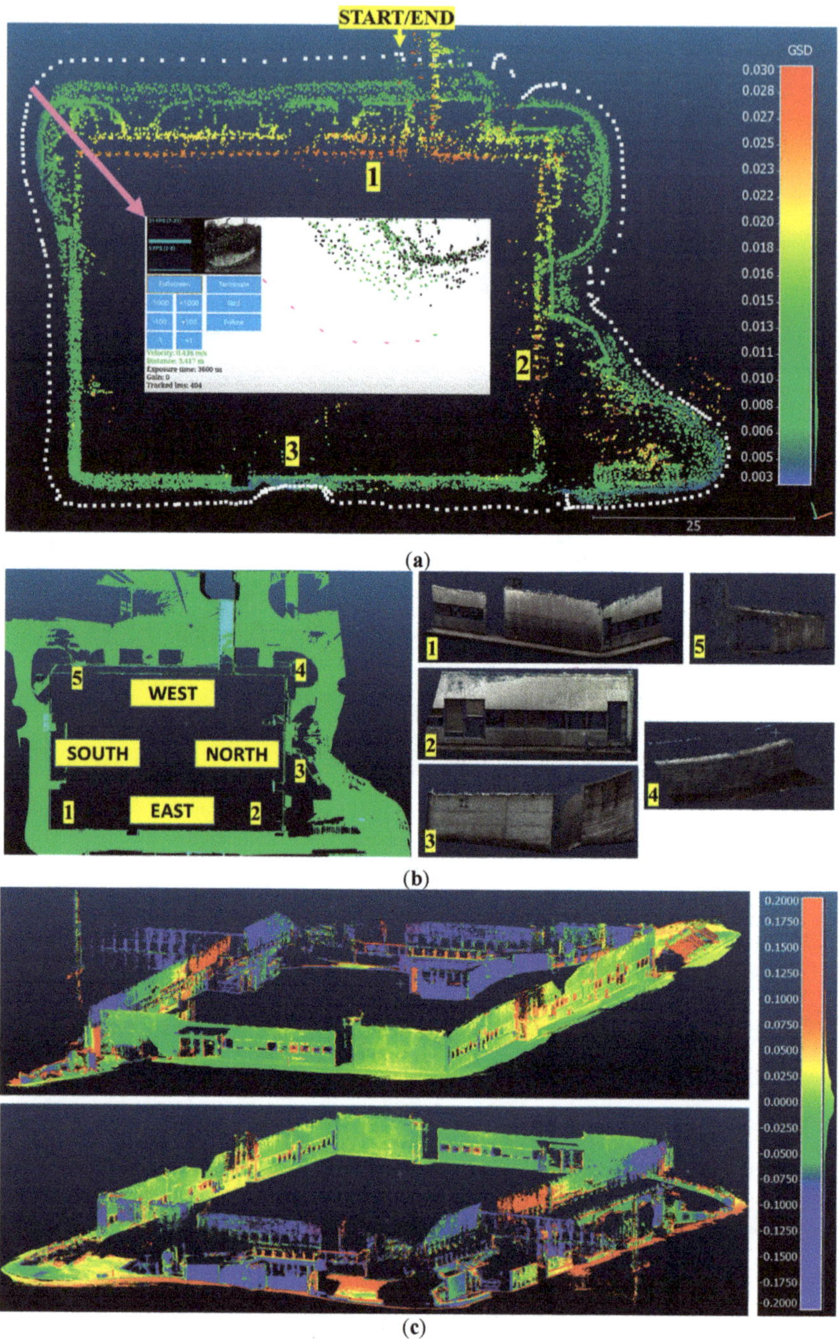

Figure 10. (a) Top view of the estimated camera positions of the selected images (white squares) and derived GSD-color-coded point cloud. (b) Building areas used to align the derived dense point cloud with the laser scanning ground truth. (c) Dense point cloud obtained using the poses estimated by our system (Case 1.), with color-coded signed Euclidean distances [m] with respect to the ground truth.

The two dense points clouds were then aligned with the laser ground truth using five selected and distributed areas (Figure 10b), achieving a final alignment error of 0.02 m (case 1.) and 0.019 m (case 2.). Agisoft Metashape was used to orient the images (case 2.) and perform the dense reconstruction (cases 1. and 2.). Finally, we computed a cloud to mesh distance between the two dense point clouds and the laser ground truth. In both cases, the errors range from a few centimeters in the south and east part of the building, to some 20 cm in the north and west ones (Figures 10c and 11). In case 1., the distribution of the signed distances has a mean of 0.003 m and a standard deviation of 0.070 m (Figure 12a), with 95% of the differences bounded in the interval [−0.159, 0.166] m. The case 2. returned a slightly better error distribution (Figure 12b), with a mean value of −0.009 m, a standard deviation of 0.054 m and 95% of the differences falling in the interval [−0.144, 0.114] m. In addition to reporting the global error distribution of the building, which is significantly related to the outcome of the ICP algorithm, the local accuracy of the derived dense point cloud is also estimated using the LME (length measurement error) and RLME (relative length measurement error) [63] between seven segments (Figure 13), manually selected both on the dense point cloud obtained from the V-SLAM poses and the ground truth. The results are reported in Table 3.

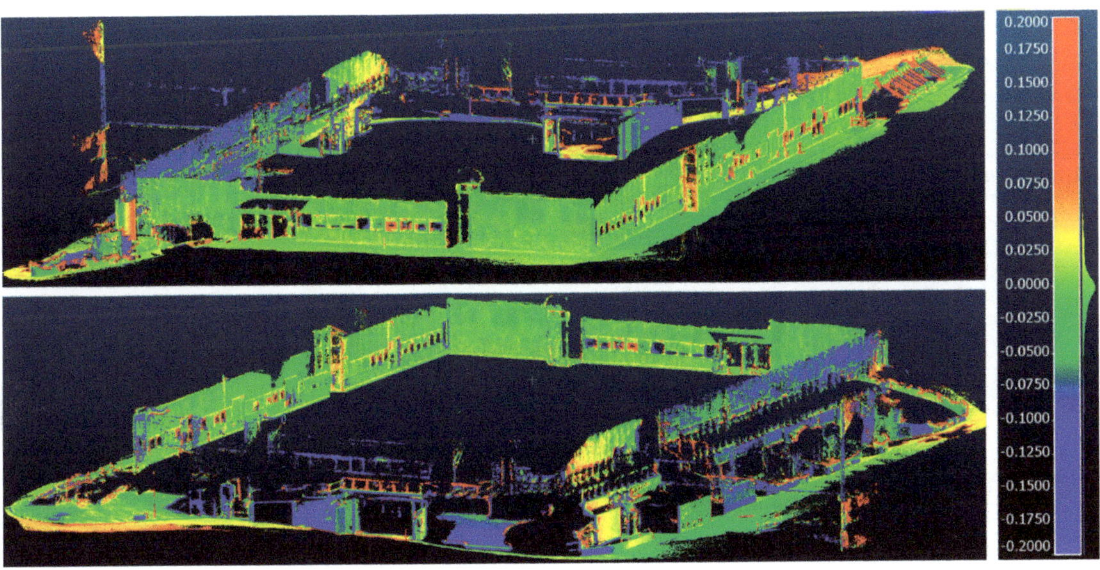

Figure 11. Dense point cloud obtained from the selected images with the standard photogrammetric pipeline (Case 2.) and color-coded signed Euclidean distances [m] with respect to the ground truth. The gaps in the lower left area correspond to zone #2 in Figure 10a.

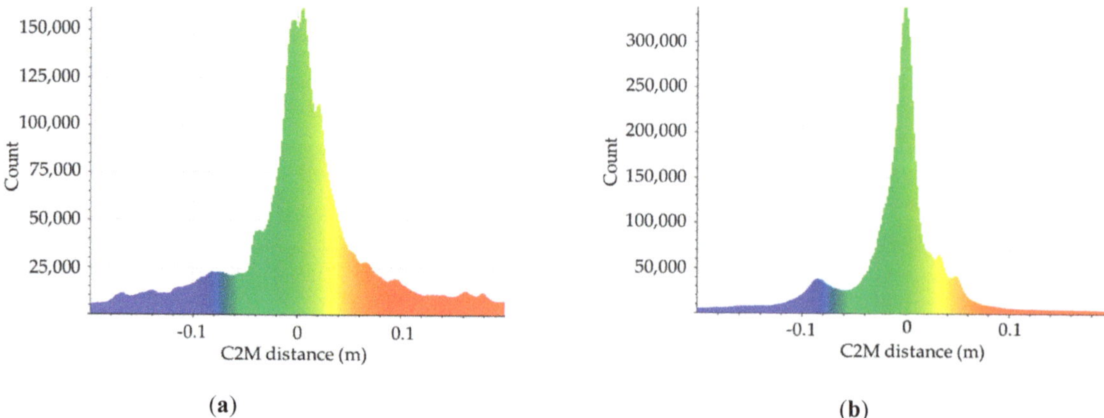

Figure 12. Histogram and distribution of the signed Euclidean distances [m] between the reference mesh and the dense point clouds obtained, respectively, with (**a**) the SLAM-based camera poses and (**b**) the standard photogrammetric pipeline.

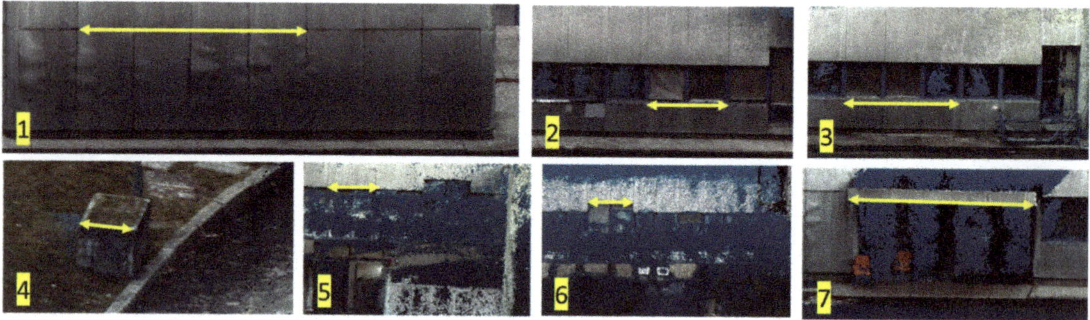

Figure 13. Segments considered in the LME and RLME analyses (Table 3).

Table 3. Measured lengths, LME and RLME between the selected segments (Figure 13).

Pair	Laser Data [m]	Dense (SLAM-Based) Cloud [m]	LME [m]	RLME [%]
1	4.7766	4.7508	0.0258	−0.5401
2	2.3843	2.3701	0.0142	−0.5956
3	3.5795	3.5854	0.0059	0.1648
4	0.4948	0.4901	0.0047	−0.9498
5	2.3888	2.3433	0.0455	−1.9047
6	2.3730	2.3840	0.011	0.4635
7	4.7647	4.7838	0.0191	0.4009
Median			**0.0142**	**−0.5401**

4. Discussion

The presented experiments showed the main advantages brought by the proposed device. Tests to validate positioning and speed feedback (Section 3.2) have shown how the system enables a precise and easy-to-understand GSD guidance. The GSD is displayed for the current position of the device but also on the final sparse point cloud. The system also detects situations of motion blur which helps an user to prevent them. Within the camera control test (Section 3.3), exploiting the tight integration with the hardware, we took advantage of the known depth of the stereo tie-points to optimize the camera auto exposure, guiding its attention to areas that matter for the acquisition. Finally, the accuracy test (Section 3.4) showed how an image acquisition properly optimized with the scene geometry avoids over or under selections (common in time-based acquisitions) and allows to achieve satisfactory 3D results in a big and challenging scenario (Figures 10c and 11). The obtained error of a few centimeters, and a RLME error around 0.5%, against the laser scanning ground truth, highlights both the quality of the image acquisition and the accuracy level reached by the developed device. The achieved results are truly inspiring if we consider the important size of the trajectory (more than 300 m), the not always collaborative surface of the building (windows, glasses and panels with poor texture-Figure 9b) and the real-time estimation of the camera poses using a simple Raspberry Pi 4. Some parts of the building, especially in the highest parts, present holes and are less accurate, but for those areas we would have required a UAV to properly record the structure with more redundancy and stronger camera network geometry. Obviously, the V-SLAM approach is meant for real-time applications therefore, considering the dimensions of the outdoor case study, we could consider satisfactory the achieved accuracy results.

5. Conclusions and Future Works

The paper presented the GuPho (Guided Photogrammetric System) prototype, a handheld, low-cost, low-powered synchronized stereo vision system that combines Visual SLAM with an optimized image acquisition and guidance system. This solution can work, theoretically, in any environment without assuming GNSS coverage or pre-calibration procedures. The combined knowledge of the system's position and the generated 3D structure of the surveyed scene have been used to design a simple but effective real-time guidance system aimed at optimizing and making image acquisitions less error-prone. The experiments showed the effectiveness of this system to enable GSD-based guidance, motion blur prevention, robust camera exposure control and an optimized image acquisition control. The accuracy of the system has been tested in a real and challenging scenario showing an error ranging from few centimeters to some twenty centimeters in the areas not properly reached during the acquisitions. These results are even more interesting if we consider that they have been achieved with a low-cost device (ca. 1000 EUR), walking at ca 30 cm/sec, processing the image pairs in real-time at 5Hz with a simple Raspberry Pi 4. Moreover, the automatic exposure adjustment, GSD and speed control proved to be very efficient and capable of adapting to unexpected environmental situations. Finally, time-constrained applications could take advantage of the already oriented images, thanks to the V-SLAM algorithm, to speed up further processing such as the estimation of dense 3D point clouds. The built map can be stored and reutilized for successive revisiting and monitoring surveys, making the device a precise and low-cost local positioning system. Once that the survey in the field is completed, full resolution photogrammetric 3D reconstructions could be carried out on more powerful resources, such as workstations or Cloud/Edge computing, with the additional advantage of speeding up the image orientation task by reusing the approximations computed by the V-SLAM algorithm. Thanks to the system modular design, the application domain of the proposed solution could be easily extended to other scenarios, e.g., structure monitoring, forestry mapping, augmented reality or supporting rescue operations in case of disasters, facilitated also by the system easy portability, relatively low-cost and competitive accuracy.

As future works, in addition to carrying out evaluation tests in a broader range of case studies covering outdoor, indoor and potentially underwater scenarios, we will add the possibility to trigger a third high resolution camera according to the image selection criteria that will consider the calibration parameters of the third camera. We will also improve the image selection by also considering the camera rotations and by detecting already acquired areas. We are also investigating how to improve, possibly taking advantage of edge computing (5G), the real-time quality feedback of the acquisition, either with denser point clouds or with augmented reality. Finally, we are very keen to explore the applicability of deep learning to make the system more aware of the content of the images, potentially boosting the real-time V-SLAM estimation, the control of the cameras (exposure and white-balance) and the guidance system. At the end, we will investigate the use of Nvidia Jetson devices, or possibly the edge computing (5G), to allow the system to take advantage of state-of-the-art image classification/segmentation/enhancement methods offered by deep learning.

Author Contributions: The article presents a research contribution that involved authors in equal measure. Conceptualization, A.T., F.M. and F.R.; methodology, A.T. and F.M.; investigation, A.T., F.M. and R.B.; software, A.T., F.M. and R.B.; validation, A.T. and F.M.; resources, A.T.; data curation, A.T.; writing–original draft preparation, A.T.; writing–review and editing, A.T., F.M., R.B. and F.R.; supervision, F.M. and F.R.; project administration, F.R.; funding acquisition, F.R. All authors have read and agreed to the published version of the manuscript.

Funding: This work is partly supported by EIT RawMaterials GmbH under Framework Partnership Agreement No. 19018 (AMICOS. Autonomous Monitoring and Control System for Mining Plants).

Institutional Review Board Statement: Not applicable.

Informed Consent Statement: Not applicable.

Data Availability Statement: Not applicable.

Conflicts of Interest: The authors declare no conflict of interest.

References

1. Riveiro, B.; Lindenbergh, R. (Eds.) *Laser Scanning: An Emerging Technology in Structural Engineering*, 1st ed.; CRC Press: Boca Raton, FL, USA, 2019. [CrossRef]
2. Puente, I.; González-Jorge, H.; Martínez-Sánchez, J.; Arias, P. Review of mobile mapping and surveying technologies. *Measurement* **2013**, *46*, 2127–2145. [CrossRef]
3. Toschi, I.; Rodríguez-Gonzálvez, P.; Remondino, F.; Minto, S.; Orlandini, S.; Fuller, A. Accuracy evaluation of a mobile mapping system with advanced statistical methods. *Int. Arch. Photogramm. Remote Sens. Spat. Inf. Sci.* **2015**, *40*, 245. [CrossRef]
4. Nocerino, E.; Rodríguez-Gonzálvez, P.; Menna, F. Introduction to mobile mapping with portable systems. In *Laser Scanning*; CRC Press: Boca Raton, FL, USA, 2019; pp. 37–52.
5. Hassan, T.; Ellum, C.; Nassar, S.; Wang, C.; El-Sheimy, N. Photogrammetry for Mobile Mapping. *GPS World* **2007**, *18*, 44–48.
6. Burkhard, J.; Cavegn, S.; Barmettler, A.; Nebiker, S. Stereovision mobile mapping: System design and performance evaluation. *Int. Arch. Photogramm. Remote Sens. Spat. Inf. Sci.* **2012**, *5*, 453–458. [CrossRef]
7. Holdener, D.; Nebiker, S.; Blaser, S. Design and Implementation of a Novel Portable 360 Stereo Camera System with Low-Cost Action Cameras. *Int. Arch. Photogramm. Remote Sens. Spat. Inf. Sci.* **2017**, *42*, 105. [CrossRef]
8. Blaser, S.; Nebiker, S.; Cavegn, S. System design, calibration and performance analysis of a novel 360 stereo panoramic mobile mapping system. *ISPRS Ann. Photogramm. Remote Sens. Spat. Inf. Sci.* **2017**, *4*, 207. [CrossRef]
9. Ortiz-Coder, P.; Sánchez-Ríos, A. An Integrated Solution for 3D Heritage Modeling Based on Videogrammetry and V-SLAM Technology. *Remote Sens.* **2020**, *12*, 1529. [CrossRef]
10. Masiero, A.; Fissore, F.; Guarnieri, A.; Pirotti, F.; Visintini, D.; Vettore, A. Performance Evaluation of Two Indoor Mapping Systems: Low-Cost UWB-Aided Photogrammetry and Backpack Laser Scanning. *Appl. Sci.* **2018**, *8*, 416. [CrossRef]
11. Al-Hamad, A.; Elsheimy, N. Smartphones Based Mobile Mapping Systems. *ISPRS Int. Arch. Photogramm. Remote Sens. Spat. Inf. Sci.* **2014**, *40*, 29–34. [CrossRef]
12. Nocerino, E.; Poiesi, F.; Locher, A.; Tefera, Y.T.; Remondino, F.; Chippendale, P.; Van Gool, L. 3D Reconstruction with a Collaborative Approach Based on Smartphones and a Cloud-Based Server. *ISPRS Int. Arch. Photogramm. Remote Sens. Spat. Inf. Sci.* **2017**, *42*, 187–194. [CrossRef]
13. Fraser, C.S. Network design considerations for non-topographic photogrammetry. *Photogramm. Eng. Remote Sens.* **1984**, *50*, 1115–1126.
14. Nocerino, E.; Menna, F.; Remondino, F.; Saleri, R. Accuracy and block deformation analysis in automatic UAV and terrestrial photogrammetry-Lesson learnt. *ISPRS Ann. Photogramm. Remote Sens. Spat. Inf. Sci.* **2013**, *2*, W1. [CrossRef]
15. Nocerino, E.; Menna, F.; Remondino, F. Accuracy of typical photogrammetric networks in cultural heritage 3D modeling projects. *Int. Arch. Photogramm. Remote Sens. Spat. Inf. Sci.* **2014**, *45*, 465–472. [CrossRef]
16. Remondino, F.; Nocerino, E.; Toschi, I.; Menna, F. A Critical Review of Automated Photogrammetricprocessing of Large Datasets. *ISPRS Int. Arch. Photogramm. Remote Sens. Spat. Inf. Sci.* **2017**, *42*, 591–599. [CrossRef]
17. Sieberth, T.; Wackrow, R.; Chandler, J.H. Motion blur disturbs—The influence of motion-blurred images in photogrammetry. *Photogramm. Rec.* **2014**, *29*, 434–453. [CrossRef]

18. Iglhaut, J.; Cabo, C.; Puliti, S.; Piermattei, L.; O'Connor, J.; Rosette, J. Structure from Motion Photogrammetry in Forestry: A Review. *Curr. For. Rep.* **2019**, *5*, 155–168. [CrossRef]

19. Ballabeni, A.; Apollonio, F.I.; Gaiani, M.; Remondino, F. Advances in Image Pre-processing to Improve Automated 3D Reconstruction. *Int. Arch. Photogramm. Remote Sens. Spat. Inf. Sci.* **2015**, *8*, 178. [CrossRef]

20. Lecca, M.; Torresani, A.; Remondino, F. Comprehensive evaluation of image enhancement for unsupervised image description and matching. *IET Image Process.* **2020**, *14*, 4329–4339. [CrossRef]

21. Kedzierski, M.; Wierzbicki, D. Radiometric quality assessment of images acquired by UAV's in various lighting and weather conditions. *Measurement* **2015**, *76*, 156–169. [CrossRef]

22. Gerke, M.; Przybilla, H.-J. Accuracy analysis of photogrammetric UAV image blocks: Influence of onboard RTK-GNSS and cross flight patterns. *Photogramm. Fernerkund. Geoinf. PFG* **2016**, *1*, 17–30. [CrossRef]

23. Štroner, M.; Urban, R.; Reindl, T.; Seidl, J.; Brouček, J. Evaluation of the Georeferencing Accuracy of a Photogrammetric Model Using a Quadrocopter with Onboard GNSS RTK. *Sensors* **2020**, *20*, 2318. [CrossRef] [PubMed]

24. Nex, F.; Duarte, D.; Steenbeek, A.; Kerle, N. Towards real-time building damage mapping with low-cost UAV solutions. *Remote Sens.* **2019**, *11*, 287. [CrossRef]

25. Sahinoglu, Z. *Ultra-Wideband Positioning Systems*; Cambridge University Press: Cambridge, UK, 2008.

26. Shim, I.; Lee, J.-Y.; Kweon, I.S. Auto-adjusting camera exposure for outdoor robotics using gradient information. In Proceedings of the 2014 IEEE/RSJ International Conference on Intelligent Robots and Systems, Chicago, IL, USA, 14–18 September 2014; pp. 1011–1017. [CrossRef]

27. Zhang, Z.; Forster, C.; Scaramuzza, D. Active exposure control for robust visual odometry in HDR environments. In Proceedings of the 2017 IEEE International Conference on Robotics and Automation (ICRA), Singapore, 29 May–3 June 2017; pp. 3894–3901. [CrossRef]

28. Taketomi, T.; Uchiyama, H.; Ikeda, S. Visual SLAM algorithms: A survey from 2010 to 2016. *IPSJ Trans. Comput. Vis. Appl.* **2017**, *9*, 16. [CrossRef]

29. Younes, G.; Asmar, D.; Shammas, E.; Zelek, J. Keyframe-based monocular SLAM: Design, survey, and future directions. *Robot. Auton. Syst.* **2017**, *98*, 67–88. [CrossRef]

30. Cadena, C.; Carlone, L.; Carrillo, H.; Latif, Y.; Scaramuzza, D.; Neira, J.; Reid, I.; Leonard, J.J. Past, Present, and Future of Simultaneous Localization and Mapping: Towards the Robust-Perception Age. *IEEE Trans. Robot.* **2016**, *32*, 1309–1332. [CrossRef]

31. Sualeh, M.; Kim, G.-W. Simultaneous Localization and Mapping in the Epoch of Semantics: A Survey. *Int. J. Control Autom. Syst.* **2019**, *17*, 729–742. [CrossRef]

32. Chen, S.; Li, Y.; Kwok, N.M. Active vision in robotic systems: A survey of recent developments. *Int. J. Robot. Res.* **2011**, *30*, 1343–1377. [CrossRef]

33. Sünderhauf, N.; Brock, O.; Scheirer, W.; Hadsell, R.; Fox, D.; Leitner, J.; Upcroft, B.; Abbeel, P.; Burgard, W.; Milford, M.; et al. The Limits and Potentials of Deep Learning for Robotics. *Int. J. Robot. Res.* **2018**, *37*, 405–420. [CrossRef]

34. Karam, S.; Vosselman, G.; Peter, M.; Hosseinyalamdary, S.; Lehtola, V. Design, Calibration, and Evaluation of a Backpack Indoor Mobile Mapping System. *Remote Sens.* **2019**, *11*, 905. [CrossRef]

35. Lauterbach, H.A.; Borrmann, D.; Heß, R.; Eck, D.; Schilling, K.; Nüchter, A. Evaluation of a Backpack-Mounted 3D Mobile Scanning System. *Remote Sens.* **2015**, *7*, 13753–13781. [CrossRef]

36. Kalisperakis, I.; Mandilaras, T.; El Saer, A.; Stamatopoulou, P.; Stentoumis, C.; Bourou, S.; Grammatikopoulos, L. A Modular Mobile Mapping Platform for Complex Indoor and Outdoor Environments. *ISPRS Int. Arch. Photogramm. Remote Sens. Spat. Inf. Sci.* **2020**, *43*, 243–250. [CrossRef]

37. Davison, A.J.; Reid, I.; Molton, N.D.; Stasse, O. MonoSLAM: Real-Time Single Camera SLAM. *IEEE Trans. Pattern Anal. Mach. Intell.* **2007**, *29*, 1052–1067. [CrossRef] [PubMed]

38. Klein, G.; Murray, D. Parallel Tracking and Mapping on a camera phone. In Proceedings of the 2009 8th IEEE International Symposium on Mixed and Augmented Reality, Orlando, FL, USA, 19–22 October 2009; pp. 83–86. [CrossRef]

39. Newcombe, R.A.; Lovegrove, S.J.; Davison, A.J. DTAM: Dense tracking and mapping in real-time. In Proceedings of the 2011 International Conference on Computer Vision, Barcelona, Spain, 6–13 November 2011; pp. 2320–2327. [CrossRef]

40. Engel, J.; Schöps, T.; Cremers, D. LSD-SLAM: Large-Scale Direct Monocular SLAM. In *Computer Vision—ECCV 2014*; Fleet, D., Pajdla, T., Schiele, B., Tuytelaars, T., Eds.; Springer: Cham, Switzerland, 2014; Volume 8690, pp. 834–849. [CrossRef]

41. Mur-Artal, R.; Montiel, J.M.M.; Tardos, J.D. ORB-SLAM: A Versatile and Accurate Monocular SLAM System. *IEEE Trans. Robot.* **2015**, *31*, 1147–1163. [CrossRef]

42. Mur-Artal, R.; Tardos, J.D. ORB-SLAM2: An Open-Source SLAM System for Monocular, Stereo and RGB-D Cameras. *IEEE Trans. Robot.* **2017**, *33*, 1255–1262. [CrossRef]

43. Sumikura, S.; Shibuya, M.; Sakurada, K. OpenVSLAM: A Versatile Visual SLAM Framework. In Proceedings of the 27th ACM International Conference on Multimedia, Nice, France, 21–25 October 2019; ACM Press: New York, NY, USA; pp. 2292–2295. [CrossRef]

44. Campos, C.; Elvira, R.; Rodríguez, J.J.G.; Montiel, J.M.M.; Tardós, J.D. ORB-SLAM3: An Accurate Open-Source Library for Visual, Visual-Inertial and Multi-Map SLAM. ArXiv200711898 Cs. Available online: http://arxiv.org/abs/2007.11898 (accessed on 12 March 2021).

45. Snavely, N.; Seitz, S.M.; Szeliski, R. Photo tourism: Exploring photo collections in 3D. In *ACM Siggraph 2006 Papers*; ACM Press: New York, NY, USA, 2006; pp. 835–846.
46. Schonberger, J.L.; Frahm, J.-M. Structure-from-motion revisited. In Proceedings of the IEEE Conference on Computer Vision and Pattern Recognition, Las Vegas, NV, USA, 27–30 June 2016; pp. 4104–4113.
47. Trajković, M.; Hedley, M. Fast corner detection. *Image Vis. Comput.* **1998**, *16*, 75–87. [CrossRef]
48. Calonder, M.; Lepetit, V.; Ozuysal, M.; Trzcinski, T.; Strecha, C.; Fua, P. BRIEF: Computing a Local Binary Descriptor Very Fast. *IEEE Trans. Pattern Anal. Mach. Intell.* **2012**, *34*, 1281–1298. [CrossRef]
49. Rublee, E.; Rabaud, V.; Konolige, K.; Bradski, G. ORB: An efficient alternative to SIFT or SURF. In Proceedings of the 2011 International Conference on Computer Vision, Barcellona, Spain, 6–13 November 2011; pp. 2564–2571.
50. Kummerle, R.; Grisetti, G.; Strasdat, H.; Konolige, K.; Burgard, W. G$_2$O: A general framework for graph optimization. In Proceedings of the 2011 IEEE International Conference on Robotics and Automation, Shanghai, China, 9–13 May 2011; pp. 3607–3613. [CrossRef]
51. Galvez-López, D.; Tardos, J.D. Bags of Binary Words for Fast Place Recognition in Image Sequences. *IEEE Trans. Robot.* **2012**, *28*, 1188–1197. [CrossRef]
52. Torresani, A.; Remondino, F. Videogrammetry vs Photogrammetry for heritage 3D reconstruction. *ISPRS Int. Arch. Photogramm. Remote Sens. Spat. Inf. Sci.* **2019**, *XLII-2/W15*, 1157–1162. Available online: http://hdl.handle.net/11582/319626 (accessed on 1 June 2021).
53. Perfetti, L.; Polari, C.; Fassi, F.; Troisi, S.; Baiocchi, V.; del Pizzo, S.; Giannone, F.; Barazzetti, L.; Previtali, M.; Roncoroni, F. Fisheye Photogrammetry to Survey Narrow Spaces in Architecture and a Hypogea Environment. In *Latest Developments in Reality-Based 3D Surveying Model*; MDPI: Basel, Switzeralnd, 2018; pp. 3–28.
54. Luhmann, T.; Robson, S.; Kyle, S.; Boehm, J. *Close-Range Photogrammetry and 3D Imaging*; Walter de Gruyter: Berlin, Germany, 2013.
55. Tong, H.; Li, M.; Zhang, H.; Zhang, C. Blur detection for digital images using wavelet transform. In *Proceedings of the 2004 IEEE International Conference on Multimedia and Expo (ICME) (IEEE Cat. No.04TH8763)*; IEEE: Toulouse, France, 2004; Volume 1, pp. 17–20. [CrossRef]
56. Ji, H.; Liu, C. Motion blur identification from image gradients. In Proceedings of the 2008 IEEE Conference on Computer Vision and Pattern Recognition, Anchorage, AK, USA, 23–28 June 2008; pp. 1–8. [CrossRef]
57. Pertuz, S.; Puig, D.; Garcia, M.A. Analysis of focus measure operators for shape-from-focus. *Pattern Recognit.* **2013**, *46*, 1415–1432. [CrossRef]
58. Krizhevsky, A.; Sutskever, I.; Hinton, G.E. ImageNet classification with deep convolutional neural networks. *Commun. ACM* **2012**, *60*, 84–90. [CrossRef]
59. Hinton, G.; Vinyals, O.; Dean, J. Distilling the Knowledge in a Neural Network. *arXiv* **2015**, arXiv:1503.02531.
60. Nourani-Vatani, N.; Roberts, J. Automatic camera exposure control. In *Proceedings of the Australasian Conference on Robotics and Automation 2007*; Dunbabin, M., Srinivasan, M., Eds.; Australian Robotics and Automation Association Inc.: Sydney, Australia, 2007; pp. 1–6.
61. Menna, F.; Nocerino, E. Optical aberrations in underwater photogrammetry with flat and hemispherical dome ports. In *Videometrics, Range Imaging, and Applications XIV*; International Society for Optics and Photonics: Bellingham, WA, USA, 2017; Volume 10332, p. 1033205. [CrossRef]
62. Cignoni, P.; Callieri, M.; Corsini, M.; Dellepiane, M.; Ganovelli, F.; Ranzuglia, G. MeshLab: An Open-Source Mesh Processing Tool. In *Eurographics Italian Chapter Conference*; CNR: Pisa, Italy, 2008; pp. 129–136.
63. Nocerino, E.; Menna, F.; Remondino, F.; Toschi, I.; Rodríguez-Gonzálvez, P. Investigation of indoor and outdoor performance of two portable mobile mapping systems. SPIE Proc. In Proceedings of the Videometrics, Range Imaging, and Applications XIV, Munich, Germany, 26 June 2017; Volume 10332. [CrossRef]

Article

Improved Point–Line Visual–Inertial Odometry System Using Helmert Variance Component Estimation

Bo Xu [1], **Yu Chen [1],***, **Shoujian Zhang [1]** and **Jingrong Wang [2]**

[1] School of Geodesy and Geomatics, Wuhan University, Wuhan 430079, China; boxu1995@whu.edu.cn (B.X.); shjzhang@sgg.whu.edu.cn (S.Z.)
[2] GNSS Research Center, Wuhan University, Wuhan 430079, China; wangjingrong@whu.edu.cn
* Correspondence: chenyuphd@whu.edu.cn

Received: 11 August 2020; Accepted: 5 September 2020; Published: 7 September 2020

Abstract: Mobile platform visual image sequence inevitably has large areas with various types of weak textures, which affect the acquisition of accurate pose in the subsequent platform moving process. The visual–inertial odometry (VIO) with point features and line features as visual information shows a good performance in weak texture environments, which can solve these problems to a certain extent. However, the extraction and matching of line features are time consuming, and reasonable weights between the point and line features are hard to estimate, which makes it difficult to accurately track the pose of the platform in real time. In order to overcome the deficiency, an improved effective point–line visual–inertial odometry system is proposed in this paper, which makes use of geometric information of line features and combines with pixel correlation coefficient to match the line features. Furthermore, this system uses the Helmert variance component estimation method to adjust weights between point features and line features. Comprehensive experimental results on the two datasets of EuRoc MAV and PennCOSYVIO demonstrate that the point–line visual–inertial odometry system developed in this paper achieved significant improvements in both localization accuracy and efficiency compared with several state-of-the-art VIO systems.

Keywords: visual–inertial odometry; Helmert variance component estimation; line feature matching method; correlation coefficient; point and line features

---

1. Introduction

Simultaneous localization and mapping (SLAM) has become a key technology in autonomous driving and autonomous robot navigation, which has attracted widespread attention from academia and industry [1]. Visual SLAM technology, using an optical lens as a sensor, has the characteristics of low power consumption and small size, and is widely used in indoor environment positioning and navigation. However, visual SLAM has higher requirements for observation conditions. When the movement speed is fast or the illumination conditions are poor, the tracked point features are easily lost, resulting in larger positioning errors. In order to improve the reliability and accuracy of the visual SLAM system, fusing inertial navigation data into the visual SLAM system can significantly improve the positioning accuracy and reliability, which has become a research hotspot.

Visual–inertial odometry (VIO) uses visual and inertial navigation data for integrated navigation, which has broad application prospects and is studied worldwide [2,3]. The earliest VIO systems are mainly based on filtering technology [4,5] by using the integral of inertial measurement unit (IMU) measurement information to predict the state variables of the motion carrier, which further updates the state variables with visual information, so as to realize the tightly coupled approaches of vision and IMU information. However, with the linearization point of the nonlinear measurement model and the state transition model fixed in the filtering process, the linearization process may pose a large error with an unreasonable initial value. Thus, most scholars adopt the method of graph optimization [6,7] and use iterative methods to achieve higher precision parameter estimation [8]. For example, the OKVIS [9] system uses tightly coupled approaches to optimize the visual constraints of feature points and the preintegration constraints of IMU, and adopts optimization strategy based on keyframe and "first-in first-out" sliding window method by marginalizing the measurements from the oldest state. The VINS [10] system is a monocular visual–inertial SLAM scheme, which uses a sliding-window-based approach to construct the tightly coupled optimization of IMU preintegration and visual measurement information. In the sliding window, the oldest frame and the latest frame are selectively marginalized to maintain the optimized state variables and achieve a good optimization effect.

At present, mainstream VIO systems generally use point features as visual observations. For example, the VINS system is designed to detect Shi–Tomasi corner points [11], which uses the Kanade–Lucas–Tomasi (KLT) sparse optical flow method for tracking [12]. The S-MSCKF system [13] is designed to detect feature from accelerated segment test (FAST) corner points [14], using the KLT sparse optical flow [12] method for tracking. The OKVIS [9] system is designed to detect Harris [15] corner points, and uses binary robust invariant scalable keypoints (BRISK) [16] to match and track feature points. In most scenarios, the number of corner points are large and stable, which can ensure positioning performance. However, in weak texture environments and scenes where the illumination changes significantly, the point features always have less visual measurement information or have large measurement errors [17,18]. In order to present relief from the insufficient point feature performance, the line features that can provide structured information are introduced into the VIO system [19]. The simplest way is to use the two endpoints of the line to represent the 3D spatial line [20,21]. The 3D spatial line represented by the endpoints requires six parameters, while the 3D spatial line only has four degrees-of-freedom (DoFs); thus this representation will further cause a rank deficit problem of the equation and add additional computational burden. Bartoli and Sturm [22] proposed an orthogonal representation of line features by using four parameters to represent the 3D spatial line, in which the three-dimensional vector is related to the rotation of the line around three axes, and the last parameter represents the vertical distance from the origin to the spatial line [23]. This representation method has good numerical stability. Based on the line feature representation, He et al. [24] proposed a tightly coupled monocular point–line visual–inertial odometry (PL–VIO), which uses point and line measurement information and IMU measurement information to continuously estimate the state of the moving platform, and the state variables are optimized by the sliding window method, which ensures the accuracy and in the meantime guarantees an appropriate number of optimization variables, thereby improving the efficiency of optimization. Wen et al. [25] proposed a tightly coupled stereo point–line visual–inertial odometry (PLS–VIO), which uses stereo point–line features and IMU measurement information for tightly coupled optimization. Compared with the monocular VIO system, the stereo VIO system has higher stability and accuracy, while the time consumption is greatly increased.

In a VIO system that uses point features and line features at the same time, the traditional line feature matching method using line binary descriptors (LBD) [26] is time consuming, which reduces the real-time performance of the entire VIO system. At the same time, in the VIO system, it is difficult to provide reasonable and reliable weights of point and line features, and these two points are the key to getting good performance of the point–line coupled VIO system.

Line features have geometric information and good pixel level information. By using these two kinds of information, line features can be matched. Helmert variance component estimation (HVCE) [27] can determine the weights of different types of observations, and has been applied in many different fields including inertial navigation system (INS) and global navigation satellite system (GNSS) fusion positioning [28,29], global positioning system (GPS) and BeiDou navigation satellite system

(BDS) pseudorange differential positioning [30], and other fields, which demonstrate the effectiveness of Helmert variance component estimation.

Based on this discussion, at the front end of the point–line VIO system, the line feature matching speed is slow; at the back end, when performing tightly coupled optimization of IMU observation, point feature observation, and line feature observation, it is difficult to determine a more reasonable point–line weight. Contributions described in this article follow:

- Aiming to solve the time-consuming problem of line feature matching, this paper comprehensively uses geometric information such as the position and angle of the line feature, as well as the pixel gray information around the line feature, and uses the correlation coefficient combined with the geometric information to match the line feature.
- Aiming to deal with the problem of difficulty in determining appropriate weights for line feature and point feature observations, this paper uses the Helmert variance component estimation (HVCE) method in the sliding window optimization based on the orthogonal representation of line features to assign more reasonable weights of point and line features.
- This article compares the improved point–line VIO system (IPL–VIO, improved PL–VIO) with OKVIS–Mono [9], VINS–Mono [10], PL–VIO [24] systems, and runs EuRoc MAV [31] and PennCOSYVIO [32] datasets. We comprehensively analyze the performance of the proposed method and other classic methods on different datasets.

The organization of this paper is as follows. After a comprehensive introduction in Section 1, the mathematical model is introduced in Section 2. The numerical experiments are conducted in Section 3 and the results are discussed in Section 4. Finally, conclusions and recommendations are given in Section 5.

2. Mathematical Formulation

In general, the VIO system is divided into two modules: the front end and the back end. The front end is designed for the processing of visual measurement information, the preintegration of IMU measurement information [8], and calculates the initial poses. The back end is designed for data fusion and optimization. The front end of PL–VIO [24] adds line feature measurement information in addition to the original point feature measurement information, which improves the robustness of the algorithm. On the basis of PL–VIO, in order to reduce the front-end running time, the matching algorithm of the line feature is improved. In order to improve the accuracy of visual information in the overall optimization, we adopt the method of Helmert variance component estimation to better determine the prior weights of point and line information.

Figure 1 shows the algorithm pipeline. At the front end, we improved the line feature matching algorithm, as is shown in the red box. Simultaneously, as shown again in the red box at the back end, before entering the sliding window optimization, we use the Helmert variance component estimation algorithm to estimate the weights of point features and line features. Finally, we add visual information and IMU measurement information to the sliding window for optimization.

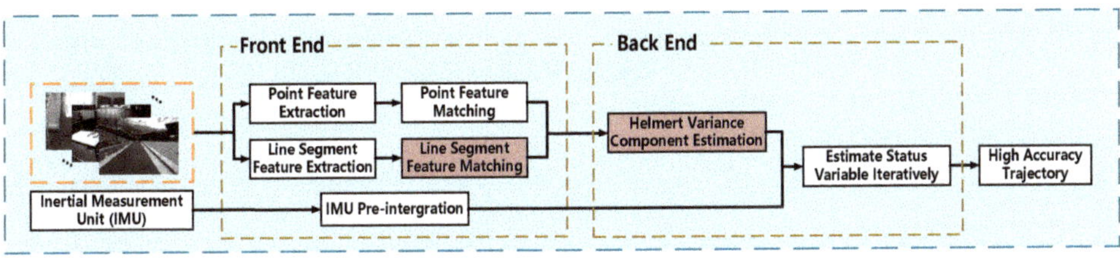

Figure 1. Overview of our improved point–line visual–inertial odometry (IPL–VIO) system.

2.1. Notations

Figure 2 [24] shows the basic principle of the point–line coupled visual–inertial odometry, and stipulates the following notation. The visual–inertial odometry uses the extracted point features and line features as visual observation values, and couples IMU measurement information for integrated navigation; c_i and b_i represent camera frame and IMU body frame at time $t = i$; f_j and L_j represent a point feature and a line feature in the world coordinate system. The variable $z_{f_j}^{c_i}$ is the jth point feature observed by ith camera frame, $z_{L_j}^{c_i}$ is the jth line feature observed by ith camera frame; they compose visual observations, $z_{b_i b_j}$ represents a preintegrated IMU measurement between two keyframes; q_{bc} and p_{bc} are the extrinsic parameters between the camera frame and the body frame.

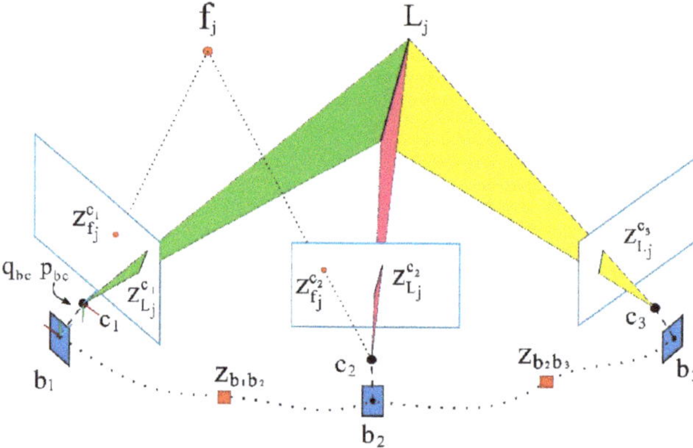

Figure 2. An illustration of visual–inertial odometry. Visual observations: point observations and line measurements. IMU observations: inertial measurement unit measurements.

2.2. Improved Line Feature Matching Algorithm

In general, most line feature matching algorithms use LBD [26] to match line features, which need to describe the line features, and the matching of the descriptors would take a certain amount of time, hugely increasing the burden of calculations.

Since the line features contain rich geometric and texture characteristics, we comprehensively use the angle, position, and pixel properties of the line features to match the line features. The algorithm can increase the matching speed of the line features. The specific algorithm follows:

(1) According to the midpoint coordinates of the line features, narrow the matching range by extracting line features of the left and right image, and the two endpoints of the line features are extracted by the line segment detector (LSD) algorithm [33]. Then the left image is divided into m × n grids and the line features extracted from the left image are mapped into different grids according to the midpoint coordinates, as shown in Figure 3. When the midpoint coordinates of the line features in the right image fall into the corresponding grid of the left image, then all line features of the left image and the right image in the same grid are obtained as candidate line features. We denote candidate line features in the left image as $\{P_1, P_2, \dots P_n\}$ and in right image as $\{Q_1, Q_2, \dots Q_n\}$.

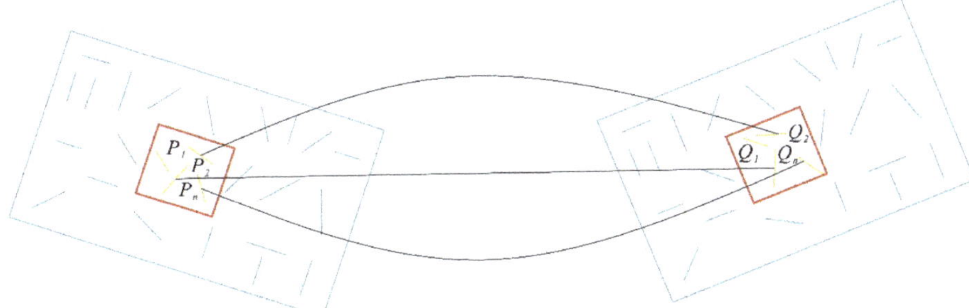

Figure 3. Using geometric information to match line features. The initial matching is performed by the pixel coordinates of the line features. The red box represents the grid divided in the image, and the yellow lines represent the line features that fall into the same grid of the two matching images.

(2) According to the correlation coefficient of the pixels in the surrounding area of the line features, the matching line features are determined. We match the candidate line feature P_i in $\{P_1, P_2, \ldots P_n\}$ with the line features $\{Q_1, Q_2, \ldots Q_n\}$, and the correlation coefficient of single pixel in the matching line is calculated using Formula (1) [34]. The respective correlation coefficients between P_i and $\{Q_1, Q_2, \ldots Q_n\}$ are calculated according to averaging the correlation coefficient of each pixel in the corresponding line features, and the correlation coefficients of the line features are sorted from small to large. If the correlation coefficient between P_i and Q_j is the largest, the respective correlation coefficients between Q_j and $\{P_1, P_2, \ldots P_n\}$ are calculated as well. If the correlation coefficient between Q_j and P_i is also the largest, P_i and Q_j are considered to be a pair of matching lines.

$$\rho(c,r,c',r') = \frac{\sum_{i=1}^{m}\sum_{j=1}^{n} g_{i+r,j+c}\cdot g'_{i+r',j+c'} - \frac{1}{m\cdot n}\left(\sum_{i=1}^{m}\sum_{j=1}^{n} g_{i+r,j+c}\right)\left(\sum_{i=1}^{m}\sum_{j=1}^{n} g'_{i+r',j+c'}\right)}{\sqrt{\left[\sum_{i=1}^{m}\sum_{j=1}^{n} g^2_{i+r,j+c} - \frac{1}{m\cdot n}\left(\sum_{i=1}^{m}\sum_{j=1}^{n} g_{i+r,j+c}\right)^2\right]\left[\sum_{i=1}^{m}\sum_{j=1}^{n} g'^2_{i+r',j+c'} - \frac{1}{m\cdot n}\left(\sum_{i=1}^{m}\sum_{j=1}^{n} g'_{i+r',j+c'}\right)^2\right]}}$$

$$(1)$$

where $c, r \in l_1$, $c', r' \in l_2$, (c, r) are the pixel coordinates of line l_1 on left image; (c', r') are the pixel coordinates of line l_2 on right image; m, n are the matching window size; $g_{i,j}$ is the gray value at (i, j) on left image; $g'_{i,j}$ is the gray value at (i, j) on right image; and $\rho(c, r, c', r')$ is the correlation coefficient.

(3) According to the rotation consistency of the line feature angles of the matched images, mismatches are eliminated. If the matching images are rotated, the angle changes of all matching line features should be consistent, which means the line feature rotation angles of the matching images have global consistency. If there is a rotation angle obviously inconsistent with the rotation angles of other matching line features, the matching pair may be seen as a mismatch and should be eliminated. This paper establishes a statistical histogram from 0 to 360 degrees in a unit of 1 degree. Through the histogram, the angle changes of matching line features are counted and the group with the largest number of histograms is retained. Line feature matching pairs that fall into other groups are considered to be mismatches and are eliminated.

2.3. Tightly Coupled VIO System

The VIO system in this paper uses point features, line features and IMU measurement information to optimize in the sliding window. In the optimization process, reasonable weights of different measurement information need to be given. Generally, the IMU measurements adopt the form of preintegration to construct the observation constraints, and the weight matrix of the IMU observation is recursively obtained, with the point features and the line features assigned prior weight matrices. Since the point feature and the line feature express different visual measurement information, the given prior weight matrices may be unreasonable to a certain extent. We use the Helmert variance component estimation method to obtain the post-test estimation of the prior weight matrices to better determine the contribution of visual measurement information to the overall optimization.

In order to better explain the improved algorithm in this article, the basic principles of tight coupling in the VIO system will be introduced in the following section, according to the basic principles of IMU error model, point feature error model, line feature error model, and Helmert variance component estimation.

2.3.1. Basic Principles of Tightly Coupled VIO System

In order to ensure accuracy and take into account efficiency at the same time, the sliding window algorithm is used to optimize state variables at the back end of the VIO system. Define the variable optimized in the sliding window at time t as [24]:

$$
\begin{aligned}
X &= [x_n, x_{n+1}, \ldots, x_{n+N}, \lambda_m, \lambda_{m+1}, \ldots, \lambda_{m+M}, l_k, l_{k+1}, \ldots l_{k+K}] \\
x_i &= [\mathbf{p}_{wb_i}, \mathbf{q}_{wb_i}, \mathbf{v}_i^w, \mathbf{b}_a^{b_i}, \mathbf{b}_g^{b_i}]T, \qquad\qquad i \in [n, n+N]
\end{aligned}
\tag{2}
$$

where x_i describes the ith IMU body state; $\mathbf{p}_{wb_i}$, $\mathbf{v}_i^w$, $\mathbf{q}_{wb_i}$ describe the position, velocity, and orientation of the IMU body in the world frame; $\mathbf{b}_a^{b_i}$, $\mathbf{b}_g^{b_i}$ describe the acceleration bias and angular velocity bias. We only use one variable, the inverse depth λ_k, to parameterize the kth point landmark from its first observed keyframe. The variable l_s is the orthonormal representation of the sth line feature in the world frame. Subscripts n, m, and k are the start indexes of the body states, point landmarks, and line landmarks, respectively. N is the number of keyframes in the sliding window. M and K are the numbers of point landmarks and line landmarks observed by all keyframes in the sliding window.

We optimize all the state variables in the sliding window by minimizing the sum of cost terms from all the measurement residuals [24]:

$$
\begin{aligned}
\min \rho(\|\mathbf{r}_p - \mathbf{J}_p X\|_{\Sigma_P}^2) + &\sum_{i \in B} \rho(\|\mathbf{r}_b(z_{b_i b_{i+1}}, X)\|_{\Sigma_{b_i b_{i+1}}}^2) + \\
&\sum_{(i,j) \in F} \rho(\|\mathbf{r}_f(z_{\mathbf{f}_j}^{c_i}, X)\|_{\Sigma_{\mathbf{f}_j}^{c_i}}^2) + \sum_{(i,l) \in L} \rho(\|\mathbf{r}_l(z_{L_i}^{c_i}, X)\|_{\Sigma_{L_i}^{c_i}}^2)
\end{aligned}
\tag{3}
$$

where $\{\mathbf{r}_p, \mathbf{J}_p\}$ is prior information after marginalizing out one frame in the sliding window, and $\mathbf{J}_p$ is the prior Jacobian matrix from the resulting Hessian matrix after the previous optimization. The variable $\mathbf{r}_b(z_{b_i b_{i+1}}, X)$ is an IMU measurement residual between the body state x_i and x_{i+1}; B is the set of all preintegrated IMU measurements in the sliding window; $\mathbf{r}_f(z_{\mathbf{f}_j}^{c_i}, X)$ and $\mathbf{r}_l(z_{L_i}^{c_i}, X)$ are the point feature reprojection residual and line feature reprojection residual, respectively. F and L are the sets of point features and line features observed by camera frames. The Cauchy robust function ρ is used to suppress outliers.

We express the abovementioned nonlinear optimization process in the form of a factor graph [35]. As shown in Figure 4, the nodes represent the variables to be optimized; in the VIO system they are the visual features and the state variables of the IMU body. The edges represent the visual constraints, IMU preintegration constraints, and prior constraints. Through the constraint information of the edges, the state variables of the nodes are optimized.

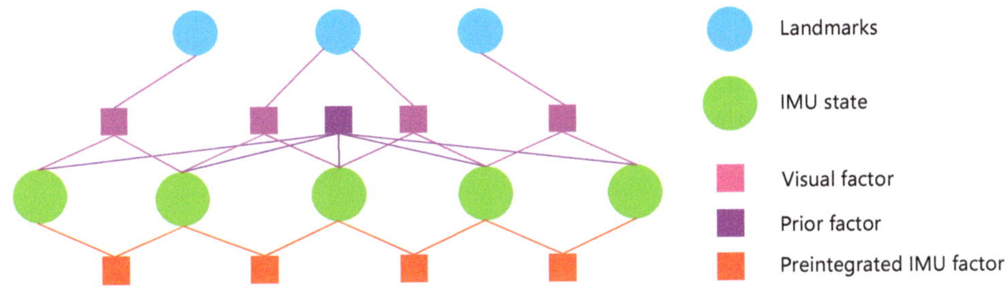

Figure 4. Optimization of a factor graph in the VIO system. Pink squares represent visual factors, purple squares represent prior factors, red squares represent IMU preintegration factors, blue nodes represent visual feature state variables to be optimized, and green nodes represent IMU body state variables to be optimized.

2.3.2. IMU Measurement Model

The IMU original observation values are preintegrated between two consecutive camera observation frames of $\mathbf{b}_i$ and $\mathbf{b}_j$, and an IMU measurement error model constructed through the preintegration [24]:

$$
\mathbf{r}_b(\mathbf{z}_{b_i b_j}, X) = \begin{bmatrix} \mathbf{r}_p \\ \mathbf{r}_\theta \\ \mathbf{r}_v \\ \mathbf{r}_{ba} \\ \mathbf{r}_{bg} \end{bmatrix} = \begin{bmatrix} \mathbf{R}_{b_i w}(\mathbf{P}_{wb_j} - \mathbf{P}_{wb_i} - \mathbf{v}_i^w \Delta t + \frac{1}{2}\mathbf{g}^w \Delta t^2) - \hat{\alpha}_{b_i b_j} \\ 2\left[\hat{\mathbf{q}}_{b_j b_i} \otimes (\mathbf{q}_{b_i w} \otimes \mathbf{q}_{wb_j})\right]_{xyz} \\ \mathbf{R}_{b_i w}(\mathbf{v}_j^w - v_i^w + \mathbf{g}^w \Delta t) - \hat{\beta}_{b_i b_j} \\ \mathbf{b}_a^{b_j} - \mathbf{b}_a^{b_i} \\ \mathbf{b}_g^{b_j} - \mathbf{b}_g^{b_i} \end{bmatrix}_{15\times 1}
\tag{4}
$$

where $\mathbf{z}_{b_i b_j} = \left[\hat{\alpha}_{b_i b_j}, \hat{\beta}_{b_i b_j}, \hat{\mathbf{q}}_{b_i b_j}\right]$ is the preintegrated measurement value of the IMU [8]; $[\cdot]_{xyz}$ extracts the real part of a quaternion, which is used to approximate the three-dimensional rotation error.

2.3.3. Point Feature Measurement Model

For a point feature, the distance from the projection point to the observation point, that is, the reprojection error, is used to construct the point feature error model. The normalized image plane coordinate of the kth point on the c_jth frame is $z_{\mathbf{f}_k}^{c_j} = [u_{\mathbf{f}_k}^{c_j}, v_{\mathbf{f}_k}^{c_j}, 1]^T$, the reprojection error is defined as [24]:

$$
\mathbf{r}_f(z_{\mathbf{f}_k}^{c_i}, X) = \begin{bmatrix} \frac{x^{c_j}}{z^{c_j}} - u_{\mathbf{f}_k}^{c_j} \\ \frac{y^{c_j}}{z^{c_j}} - v_{\mathbf{f}_k}^{c_j} \end{bmatrix}
\tag{5}
$$

where $z_{\mathbf{f}_k}^{c_j} = [u_{\mathbf{f}_k}^{c_j}, v_{\mathbf{f}_k}^{c_j}, 1]^T$ indicates the point on the normalized image plane that is observed by the camera frame c_i and $[x^{c_j}, y^{c_j}, z^{c_j}]$ indicates the point transformed into the camera frame c_i.

2.3.4. Line Feature Measurement Model

The reprojection error of a line feature is defined as the distance from the endpoints to the projection line. For a pinhole model camera, a 3D spatial line $L = [\mathbf{n}, \mathbf{d}]^T$ is projected to the camera image plane by the following formula [24]:

$$
\mathbf{l} = \begin{bmatrix} l_1 \\ l_2 \\ l_3 \end{bmatrix} = K\mathbf{n}^c = \begin{bmatrix} f_y & 0 & 0 \\ 0 & f_x & 0 \\ -f_y c_x & -f_x c_y & f_x f_y \end{bmatrix}\mathbf{n}^c
\tag{6}
$$

where a 3D spatial line is represented by the normal vector $\mathbf{n}$ and the direction vector $\mathbf{d}$, K is the projection matrix for a line feature. According to the projection of a line (Equation (6)), the normal vector of a 3D spatial line is projected to the normalized plane, which is the projection line of a 3D spatial line.

The reprojection error of the line feature in camera frame c_i is defined as (7) [24]:

$$\mathbf{r}_l(\mathbf{z}_{L_i}^{c_i}, X) = \left[\begin{array}{c} d(\mathbf{s}_l^{c_i}, \mathbf{l}_l^{c_i}) \\ d(\mathbf{e}_l^{c_i}, \mathbf{l}_l^{c_i}) \end{array} \right] \tag{7}$$

where $d(\mathbf{s}, \mathbf{l})$ indicates the distance function from endpoint $\mathbf{s}$ to the projection line $\mathbf{l}$.

2.4. Basic Principle of Helmert Variance Component Estimation

Perform the first-order Taylor expansion of the point feature error formula (5) and the line feature error Formula (7) to obtain:

$$\mathbf{r}_f(\mathbf{z}_{f_k}^{c_i}, X) \approx \mathbf{r}_f(\mathbf{z}_{f_k}^{c_i}, X_0) + \mathbf{J}_f \Delta x \tag{8}$$

$$\mathbf{r}_l(\mathbf{z}_{L_i}^{c_i}, X) \approx \mathbf{r}_l(\mathbf{z}_{L_i}^{c_i}, X_0) + \mathbf{J}_l \Delta x \tag{9}$$

where $\mathbf{r}_f(\mathbf{z}_{f_k}^{c_i}, X_0)$ and $\mathbf{r}_l(\mathbf{z}_{L_i}^{c_i}, X_0)$ are the values of the point feature error model and the line feature error model at the state variable X_0, respectively, $\mathbf{J}_f$ and $\mathbf{J}_l$ are the corresponding Jacobian matrices.

The constructed least squares optimization is:

$$\mathbf{H}\Delta x = \mathbf{b} \tag{10}$$

$$\mathbf{H} = \mathbf{J}_f^T \mathbf{P}_f \mathbf{J}_f + \mathbf{J}_l^T \mathbf{P}_l \mathbf{J}_l = \mathbf{H}_f + \mathbf{H}_l \tag{11}$$

$$\mathbf{b} = -\mathbf{J}_f^T \mathbf{P}_f \mathbf{r}_f - \mathbf{J}_l^T \mathbf{P}_l \mathbf{r}_l = \mathbf{b}_f + \mathbf{b}_l \tag{12}$$

where $\mathbf{P}_f$ and $\mathbf{P}_l$ are the weight matrices corresponding to the point feature observations and the line feature observations, respectively.

In general, during the first optimization, the weights of the point feature observations and the line feature observations are inappropriate, or the corresponding unit weight variances are not equal. Let the unit weight variance of the point feature and the line feature observations be σ_f^2, σ_l^2, the corresponding relationship between covariance matrix and the weight matrix is:

$$\Sigma_f = \sigma_f^2 \mathbf{P}_f^{-1} \tag{13}$$

$$\Sigma_l = \sigma_l^2 \mathbf{P}_l^{-1} \tag{14}$$

where Σ_f and Σ_l are the covariance matrices of the point and line features.

Using the rigorous formula of Helmert variance component estimation, we get:

$$\begin{aligned} E(\mathbf{r}_f^T \mathbf{P}_f \mathbf{r}_f) &= \sigma_f^2 \left\{ tr(\mathbf{H}^{-1}\mathbf{H}_f\mathbf{H}^{-1}\mathbf{H}_f) - 2tr(\mathbf{H}^{-1}\mathbf{H}_f) + n_1 \right\} \\ &+ \sigma_l^2 \, tr(\mathbf{H}^{-1}\mathbf{H}_f\mathbf{H}^{-1}\mathbf{H}_l) \end{aligned} \tag{15}$$

$$\begin{aligned} E(\mathbf{r}_l^T \mathbf{P}_l \mathbf{r}_l) &= \sigma_l^2 \left\{ tr(\mathbf{H}^{-1}\mathbf{H}_l\mathbf{H}^{-1}\mathbf{H}_l) - 2tr(\mathbf{H}^{-1}\mathbf{H}_l) + n_2 \right\} \\ &+ \sigma_f^2 \, tr(\mathbf{H}^{-1}\mathbf{H}_f\mathbf{H}^{-1}\mathbf{H}_l) \end{aligned} \tag{16}$$

where n_1, n_2 are the number of observations of point feature and line feature.

After combining the formulas we get:

$$S = \left[\begin{array}{cc} tr(\mathbf{H}^{-1}\mathbf{H}_f\mathbf{H}^{-1}\mathbf{H}_f) - 2tr(\mathbf{H}^{-1}\mathbf{H}_f) + n_1 & tr(\mathbf{H}^{-1}\mathbf{H}_f\mathbf{H}^{-1}\mathbf{H}_l) \\ tr(\mathbf{H}^{-1}\mathbf{H}_f\mathbf{H}^{-1}\mathbf{H}_l) & tr(\mathbf{H}^{-1}\mathbf{H}_l\mathbf{H}^{-1}\mathbf{H}_l) - 2tr(\mathbf{H}^{-1}\mathbf{H}_l) + n_2 \end{array} \right] \tag{17}$$

$$W = \begin{bmatrix} \mathbf{r}_f^T \mathbf{P}_f \mathbf{r}_f \\ \mathbf{r}_l^T \mathbf{P}_l \mathbf{r}_l \end{bmatrix} \hat{\theta} = \begin{bmatrix} \hat{\sigma}_f^2 \\ \hat{\sigma}_l^2 \end{bmatrix} \tag{18}$$

$$S\hat{\theta} = W \tag{19}$$

$$\hat{\theta} = S^{-1}W \tag{20}$$

We take the post-test unit weight variance $\hat{\sigma}_f^2$ of the point feature as the unit weight variance, then the post-test weights of the point feature and the line feature are:

$$\hat{\mathbf{P}}_f = \mathbf{P}_f \tag{21}$$

$$\hat{\mathbf{P}}_l = \frac{\hat{\sigma}_f^2}{\hat{\sigma}_l^2} \mathbf{P}_l \tag{22}$$

In sliding window optimization, in order to improve the efficiency of optimization, we ignore the trace part in the coefficient matrix S.

3. Experimental Results

We performed two improvements to the IPL–VIO system: the front-end line feature matching method and the back-end Helmert variance component estimation. In order to evaluate the performance of the algorithm in this paper, we used the EuRoc MAV [31] and PennCOSYVIO [32] datasets for verification.

We compared the IPL–VIO proposed in this paper with OKVIS–Mono [9], VINS–Mono [10], and PL–VIO [24] to verify the effectiveness of the method. OKVIS is a VIO system which can work with monocular or stereo modes. It uses a sliding window optimization algorithm to tightly couple visual point features and IMU measurements. VINS–Mono is a monocular visual inertial SLAM system that uses visual point features to assist in optimizing the IMU state. It uses a sliding window method for tightly coupling optimization and has closed-loop detection. PL–VIO is a monocular VIO system that uses a sliding window algorithm to tightly couple and optimize visual points, line features, and IMU measurement. Since the IPL–VIO in this article is a monocular VIO system, we compared it with the OKVIS in monocular mode and VINS–Mono without loop closure.

All the experiments were performed in the Ubuntu 16.04 system by an Intel Core i7-9750H CPU with 2.60 GHz and 8 GB RAM, on the ROS Kinetic [36].

3.1. Experimental Data Introduction

The EuRoc microaerial vehicle (MAV) datasets were collected by an MAV containing two scenes, a machine hall at ETH Zürich and an ordinary room, as shown in Figure 5. The datasets contain stereo images from a global shutter camera at 20 FPS and synchronized IMU measurements at 200 Hz [31]. Each dataset provides a groundtruth trajectory given by the VICON motion capture system. The datasets also provide all the extrinsic and intrinsic parameters. In our experiments, we only used the images from the left camera.

The PennCOSYVIO dataset contains images and synchronized IMU measurements that are collected with handheld equipment, including indoor and outdoor scenes of a glass building, as shown in Figure 5 [32]. Challenging factors include illumination changes, rapid rotations, and repetitive structures. The dataset also contains all the intrinsic and extrinsic parameters as well as the groundtruth trajectory.

We used the open source accuracy evaluation tool evo (https://michaelgrupp.github.io/evo/) to evaluate the accuracy of the EuRoc MAV datasets. We used absolute pose error (APE) as the error evaluation standard. For better comparison and analysis, we compared the rotation and translation parts of the trajectory and the groundtruth, respectively. Meanwhile, the tool provides a visualization of the comparison results, thereby the accuracy of the results can be analyzed more intuitively.

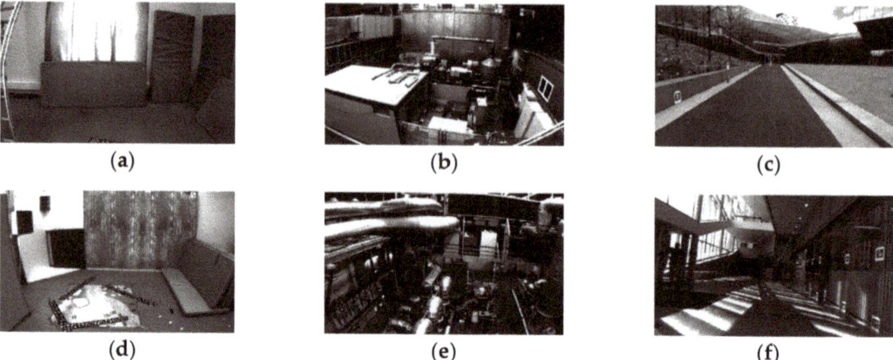

Figure 5. EuRoc MAV datasets and PennCOSYVIO dataset scenes. Images (**a,d**) are the room scenes of the V1_01_easy dataset in the EuRoc MAV datasets. Images (**b,e**) are the machine hall scenes of the MH_05_difficult dataset in the EuRoc MAV datasets. Images (**c,f**) are the indoor and outdoor scenes of PennCOSYVIO dataset.

The PennCOSYVIO dataset is equipped with accuracy assessment tools (https://daniilidis-group. github.io/penncosyvio/). We used absolute pose error (APE) and relative pose error (RPE) as the evaluation criteria for errors. For RPE, it expresses the errors in percentages by dividing the value with the path length [32]. The creator of PennCOSYVIO cautiously selected the evaluation parameters, so their tool is suited for evaluating VIO approaches in this dataset. Therefore, we adopted this evaluation tool in our experiments.

3.2. Experimental Analysis of the Improved Line Feature Matching Algorithm

We compared the proposed line feature matching method with the LBD descriptor matching method. Figure 6 shows the line feature matching effect of the LBD descriptor matching method and the method proposed in this paper. Figure 7 shows the trajectory errors of two methods running on EuRoc MAV's MH_02_easy dataset and V1_03_difficult dataset. We comprehensively used geometric information such as the position and angle of the line features, as well as the pixel gray information around the line feature to match the corresponding line feature. It can be seen that the accuracy of the improved algorithm is equivalent to the descriptor matching method.

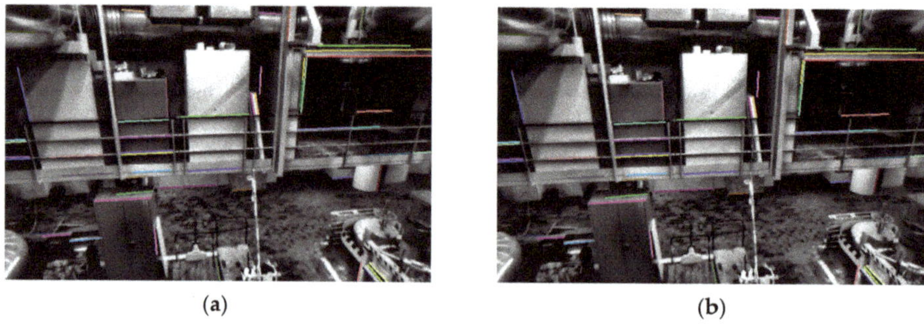

Figure 6. *Cont.*

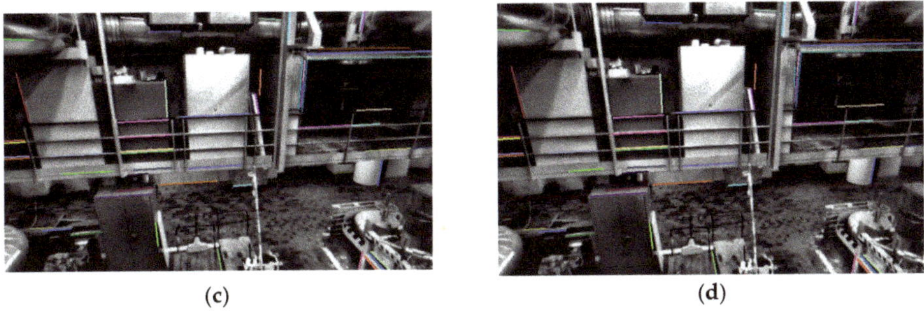

Figure 6. Comparison of the matching effect between line binary descriptors (LBD) matching method and improved matching method on the MH_02_easy dataset. Images (**a**,**b**) are LBD descriptor matching scenes—the line of the same color is the corresponding matching line; (**c**,**d**) are the matching scenes of the improved matching method, and the line of the same color is the corresponding matching line.

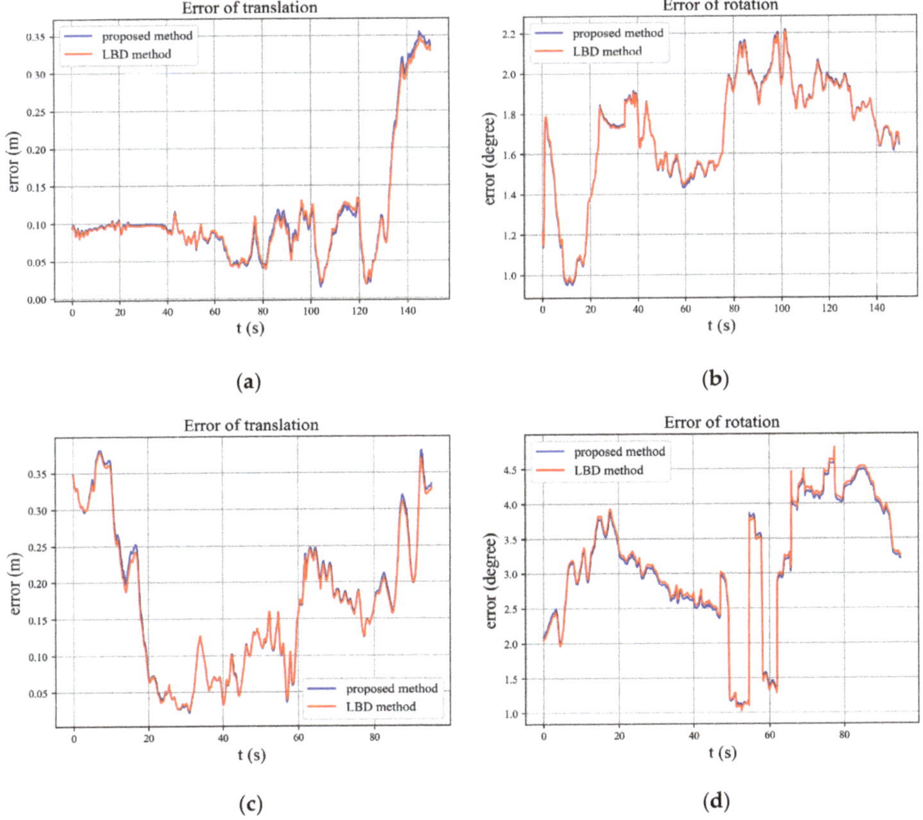

Figure 7. Comparison of improved matching method and LBD descriptor matching method: (**a**,**c**) compare translation errors on the MH_02_easy and V1_03_difficult datasets; (**b**,**d**) compare rotation errors on the MH_02_easy and V1_03_difficult datasets.

We counted the trajectory error and time of the two methods after running the MH_02_easy and V1_03_difficult dataset of EuRoc MAV; the root mean square error (RMSE) of APE is used to evaluate the translation error and rotation error, respectively, and the time is the average time of the different algorithms running the datasets, as shown in Table 1. It can be seen that running the MH_02_easy dataset by using the LBD descriptor matching algorithm, the errors of the translation part and rotation part are 0.13057 m and 1.73778 degrees; using the matching algorithm proposed in this article, the errors of the translation part and rotation part are 0.13253 m and 1.73950 degrees. Although the accuracy has decreased, it is very limited. When running the V1_03_difficult dataset, using the LBD descriptor matching algorithm to run the dataset, the errors of the translation part and rotation part are 0.19490 m and 3.31055 degrees; using the matching algorithm proposed in this paper, the errors of the translation part and rotation part are 0.19792 m and 3.27675 degrees. The accuracy of the translation part is slightly decreased, but the accuracy of the rotation part is slightly increased, and the overall accuracy is equivalent.

Table 1. Comparison of LBD descriptor matching method and the matching method we proposed.

Algorithm	Translation Error (m)		Rotation Error (°)		Time (ms)	
	LBD	**Proposed**	**LBD**	**Proposed**	**LBD**	**Proposed**
MH_02_easy	0.13057	0.13253	1.73778	1.73950	74	15
V1_03_difficult	0.19490	0.19792	3.31055	3.27675	37	10

Using the improved line feature matching method and the LBD descriptor matching method, the final trajectory accuracy is equivalent. However, when comparing the running time for the MH_02_easy dataset, the LBD descriptor matching takes an average of 74 ms per frame, and the method described in this paper takes 15 ms; it can be seen that the running time is 20% that of the LBD descriptor matching method; for the V1_03_difficult dataset, LBD descriptor matching takes an average of 37 ms per frame, the method described in this paper takes 10 ms, and the running time is 27% of the LBD descriptor matching method. It can be seen that the method proposed in this article can effectively speed up the line feature matching.

3.3. Experimental Analysis of Helmert Variance Component Estimation

We ran OKVIS–Mono, VINS–Mono, PL–VIO and IPL–VIO systems on the EuRoc MAV datasets to evaluate the accuracy. Table 2 shows the trajectories' root mean square error (RMSE) of the translation part (m) and rotation part (degrees) of the four systems, the numbers in bold representing the estimated trajectory are more close to the groundtruth. Simultaneously, we made statistics of the histogram, which can be seen in Figure 8. As shown in Table 2, in terms of translation, the IPL–VIO system has higher accuracy than other systems on MH_02_easy, MH_05_difficult, V1_03_difficult, V2_01_easy, and V2_02_medium. In terms of rotation, the IPL–VIO system has higher accuracy on MH_02_easy, MH_04_difficult, V1_03_difficult, V2_01_easy, and V2_02_medium.

Table 2. The root mean square error (RMSE) results on several EuRoc MAV datasets.

Seq.	OKVIS–Mono		VINS–Mono		PL–VIO		IPL–VIO	
	Trans (m)	Rot (°)	Trans (m)	Rot (°)	Trans (m)	Rot (°)	Trans (m)	Rot (°)
MH_02_easy	0.30655	3.92590	0.17143	2.30959	0.13057	1.74408	**0.11534**	**1.47136**
MH_03_medium	0.33372	3.30597	**0.19401**	**1.64611**	0.26095	1.70340	0.25248	1.96238
MH_04_difficult	0.38942	2.28610	**0.34633**	1.49141	0.35759	1.64553	0.36427	**1.15279**
MH_05_difficult	0.46736	2.37892	0.29151	**0.71333**	0.24446	1.07200	**0.19262**	1.25478
V1_01_easy	0.08982	5.83328	0.08683	6.33691	**0.07792**	**5.82240**	0.08778	5.85792
V1_03_difficult	0.27364	5.58748	0.20710	6.20628	0.19489	3.20856	**0.18983**	**3.09684**
V2_01_easy	0.13543	2.21792	0.08162	2.03056	0.08432	2.06150	**0.07394**	**1.89420**
V2_02_medium	0.19826	4.85181	0.15685	4.34073	0.14284	2.97881	**0.11158**	**2.60868**

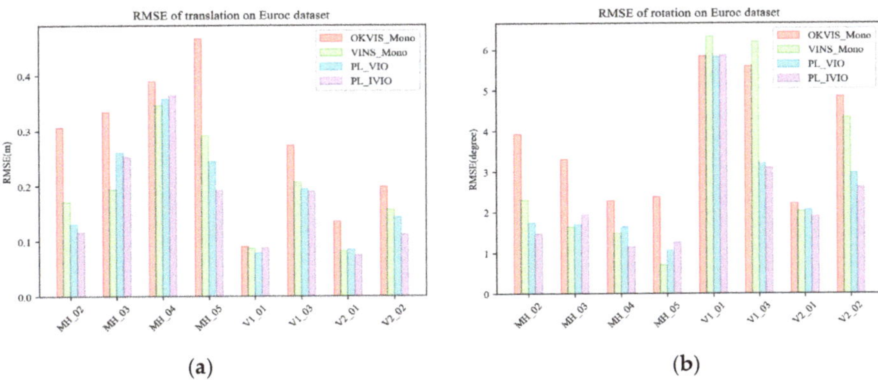

(a)   **(b)**

Figure 8. RMSEs for OKVIS–Mono, Vins–Mono without loop closure, PL–VIO, and the proposed IPL–VIO using the EuRoc MAV datasets. (**a**) RMSEs in translation. (**b**) RMSEs in rotation.

However, there are datasets in Table 2 whose accuracy decreases after the Helmert variance component method is used. As shown in Figure 9, in the V1_01_easy dataset, there are a large number of weak texture environments in the dataset scene, the quality of the extracted point features is relatively low. These still contain repetitive textures that make line features prone to the mismatch problem. Therefore, the RMSE of the translation part of PL–VIO is 0.07792 m and the RMSE of the rotation part is 5.82240 degrees. After using the Helmert variance component estimation, the results are susceptible to errors, resulting in a decrease in accuracy. The RMSE of the translation part of the IPL–VIO is 0.08778 m and the RMSE of the rotation part is 5.85792 degrees.

(a)   **(b)**

Figure 9. V1_01_easy visual feature extraction: (**a**) line features extraction, (**b**) point features extraction.

Another representative dataset is MH_03_medium. Compared with VINS–Mono, the accuracy of PL–VIO with added line features decreased. This is because in MH_03_medium, there are mismatches of line features, as shown in Figure 10; the line features in the scene are also relatively short and fragmented, which increase error. However, it can be seen from Table 2 that after Helmert variance component estimation, compared with PL–VIO, the accuracy of the translation part of IPL–VIO improved from 0.26095 to 0.25248 m.

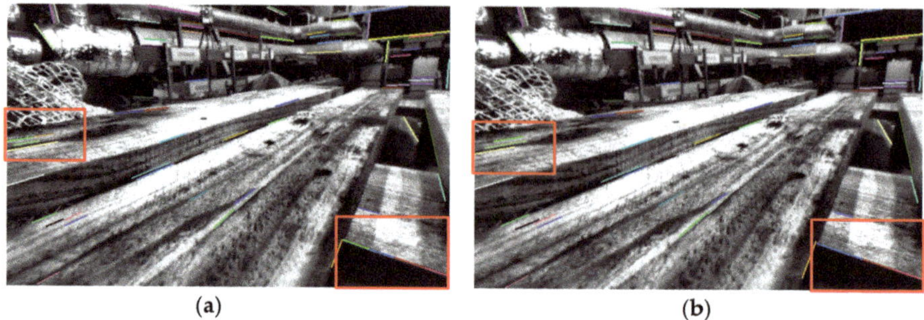

(a)   (b)

Figure 10. MH_03_medium line feature matching. The line features of the two frames at the previous time (**a**) and the next time (**b**) are matched. The line of the same color represents the corresponding matching line, and the red boxes on the left and right represent the mismatches of the line features.

In order to show a more intuitive result, we have drawn the trajectory estimation heat map of both PL–VIO and IPL–VIO in a same figure for the MH_05_difficult and V2_02_medium datasets. As shown in Figures 11 and 12, the more reddish the figure, the larger the translation error of the trajectory. It can be seen that by adjusting the weights of the point and line features, the IPL–VIO has higher accuracy than PL–VIO.

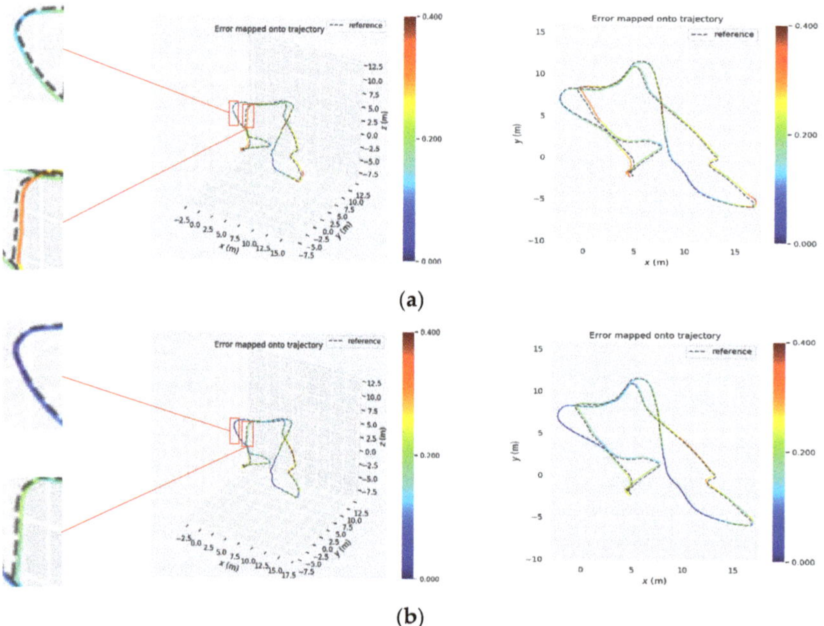

(a)

(b)

Figure 11. Comparison of trajectory translation errors between IPL–VIO and PL–VIO for the MH_05_difficult dataset: (**a**) PL–VIO trajectory error details, overall diagram, and bird's eye view; (**b**) IPL–VIO trajectory error details, overall diagram, and bird's eye view.

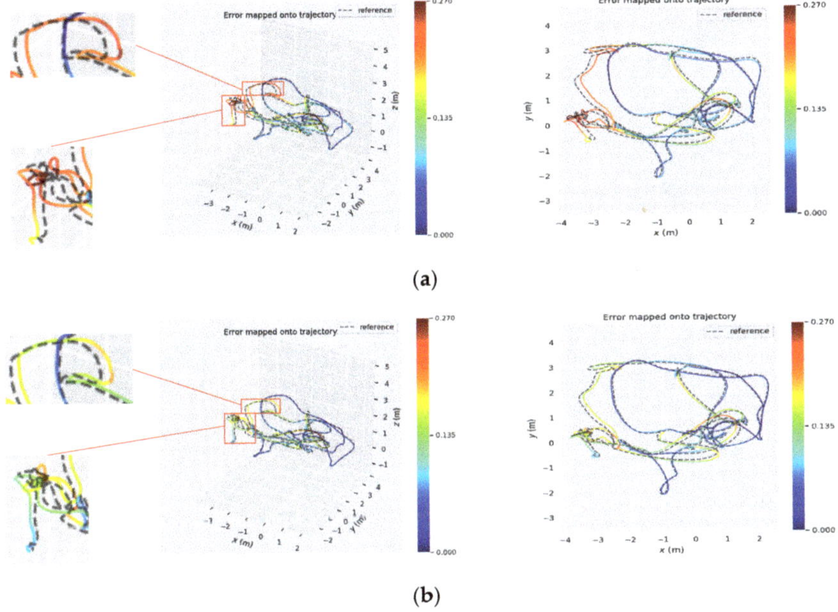

Figure 12. Comparison of trajectory translation errors between IPL–VIO and PL–VIO for the V2_02_medium dataset: (**a**) PL–VIO trajectory error details, overall diagram, and bird's eye view; (**b**) IPL–VIO trajectory error details, overall diagram, and bird's eye view.

When the carrier undergoes significant rotation changes or runs along straight lines, as shown in Figure 11a,b, using the Helmert variance component to estimate the weights of the points and lines, the trajectory accuracy can be significantly improved. From Figure 12a,b, we can see that for continuous rapid rotation changes, we can effectively improve the accuracy by adjusting the weights of point features and line features.

The PennCOSYVIO dataset contains various scenes such as obvious changes in lighting, rapid rotation, and repeated texture. For these challenges, the point and line features have different characteristics, so we used this dataset to compare and analyze the accuracy and time consumption of PL–VIO and IPL–VIO.

It can be seen from Figure 13 that the dataset contains a large number of repetitive linear textures and scenes with changes in light, illumination, and darkness, which can fully verify the method proposed in this article. We used the Helmert variance component estimation method to weight the two visual features, and the accuracy of the trajectory can be significantly improved. As shown in Table 3, we compared APE and RPE of the trajectory after running PL–VIO and IPL–VIO. The rotation errors for the APE and RPE are expressed in degrees. The translation errors are expressed in the x, y, z axes, and the APE of translation part is expressed in meters, while the RPE of translation part is expressed in percentages. The numbers in bold, representing the estimated trajectory, are closer to the groundtruth. We can see that the trajectory accuracy has a significant improvement when compared to APE and RPE.

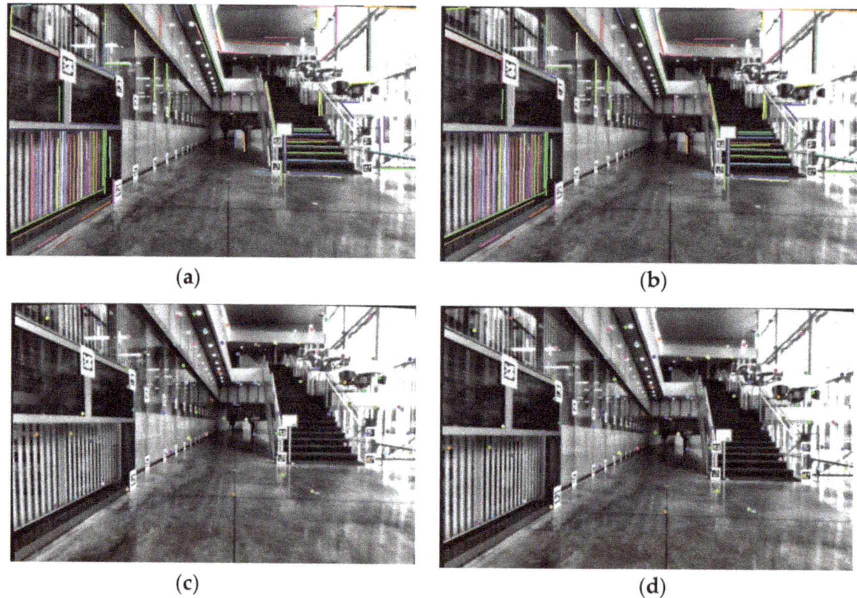

(a)   (b)

(c)   (d)

Figure 13. Point and line features matching in the PennCOSYVIO dataset: (**a,b**) are the matching of line features, and the line of the same color is the matched line feature; (**c,d**) are the matching of point features, the point of the same color is tracked by the optical flow [12].

Table 3. Absolute and relative pose error (APE and RPE) of the trajectory by running PL–VIO and IPL–VIO on the PennCOSYVIO dataset.

Algorithm	APE				RPE			
	x (m)	y (m)	z (m)	rot (°)	x (%)	y (%)	z (%)	rot (°)
PL–VIO	0.406	0.169	1.006	2.3756	2.561	1.221	5.323	1.8276
IPL–VIO	**0.371**	**0.137**	**0.911**	**2.2657**	**2.401**	1.254	**4.827**	**1.7983**

Table 4 shows the time consumption of each module in IPL–VIO. It can be seen that for the average time per frame of line feature extraction and matching, the original method takes 74 ms; the method proposed in this article takes 60 ms. At the back end, without the Helmert variance component estimation method, it takes 23 ms, and using the Helmert variance component estimation method, it takes 24 ms. Thus, the time increase is negligible.

Table 4. The running time of each module of PL–VIO and IPL–VIO.

Module	Operation	Algorithm Times (ms)	
		PL–VIO	IPL–VIO
front end	Point feature detection and matching	18	18
	Line feature detection and matching	74	60
	IMU forward propagation	1	1
back end	Nonlinear optimization	23	24

4. Discussion

In this paper, an improved point line coupled VIO system (IPL–VIO) was proposed. IPL–VIO has two main improvements. Firstly, geometric information such as the position and angle of the line feature and the gray information of the pixels around the line features were explored. We comprehensively used the geometric information and correlation coefficient to match the line features. Secondly, the Helmert variance component estimation method was introduced in the sliding window optimization, which ensured that more reasonable weights can be assigned for point features and line features. Compared with point features, line features are high-dimensional visual feature information that contain structured and geometric information, but matching line features is more time consuming. Thus, our proposed line feature matching method can shorten the matching time without any loss of accuracy. In addition, in the sliding window optimization, we used the Helmert variance component estimation method to determine more reasonable posterior weights for point features and line features, and improved the accuracy of visual information in the VIO system.

In order to verify the effectiveness of the proposed IPL–VIO system, a series of experiments were conducted. The improved line feature matching method was compared with the traditional LBD descriptor matching method, and the EuRoc MAV datasets were used for verification. As is shown, the improved matching method had the same accuracy as the traditional method, but reduced the running time to about a quarter of the traditional one. We compared and analyzed IPL–VIO with the current mainstream VIO systems: OKVIS–Mono, VINS–Mono, and PL–VIO. The test results on the EuRoc MAV datasets showed that the proposed IPL–VIO system performed well on most datasets when compared to other systems. There are also datasets with reduced accuracy, such as the V1_01_easy dataset, where there are a large number of weak texture and repetitive texture environments in the dataset scenes; the quality of point features and line features is both poor, after adjusting the weights, and the accuracy of the trajectory decreased. From the error heat map of the trajectory, it can be seen that the trajectory accuracy of IPL–VIO can be improved whether it is smooth running or exhibiting continuous large-angle rotation. We also compared and analyzed the proposed IPL–VIO system and the PL–VIO system on the PennCOSYVIO dataset, which contains challenging scenes such as significant changes in lighting, large-angle rotation, and repeated textures. It was seen that the IPL–VIO system can improve the final trajectory accuracy after readjusting the point-line weights with the Helmert variance component estimation method. Furthermore, we assessed the speed of each module of IPL–VIO and PL–VIO. The improved line feature matching method can reduce the time consumption of the front end, and the Helmert variance component estimation method added in the back end was effective for the back end; the increase load was quite limited and almost negligible, which proved the effectiveness of the proposed IPL–VIO system.

The algorithm in this paper improved the basis of PL–VIO. Therefore, in Tables 2–4, we indicate the results of a comprehensive comparison of PL–VIO and IPL–VIO. As is shown in Table 2, IPL–VIO had higher accuracy than PL–VIO in most datasets, which shows that the algorithm in this paper has better performance in different scenarios. As can be seen from Table 3, the error in the x, y, z three-axis direction of IPL–VIO was almost small compared with PL–VIO. It can be seen from Table 4 that the method proposed in this paper shortened the matching time of line features and leaves more time for the operation of other modules.

5. Conclusions

This paper proposes an improved point–line VIO system IPL–VIO. The IPL–VIO system has two main improvement modules: the front end and the back end. In the front-end module, an improved line feature matching algorithm is proposed, which comprehensively uses the geometric information and the pixel gray information of the line feature to match. In the back-end module, we use the Helmert variance component estimation method to determinate the weights of the point features and line features. We compared IPL–VIO with OKVIS–Mono [9], VINS–Mono [10], and PL–VIO [24], and verified the effectiveness of the algorithm on the EuRoc MAV [31] and PennCOSYVIO [32] datasets. According to the analysis and results, there are two further conclusions:

1. Compared with traditional line feature matching methods using LBD descriptors, using geometric information and pixel gray information to match has the same accuracy as the traditional method, but reduces the running time to about a quarter of the traditional one.
2. By using the Helmert variance component estimation method to determine more reasonable posterior weights for point features and line features, this method can improve the accuracy of visual information in the VIO system. The final trajectory accuracy is improved and time consumption is almost negligible.

We also look forward to the next work. At the back end, we use the simplified formula of the Helmert variance component estimation method, which introduces a certain degree of error. In the future, we would like to study how to improve the accuracy of weight determination without increasing the back-end overhead. We only use the Helmert variance component estimation method to estimate the weights of visual features; in the future, we will try to figure out how to better determine the weights of visual information and IMU information.

Author Contributions: B.X. and Y.C. conceived and designed the algorithm; B.X. performed the experiments, analyzed the data, and drafted the paper; J.W. contributed analysis tools; Y.C. and S.Z. revised the manuscript. All authors have read and agreed to the published version of the manuscript.

Funding: This research was funded by the National Key R&D Program of China (Grant No. 2017YFC0803801) and the National Key R&D Program of China (Grant No. 2016YFB0501803). We owe great appreciation to the anonymous reviewers for their critical, helpful, and constructive comments and suggestions.

Conflicts of Interest: The authors declare no conflict of interest.

References

1. Fuentes-Pacheco, J.; Ruiz-Ascencio, J.; Rendón-Mancha, J.M. Visual simultaneous localization and mapping: A survey. *Artif. Intell. Rev.* **2015**, *43*, 55–81. [CrossRef]
2. Kelly, J.; Saripalli, S.; Sukhatme, G.S. Combined Visual and Inertial Navigation for an Unmanned Aerial Vehicle. In Proceedings of the Field and Service Robotics, Chamonix, France, 9–12 July 2007; pp. 255–264.
3. Bloesch, M.; Burri, M.; Omari, S.; Hutter, M.; Siegwart, R. Iterated extended Kalman filter based visual-inertial odometry using direct photometric feedback. *Int. J. Robot. Res.* **2017**, *36*, 1053–1072. [CrossRef]
4. Jones, E.S.; Soatto, S. Visual-inertial navigation, mapping and localization: A scalable real-time causal approach. *Int. J. Robot. Res.* **2011**, *30*, 407–430. [CrossRef]
5. Bloesch, M.; Omari, S.; Hutter, M.; Siegwart, R. Robust visual inertial odometry using a direct EKF-based approach. In Proceedings of the 2015 IEEE/RSJ International Conference on Intelligent Robots and Systems (IROS), Hamburg, Germany, 28 September–2 October 2015; pp. 298–304.
6. Kasyanov, A.; Engelmann, F.; Stückler, J.; Leibe, B. Keyframe-based visual-inertial online SLAM with relocalization. In Proceedings of the 2017 IEEE/RSJ International Conference on Intelligent Robots and Systems (IROS), Vancouver, BC, Canada, 24–28 September 2017; pp. 6662–6669.
7. Usenko, V.; Engel, J.; Stückler, J.; Cremers, D. Direct visual-inertial odometry with stereo cameras. In Proceedings of the 2016 IEEE International Conference on Robotics and Automation (ICRA), Stockholm, Sweden, 16–21 May 2016; pp. 1885–1892.
8. Forster, C.; Carlone, L.; Dellaert, F.; Scaramuzza, D. On-Manifold Preintegration for Real-Time Visual–Inertial Odometry. *IEEE Trans. Robot.* **2017**, *33*, 1–21. [CrossRef]
9. Leutenegger, S.; Lynen, S.; Bosse, M.; Siegwart, R.; Furgale, P. Keyframe-based visual–inertial odometry using nonlinear optimization. *Int. J. Robot. Res.* **2015**, *34*, 314–334. [CrossRef]
10. Qin, T.; Li, P.; Shen, S. Vins-Mono: A Robust and Versatile Monocular Visual-Inertial State Estimator. *IEEE Trans. Robot.* **2018**, *34*, 1004–1020. [CrossRef]
11. Shi, J.; Tomasi, C. Good features to track. In Proceedings of the 1994 Proceedings of IEEE Conference on Computer Vision and Pattern Recognition, Seattle, WA, USA, 21–23 June 1994; pp. 593–600.

12. Lucas, B.D.; Kanade, T. An iterative image registration technique with an application to stereo vision. In Proceedings of the 7th International Joint Conference on artificial Intelligence (IJCAI), Vancouver, BC, Canada, 24–28 August 1981.
13. Sun, K.; Mohta, K.; Pfrommer, B.; Watterson, M.; Liu, S.; Mulgaonkar, Y.; Taylor, C.J.; Kumar, V. Robust Stereo Visual Inertial Odometry for Fast Autonomous Flight. *IEEE Robot. Autom. Lett.* **2017**, *3*, 965–972. [CrossRef]
14. Trajković, M.; Hedley, M. Fast corner detection. *Image Vis. Comput.* **1998**, *16*, 75–87. [CrossRef]
15. Harris, C.G.; Stephens, M. A Combined Corner and Edge Detector. In Proceedings of the Alvey Vision Conference, AVC 1988, Manchester, UK, 31 August–2 September 1988; Taylor, C.J., Ed.; Alvey Vision Club: Manchester, UK, 1988; pp. 1–6.
16. Leutenegger, S.; Chli, M.; Siegwart, R.Y. BRISK: Binary robust invariant scalable keypoints. In Proceedings of the 2011 International Conference on Computer Vision, Barcelona, Spain, 6–13 November 2011; pp. 2548–2555.
17. Kong, X.; Wu, W.; Zhang, L.; Wang, Y. Tightly-coupled stereo visual-inertial navigation using point and line features. *Sensors* **2015**, *15*, 12816–12833. [CrossRef]
18. Kottas, D.G.; Roumeliotis, S.I. Efficient and consistent vision-aided inertial navigation using line observations. In Proceedings of the 2013 IEEE International Conference on Robotics and Automation, Karlsruhe, Germany, 6–10 May 2013; pp. 1540–1547.
19. Zhang, G.; Lee, J.H.; Lim, J.; Suh, I.H. Building a 3-D line-based map using stereo SLAM. *IEEE Trans. Robot.* **2015**, *31*, 1364–1377. [CrossRef]
20. Pumarola, A.; Vakhitov, A.; Agudo, A.; Sanfeliu, A.; Moreno-Noguer, F. PL-SLAM: Real-time monocular visual SLAM with points and lines. In Proceedings of the 2017 IEEE International Conference on Robotics and Automation (ICRA), Singapore, 29 May–3 June 2017; pp. 4503–4508.
21. Gomez-Ojeda, R.; Moreno, F.-A.; Zuñiga-Noël, D.; Scaramuzza, D.; Gonzalez-Jimenez, J. PL-SLAM: A stereo SLAM system through the combination of points and line segments. *IEEE Trans. Robot.* **2019**, *35*, 734–746. [CrossRef]
22. Bartoli, A.; Sturm, P. The 3D line motion matrix and alignment of line reconstructions. *Int. J. Comput. Vis.* **2004**, *57*, 159–178. [CrossRef]
23. Zuo, X.; Xie, X.; Liu, Y.; Huang, G. Robust visual SLAM with point and line features. In Proceedings of the 2017 IEEE/RSJ International Conference on Intelligent Robots and Systems (IROS), Vancouver, BC, Canada, 24–28 September 2017; pp. 1775–1782.
24. He, Y.; Zhao, J.; Guo, Y.; He, W.; Yuan, K. Pl-VIO: Tightly-Coupled Monocular Visual–Inertial Odometry Using Point and Line Features. *Sensors* **2018**, *18*, 1159. [CrossRef] [PubMed]
25. Wen, H.; Tian, J.; Li, D. PLS-VIO: Stereo Vision-inertial Odometry Based on Point and Line Features. In Proceedings of the 2020 International Conference on High Performance Big Data and Intelligent Systems (HPBD&IS), Shenzhen, China, 23 May 2020; pp. 1–7.
26. Zhang, L.; Koch, R. An efficient and robust line segment matching approach based on LBD descriptor and pairwise geometric consistency. *J. Vis. Commun. Image Represent.* **2013**, *24*, 794–805. [CrossRef]
27. Yu, Z. A universal formula of maximum likelihood estimation of variance-covariance components. *J. Geod.* **1996**, *70*, 233–240. [CrossRef]
28. Zhang, P.; Tu, R.; Gao, Y.; Zhang, R.; Liu, N. Improving the performance of multi-GNSS time and frequency transfer using robust helmert variance component estimation. *Sensors* **2018**, *18*, 2878. [CrossRef]
29. Gao, Z.; Shen, W.; Zhang, H.; Ge, M.; Niu, X. Application of Helmert variance component based adaptive Kalman filter in multi-GNSS PPP/INS tightly coupled integration. *Remote Sens.* **2016**, *8*, 553. [CrossRef]
30. Deng, J.; Zhao, X.; Zhang, A.; Ke, F. A robust method for GPS/BDS pseudorange differential positioning based on the helmert variance component estimation. *J. Sens.* **2017**, *2017*, 1–8. [CrossRef]
31. Burri, M.; Nikolic, J.; Gohl, P.; Schneider, T.; Rehder, J.; Omari, S.; Achtelik, M.W.; Siegwart, R. The EuRoC micro aerial vehicle datasets. *Int. J. Robot. Res.* **2016**, *35*, 1157–1163. [CrossRef]
32. Pfrommer, B.; Sanket, N.; Daniilidis, K.; Cleveland, J. PennCOSYVIO: A Challenging Visual Inertial Odometry Benchmark. In Proceedings of the 2017 IEEE International Conference on Robotics and Automation (ICRA), Singapore, 29 May–3 June 2017; pp. 3847–3854.
33. Von Gioi, R.G.; Jakubowicz, J.; Morel, J.-M.; Randall, G. LSD: A fast line segment detector with a false detection control. *IEEE Trans. Pattern Anal. Mach. Intell.* **2008**, *32*, 722–732. [CrossRef]
34. Chen, Y.; Yan, L.; Xu, B.; Liu, Y. Multi-Stage Matching Approach for Mobile Platform Visual Imagery. *IEEE Access* **2019**, *7*, 160523–160535. [CrossRef]

35. Kaess, M.; Johannsson, H.; Roberts, R.; Ila, V.; Leonard, J.J.; Dellaert, F. iSAM2: Incremental smoothing and mapping using the Bayes tree. *Int. J. Robot. Res.* **2012**, *31*, 216–235. [CrossRef]
36. Quigley, M.; Conley, K.; Gerkey, B.; Faust, J.; Foote, T.; Leibs, J.; Wheeler, R.; Ng, A.Y. ROS: An Open-Source Robot Operating System. In Proceedings of the ICRA Workshop on Open Source Software, Kobe, Japan, 12–17 May 2009; ICRA: Kobe, Japan, 2009; p. 5.

remote sensing

Article

An Imaging Network Design for UGV-Based 3D Reconstruction of Buildings

Ali Hosseininaveh [1] and Fabio Remondino [2,*]

[1] Department of Photogrammetry and Remote Sensing, Faculty of Geodesy and Geomatics Engineering, K. N. Toosi University of Technology, Tehran 15433-19967, Iran; hosseininaveh@kntu.ac.ir
[2] 3D Optical Metrology (3DOM) Unit, Bruno Kessler Foundation (FBK), 38123 Trento, Italy
* Correspondence: remondino@fbk.eu

Abstract: Imaging network design is a crucial step in most image-based 3D reconstruction applications based on Structure from Motion (SfM) and multi-view stereo (MVS) methods. This paper proposes a novel photogrammetric algorithm for imaging network design for building 3D reconstruction purposes. The proposed methodology consists of two main steps: (i) the generation of candidate viewpoints and (ii) the clustering and selection of vantage viewpoints. The first step includes the identification of initial candidate viewpoints, selecting the candidate viewpoints in the optimum range, and defining viewpoint direction stages. In the second step, four challenging approaches—named façade pointing, centre pointing, hybrid, and both centre & façade pointing—are proposed. The entire methodology is implemented and evaluated in both simulation and real-world experiments. In the simulation experiment, a building and its environment are computer-generated in the ROS (Robot Operating System) Gazebo environment and a map is created by using a simulated robot and Gmapping algorithm based on a Simultaneously Localization and Mapping (SLAM) algorithm using a simulated Unmanned Ground Vehicle (UGV). In the real-world experiment, the proposed methodology is evaluated for all four approaches for a real building with two common approaches, called continuous image capturing and continuous image capturing & clustering and selection approaches. The results of both evaluations reveal that the fusion of centre & façade pointing approach is more efficient than all other approaches in terms of both accuracy and completeness criteria.

Keywords: view planning; imaging network design; building 3D modelling; path planning

Citation: Hosseininaveh, A.; Remondino, F. An Imaging Network Design for UGV-Based 3D Reconstruction of Buildings. *Remote Sens.* **2021**, *13*, 1923. https://doi.org/10.3390/rs13101923

Academic Editors: Ville Lehtola, François Goulette and Andreas Nüchter

Received: 17 March 2021
Accepted: 12 May 2021
Published: 14 May 2021

Publisher's Note: MDPI stays neutral with regard to jurisdictional claims in published maps and institutional affiliations.

1. Introduction

The 3D reconstruction of building is of interest for many companies and researchers who are working in the field of Building Information Modelling [1] or heritage documentation. Indeed, 3D modelling of buildings can be used for many applications, including accurate documentation [2], for reconstruction or repairing in the case of damage [3,4], for visualization purposes, for the generation of education resources for history and culture students and researchers [5], and for virtual tourism and for (Heritage) Building Information Modelling (H-BIM/BIM) [6,7]. Most of these applications require several necessities, which can be summarized as a fully automatic, low-cost, portable 3D modelling system which can deliver a high accurate, comprehensive, photorealistic 3D model with all details.

Image-based 3D reconstruction is one of the most feasible, accurate and fast techniques that can be used for building 3D reconstructions [8]. Images of buildings can be captured by an Unmanned Ground Vehicle (UGV), Unmanned Aerial Vehicle (UAV) or hand-held camera carried by an operator, as well as some novel approaches for capturing stereo images [9]. If a UGV equipped with a height-adjustable and pan-tilt camera is used for such task, the maximum height of the camera will be far lower than the height of the building. This restriction decreases the quality of the final model generated with the captured UGV-based images for the top parts of the building. On the other hand, using UAVs in urban areas is a challenging project due to numerous technical and operational issues, including regulatory restrictions, problems with data transfer, low payload, limited flight time to carry a high-quality camera, safety hazard for other aircrafts, impacts with people or structures, etc. [10]. These issues force users to exploit UAVs in a safe mode over buildings, decreasing the quality of the final model on façades. In an ideal situation, a combination of a UAV and a UGV can be used. In such way, a UAV flight planning technique can be used for the top parts of the building [11,12]. Similarly, path planning for UGVs should be carried out to capture images in suitable poses, in an optimal and obstacle-free way.

Imaging network design is one of the critical steps in image-based 3D reconstruction of buildings [13–16]. This step, also known in robotics as Next Best View Planning [17], aims to determine a minimum number of images that are sufficient to provide an accurate and complete digital twin of the surveyed scene. This can be done by considering the optimal range for each viewpoint and the suitable coverage and overlap between viewpoints. Although this issue has been taken into account by photogrammetry researchers [18,19] working with hand-held cameras or Unmanned Aerial Vehicles (UAV) [20], it has not yet been considered for UGVs equipped with digital cameras. Imaging network design for a UGV differs from that for UAVs as a result of the height and camera orientation constraints of the UGV.

A summary of investigations in image-based view planning for 3D reconstruction purposes was cited in [13], where research activities were classified into three main categories, including:

1. Next Best View Planning: starting from initial viewpoints, the research question is where the next viewpoints should be placed. Most of the approaches use Next Best View (NBV) methods to plan viewpoints without prior geometric information of the target object in the form of 3D model. Generally, NBV methods iteratively find the next best viewpoint based on a cost-planning function and information from previously planned viewpoints. These methods also use partial geometric information of the target object, reconstructed from planned viewpoints, to plan future sensor placements [21]. To find the next best viewpoints, one of three methods for representing the generated scanned area in the initial viewpoints is used, including triangular meshes [22] and volumetric [23] and Surfel representations [24].
2. Clustering and Selecting the Vantage Viewpoints: given a dense imaging network, clustering and selecting the vantage images is the primary goal [18]. Usually, in this category, the core functionality is performed by defining a visibility matrix between sparse surface points (rows) and the camera poses (columns), which can be estimated through a structure from motion procedure [15,16,25–27].
3. Complete Imaging Network Design (also known as model-based design): contrary to the previous methods, complete imaging network design is performed without any initial network, but an initial geometric model of the object should be available. The common approaches in this category are classified into set theory, graph theory and computational geometry [28].

Most of the previous works in this field have focused mainly on view planning for the 3D reconstruction of small industrial or cultural heritage objects using either an arm robot or a person holding a digital camera [15,19,20,29–31]. These methods follow a common workflow that includes generating a large set of initial candidate viewpoints, and then clustering and selecting a subset of vantage viewpoints through an optimization technique [32]. Candidate viewpoints are typically produced by offsetting from the surface of the object of interest [33], or on the surface of the sphere [34] or ellipsoid that encapsulates it [13].

A comparison of view planning algorithms in the complete design (the third group) and next best view planning (the first group) groups is presented in [35], where 13 state-of-the-art algorithms were compared with each other using a six-axis robotic arm manipulator equipped with a projector and two cameras mounted on a space bar and placed in front of a rotation table. All the methods were exploited to generate a complete and accurate 3D point cloud for five cultural heritage objects. The comparison was performed based on four criteria, including the number of directional measurements, digitization time, total positioning distance, and surface coverage.

Recently, view planning has been integrated into UAV applications, where large target objects, such as buildings or outdoor environments, need to be inspected or reconstructed via aerial or terrestrial photography [12,36–39]. A survey of view planning methods, also including UAV platforms, is presented in [36]. In the case of planning for 3D reconstruction purposes, methods are divided into two main groups: off-the-shelf flight planning and explore-then-exploit groups. While in the former group, commercial flight planners for UAVs use simple aerial photogrammetry imaging network constraints to plan the flight, in the latter group, having an initial flight based on off-the-shelf flight planning, a model is generated and flight view planning algorithms are used. Researchers have proposed different view planning algorithms, including complete design and next best view planning strategies, for the second view planning of the explore-then-exploit groups. For instance, [40–42] proposed on-line next best view planning for UAV with an initial 3D model. Other authors proposed different complete design view planning algorithms for 3D reconstruction of buildings using UAVs [12,21,39,43–47]. For instance, in [21], the footprint of a building is extracted from DSM generated by nadir UAV imagery. Next, a workflow including façade definition, dense camera network design, visibility analysis, and coverage-based filtering (three viewpoints for each point) were then applied to generate an optimum camera poses for acquiring façade images and generate a complete geometric 3D model of the structure. UAV imagery is in most cases not enough to obtain a highly accurate, complete and dense point cloud of a building, and terrestrial imaging should also be performed [37]. Moreover, UAV imagery needs a proper certificate of waiver or authorization in urban regions for flying.

Investigating optimum network design in the real world presents difficulties due to the diversity of parameters influencing the final result. In this article, before real-world experiments, the different proposed approaches of network design were tested in a simulation environment known as Gazebo with a simulated robot operated on ROS. ROS is an open-source middleware operating system that offers libraries and tools in the form of stacks, packages and nodes written in python or C++ to assist software developers in generating robot applications. It works based on a specific communication architecture in the form of a message passing through common topics, server and client communication in the form of request and response, and dynamic reconfiguration using services [48]. On the other hand, Gazebo provides tools to accurately and efficiently simulate different robots in complex indoor and outdoor environments [49]. To achieve ROS integration with Gazebo, a set of ROS packages provide wrappers for using Gazebo under ROS [50]. They offer the essential interfaces to simulate a robot in Gazebo using the ROS communication architecture. Researchers and companies have developed many simulated robots in ROS Gazebo and have made them freely available based on ROS licenses. For instance, Husky is a four-wheeled robot generated by the Clear Path company in both ROS Gazebo simulation and real-world scenarios [51]. Moreover, many robotics researchers have developed software packages for different robotics concepts such as navigation, localization and SLAM based on ROS rules. For example, GMapping is a Rao-Blackwellized particle filer for solving SLAM problems. Each particle carries an individual map of the environment for its poses. The weighting of each particle is performed based on the similarity between the 2D laser data and the map of the particle. An adaptive technique is used to reduce the number of particles in a Rao-Blackwellized particle filter using the movement of the robot and the most recent observations [52].

This paper aims to propose a novel photogrammetric imaging network design to automatically generate optimum poses for the 3D reconstruction of a building using terrestrial image acquisitions. The images can then be captured by either a human operator or a robot located in the designed poses and can be used in photogrammetric tools for accurate and complete 3D reconstruction purposes.

The main contribution of the article is a view planning method for the 3D reconstruction of a building from terrestrial images acquired with a UGV platform carrying a digital camera. Other contributions of the article are as follows:

(i)   It proposes a method to suggest camera poses reachable by either a robot in the form of numerical values of the poses or a human operator in the form of vectors on a metric map.

(ii) In contrast to the imaging network design methods which have been developed to generate initial viewpoints located on an ellipse or a sphere at an optimal range from the object, the initial viewpoints are here placed within the maximum and minimum optimal ranges on a two-dimensional map.

(iii) Contrary to other imaging network design methods developed for building 3D reconstruction (e.g., [21]), the presented method takes into account range-related constraints in defining the suitable range from the building. Moreover, clustering and selecting approach is accomplished using a visibility matrix defined based on a four-zone cone instead of filtering for coverage with only three rays for each point and filtering for accuracy without considering the impact of a viewpoint in dense 3D reconstruction. Additionally, in the presented method, four different definitions of viewpoints are examined to evaluate the best viewpoint directions.

(iv) To evaluate the proposed methods, a simulated environment including textured buildings in ROS Gazebo, as well as a ROS-based simulated UGV equipped with a 2D LiDAR, a DSLR camera and an IMU, are provided and are freely available in https://github.com/hosseininaveh/Moor_For_BIM (accessed on 13 May 2021). Researchers can use them for evaluating their methods.

In the following sections of this article, the novel imaging network design method is presented. Method implementation and results are provided with simulation and real experiments for façade 3D modelling purposes in Section 3. Finally, the article ends with a discussion and concluding considerations and some suggestions for future works in Sections 4 and 5.

2. Materials and Methods

The general structure of the developed methodology consists of four main stages (Figure 1):

1. A dataset is created for running the proposed algorithm for view planning, including a 2D map of the building; an initial 3D model of the building generated simply by defining a thickness for the map using the height of the building; camera calibration parameters; and, minimum distance for candidate viewpoints (in order to keep the correct Ground Sample Distance (GSD) for 3D reconstruction purposes).

2. A set of candidate viewpoints is provided by generating a grid of sample viewpoints on binary maps extracted from the 2D map and selecting some of the viewpoints located in suitable range with considering imaging network constraints. The direction of the camera for each viewpoint is calculated based on pointing towards the façade (called façade pointing) or pointing towards to the centre of the building (called centre pointing), or two directions in each viewpoint locations with centre & façade pointing. The candidate viewpoints in each pose are duplicated at two different heights (0.4 and 1.6 m).

3. The generated candidate viewpoints, the camera calibration parameters and the initial 3D model of the building are used in the process of clustering and selecting vantage viewpoints with four different approaches including: centre pointing, façade pointing, hybrid, and centre & façade pointing.

4. Given the viewpoint poses selected in the above-mentioned approaches, a set of images is captured at the designed viewpoints and processed with photogrammetric methods to generate dense 3D point clouds.

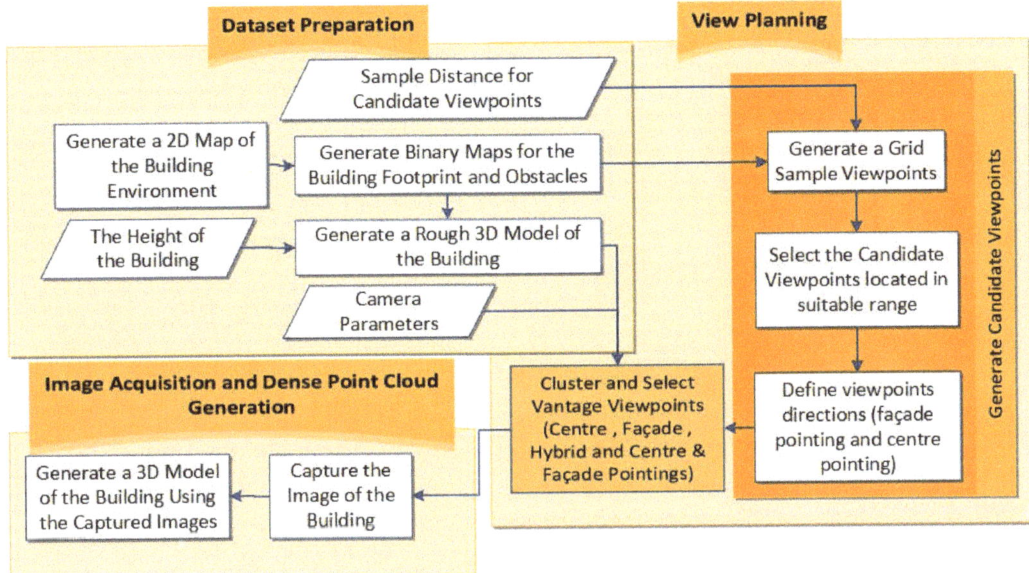

Figure 1. The proposed methodology for imaging network design for image-based 3D reconstruction of buildings.

2.1. Dataset Preparation

To run the proposed algorithm for view planning, dataset preparation is needed. The dataset includes a 2D map (with building footprint and obstacles), camera calibration parameters and a rough 3D model of the building to be surveyed. These would be the same materials required to plan a traditional photogrammetric survey. The 2D map can be generated using different methods, including classic surveying, photogrammetry, remote sensing techniques or Simultaneous Localization And Mapping (SLAM) methods. In this work, SLAM was used for the synthetic dataset and the surveying method was used for the real experiment. The rough 3D model can be provided by different techniques, including quick sparse photogrammetry [53], quick 3D modelling software [54], or simply by defining a thickness for the building footprint as walls and generating sample points on each wall with a specific sample distance. In this work, the latter method, defining a thickness for the building footprint, was used.

2.2. Generating Candidate Viewpoints

Candidate viewpoints are provided in three steps: grid sample viewpoint generation, candidate viewpoint selection, and viewpoint direction definition.

2.2.1. Generating a Grid of Sample Viewpoints

The map is converted into three binary images: (i) a binary image with only the building, (ii) a binary image with the surrounding objects known as obstacles, and (iii) a full binary image including building and obstacles. The binary images were automatically generated by running a global threshold using Otsu thresholding method [55]. The ground coordinates (in the global coordinate system) and the image coordinates of the buildings corners are used to determine the transformation parameters between the coordinate systems. These parameters are used in the last stage of the procedure for a 2D affine transformation to transfer the estimated viewpoints coordinates from the image coordinate system to the map coordinate systems. Given the full binary image, a grid of viewpoints with a specific sample distances (e.g., one metre on the ground) is generated over the map.

2.2.2. Selecting the Candidate Viewpoints Located in a Suitable Range

The viewpoints located on the building and the obstacles are removed from the grid viewpoints. Since the footprint and obstacles are black in the full binary image, this process can be carried out by simply removing points with zero grey pixel values from the grid viewpoints. Moreover, the refinement of the initial viewpoints is performed by eliminating viewpoints outside the optimal range of the camera with considering the parameters of the camera in photogrammetric imaging network constraints such as imaging scale constraint (D_{scale}^{max}), resolution (D_{Reso}^{max}), depth of field (D_{DOF}^{near}) and camera field of view (D_{FOV}^{max}, D_{FOV}^{min}). The optimum range is estimated using Equation (1) [56]. Further details of each equation are provided in Tables 1 and 2.

$$D_{max} = \min(D_{scale}^{max}, D_{Reso}^{max}, D_{FOV}^{max})$$

$$D_{min} = \max\left(D_{DOF}^{near}, D_{FOV}^{min}\right) \tag{1}$$

$$\text{Range} = D_{max} - D_{min}$$

Table 1. The equations for estimating the maximum distance from the building.

D_{max}	The Maximum Distance From The Object: $\mathrm{Min}(D_{scale}^{max}, D_{Reso}^{max}, D_{FOV}^{max})$
D_{scale}^{max}	$\dfrac{D \times f \times \sqrt{k}}{q \times S_p \times \delta}$ (mm) [56]
F	The focal length of the Camera (mm)
D	The maximum length of the object (mm)
K	The number of the images in each station
Q	The design factor (between 0.4 and 0.7)
S_p	The expected relative precision ($1/S_p$)
δ	The image measurement error (half a pixel size) (mm)
D_{Reso}^{max}	$\dfrac{f \times D_T \times \sin(\varphi)}{I_{res} \times D_t}$ (mm)
D_T	The expected minimum distance between two points in the final point cloud (mm)
D_t	The minimum distance between two recognizable points in the image (pixel)
I_{res}	The image resolution or pixel size (mm)
φ	The angle between the ray coming from the camera and the surface plane (radians)
D_{FOV}^{max}	$\dfrac{D_i \times \sin(\varphi + \alpha)}{2 \times \sin(\alpha)}$ (mm) [56]
A	The field of view of the camera:$\mathrm{atan}\left(\dfrac{0.9 \times H_i}{2 \times f}\right)$
D_i	The maximum object length to be seen in the image (mm)
H_i	The minimum image frame size (mm)

Table 2. The equations for obtaining the minimum distance from the building.

D_{min}	The minimum distance from the object: $\max(D_{DOF}^{near}, D_{FOV}^{min})$
D_{DOF}^{near}	The minimum distance from the object by considering the depth of fiel: $\dfrac{D_Z \times D_{HF}}{D_{HF} + (D_{HF} - f)}$ (mm) [56]
D_{HF}	The hyper focal distance: $\dfrac{f^2}{F_{stop} \times c}$ (mm)
F_{stop}	The F number of the camera
C	The circle of confusion ($\frac{f}{1720}$)
D_Z	The camera distance focus (D_{max} obtained from Table 1)
D_{FOV}^{min}	$\dfrac{H_O \times \sin(\varphi + \alpha)}{2 \times \sin(\alpha)}$ (mm) [56]
H_O	The height of the object (mm)

Having computed the suitable range, they should be converted into pixels. Then, a buffer is generated on the map by inverting the map of the building and subtracting two morphology operations from each other. The morphology operations are two dilations with a kernel size twice the maximum and minimum ranges. Having generated the buffer, the sample points located outside the buffer should be removed. In order to achieve redundancy in image observations in the z direction (height), two viewpoints are considered in each location with different heights (0.4 and 1.6 m) based on the height of an operator in sitting and standing states or different height level of the camera tripod on the robot.

2.2.3. Defining Viewpoint Directions

Having generated the viewpoint locations, in order to estimate the direction of the camera in each viewpoint location, three different approaches can be used:

(i)   the camera is looking at the centre of the building (centre pointing) [29]: the directions of viewpoints are generated by simply estimating the centre of the building in the binary image, estimated by computing the centroid derived from image moments on the building map [57] and defining the vector between each viewpoints locations and the estimated centre.

(ii)   the camera in each location is looking at the nearest point on the façade (façade pointing): some of the points on the façade may not be visible due to their being located behind obstacles or on the corners of complex buildings. These points are recognized by running the Harris corner detector on the map of the building for finding the corner of the buildings and recognizing the edge points located in front of big obstacles within the range buffer. Given these points, the directions of the six nearest viewpoints to these points are modified towards them.

(iii)   two directions (centre & façade pointing) are defined for any viewpoint locations: the directions of the viewpoints for both previous approaches are considered.

To estimate the directions in the form of quaternion, a vector between each viewpoint and the nearest point on the façade (in the façade pointing) or the centre of the building (in the centre pointing method) is drawn, and vector to quaternion equations are used to estimate the orientation parameters of the camera. These parameters were estimated by considering the normalized drawn vector in binary image with z value equal to zero ($\vec{a}$) and the initial camera orientation in the ground coordinate systems ($\vec{b} = [0, 0, -1]$), as follows:

$$e = \vec{a} \times \vec{b} \tag{2}$$

$$\alpha = \cos^{-1}\left(\frac{\vec{a}\,\vec{b}}{\left|\vec{a}\right| \cdot \left|\vec{b}\right|}\right) \tag{3}$$

$$q = [q_0, q_1, q_2, q_3] = [\cos(\alpha/2),\ e_x * \sin(\alpha/2),\ e_y * \sin(\alpha/2),\ e_z * \sin(\alpha/2)] \tag{4}$$

$$roll = \arctan\frac{2(q_0\,q_1 + q_2q_3)}{1 - 2(q_1^2 + q_2^2)}$$

$$pitch = \arcsin(2(q_0q_2 - q_3q_1)) \tag{5}$$

$$yaw = \arctan\frac{2(q_0\,q_3 + q_1q_2)}{1 - 2(q_2^2 + q_3^2)}$$

2.3. Clustering and Selecting Vantage Viewpoints

The initial dense viewpoints generated from the previous step are suitable for accessibility and visibility, but the number and density of these viewpoints is generally very high. Therefore, a large amount of processing time is required to generate a dense 3D point cloud from the images captured in these viewpoints. Consequently, optimum viewpoints should be chosen by clustering and selecting vantage viewpoints using a visibility matrix [16]. As this method is presented in [16], for each point of the available rough 3D model of the building, a four-zone cone with an axis aligned with the surface normal is defined (Figure 2, right). The opening angle of the cone (80 degrees) is estimated based on the maximum incidence angle for a point to be visible in an image (60 degrees). The opening angle of the cone is divided into four sections to provide the four zones of the point. A visibility matrix is created by using the four zones of each of points as rows and all viewpoints as columns (Figure 2, left). The matrix is filled with binary values by checking visibility between viewpoints and points at each zone of the cone. For this checking, the angle between the ray coming from each viewpoint and the surface normal on the point is computed and it is compared with the threshold values for each zone [16].

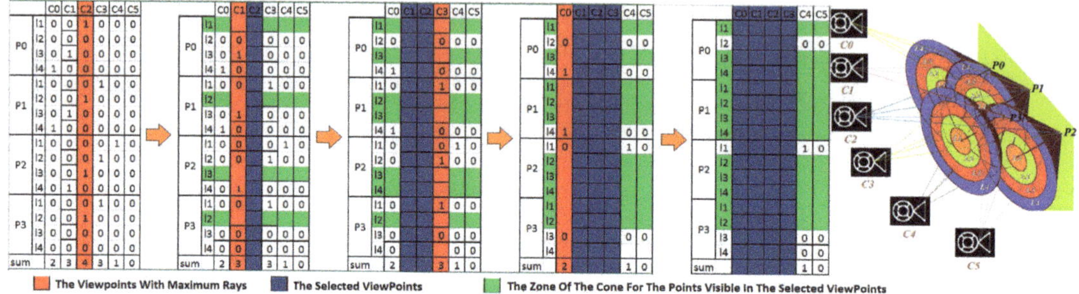

Figure 2. Visibility matrix and the procedure of clustering and selecting vantage viewpoints (**left**), the cone of points and viewpoints (**right**).

Having generated the visibility matrix, an iterative procedure is carried out to select the optimum viewpoints. In this procedure, the sum of all the columns of the obtained visibility matrix is estimated, the column with the highest sum value is selected as the optimal viewpoint, and then all of the rows with a value of 1 in that column, as well as the column itself, are removed from the visibility matrix. Finally, in any iteration of the procedure, a photogrammetric space intersection is done. Photogrammetric space intersection can be run on the common points between at least two viewpoints without any redundancy in image observations. However, when aiming to estimate the standard deviation, an extra viewpoint is needed. The procedure is repeated until completeness and accuracy criteria are satisfied. The accuracy criterion, relative precision, is obtained through running photogrammetric space intersection on all visible points (the points visible at least in three viewpoints) and the selected viewpoints and dividing the estimated standard division by the maximum length of the building. The completeness criterion is estimated through dividing the number of points that has been seen in at least three viewpoints (has only one row in the final visibility matrix) by all points in the rough 3D model. If this ratio is greater than a given threshold (e.g., 95 percent) and an accuracy criterion (a threshold given by the operator) is satisfied, the iteration is terminated. This approach is similar to the method presented in [16], but a modification is performed with respect to ignoring range-related constraints in the visibility matrix when considering these constraints in generating the candidate viewpoints step (Section 3.2).

To choose the optimum viewpoints from the initial candidate viewpoints generated with the three approaches described in the previous step, the visibility matrix approach can be run using four approaches:

Centre Pointing: the initial candidate viewpoints that point towards the centre of the building; the camera calibration parameters and the rough 3D model of the building are used in the clustering and selecting approach.

Façade Pointing: the camera calibration parameters and the rough 3D model of the building are used in the clustering and selecting approach, with the initial candidate viewpoint pointing towards façade and corners of the building.

Hybrid: both camera calibration and the rough 3D model are identical to the previous approach, whereas the initial candidate viewpoints of both previous approaches are used as inputs for the clustering and selection step.

Centre & Façade Pointing: the output of the first two approaches is assumed to be the vantage viewpoint.

2.4. Image Acquisition and Dense Point Cloud Generation

Once viewpoints have been determined, images are captured in the determined positions for all four approaches presented in the previous section. This can be performed with either a robot equipped with a digital camera using the provided numerical poses for the viewpoints or a person with a hand-held camera using GPS app on his/her smart phone or a handy GPS and the provided guide map. The images are then processed with photogrammetric methods [56–58], including (1) key point detection and matching; (2) outlier removal; (3) estimation of the camera interior and exterior parameters and the generation of a sparse point cloud; and (4) generation of a dense point cloud using multi-view dense matching. In this work, Agisoft Metashape [59] was chosen for evaluating the performance of the presented network design and viewpoint selection.

3. Results

The proposed methodology was implemented in Matlab (https://github.com/hossein inaveh/IND_UGV (accessed on 13 May 2021)) and was evaluated in both simulated and real environments, using a simulated robot developed in this work, and as later presented, respectively.

3.1. Simulation Experiments on a Building with Rectangular Footprint

To test the performances and reliability of the proposed method, ROS and Gazebo simulations were exploited with the use of a ground vehicle robot, equipped with a digital camera, in order to survey a building.

To evaluate the method proposed in Section 3, the refectory building of the K.N. Toosi University campus (Figure 3, right) was modelled in the ROS Gazebo simulation environment. A UGV equipped with a DSLR camera and a 2D range finder (Figure 3, left) was used in the simulation environment to provide a map of the scene using the GMapping algorithm, and also to capture images (Figure 4). To evaluate the performance of the proposed imaging network design algorithm, the four steps of the algorithm were followed in order to generate the 3D point cloud of the refectory building.

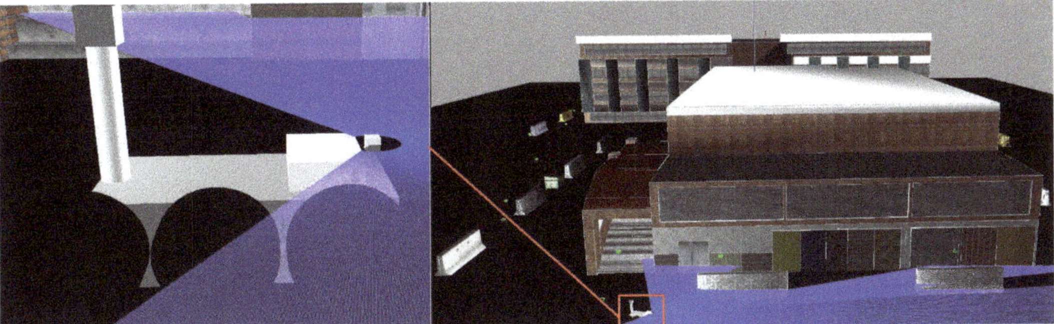

Figure 3. The ROS Gazebo simulation of the refectory buildings (**right**) and the simulated UGV/robot (**left**) moving in the scene.

Figure 4. An example of a captured (simulated) image using a camera mounted on the simulated UGV/robot (**left**) and the map of the simulated world generated with SLAM technique using a LiDAR sensor on the robot (**right**).

3.1.1. Generating Initial Candidate Viewpoints

The steps for generating initial sample viewpoints are depicted in Figure 5. Three image maps (Figure 5A–C) were extracted from the map of the building generated with the GMapping algorithm [60,61]. The coordinates of four exterior corners of the building were measured in the Gazebo model and image maps to estimate the 2D affine transformation parameters. Given the pixel size of the map on the ground (53 mm), the initial sample viewpoints were then generated over the map with one-meter sample distances, where the pixel values of the image map were not zero (green points shown in Figure 5D).

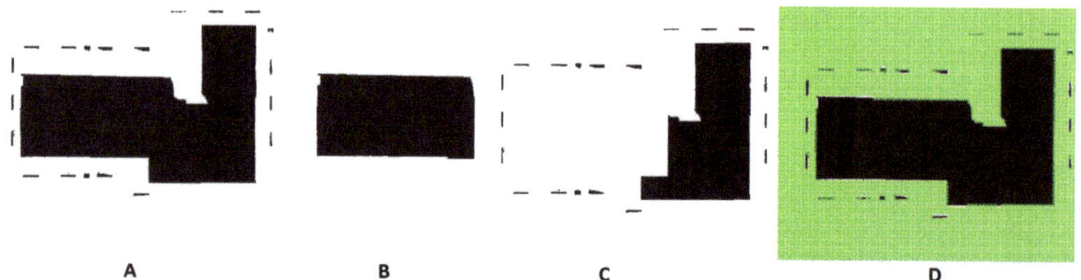

Figure 5. The map of the refectory and other buildings with their surrounding objects in the simulation space (**A**); the refectory building map (**B**); the obstacle map (**C**); the initial sample viewpoints on the map (**D**).

3.1.2. Selecting the Candidate Viewpoints Located in a Suitable Range

Given the sample viewpoints, the viewpoints located very close or very far from the building optimum range of the camera were removed using the range imaging network constraints [13] (Figure 6). Given the building dimensions (the perimeter is around 140 m) and considering a mm accuracy for the produced 3D point cloud of the building, the relative precision would be 1/14,000. The minimum range (2.56 m) and the maximum range (4.49 m) were obtained considering camera parameters as follows: focal length (18 mm), f-Stop (8), expected accuracy (1/14,000), image measurement precision (half a pixel size of the camera (0.0039 mm) and sensor size (23.5 × 15.6 mm). Having obtained the minimum and maximum range, the buffer was generated on the map, including the sample viewpoints located within a suitable range.

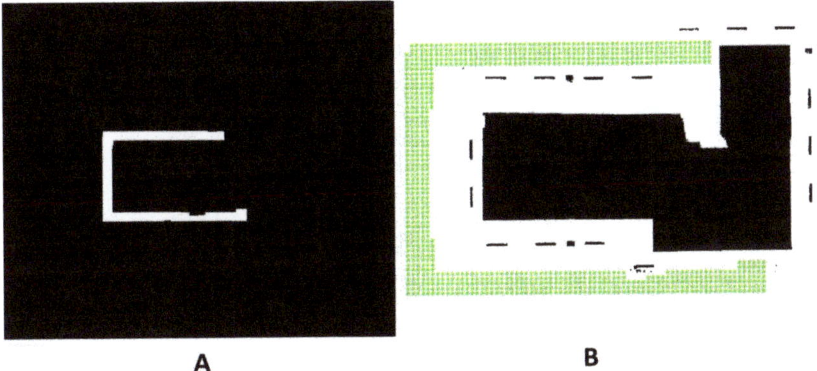

A   **B**

Figure 6. The buffer of the optimum camera range on the map (**A**); the sample points on the buffer of optimum camera range (**B**).

3.1.3. Defining Viewpoints Directions

To find the direction of each viewpoint in the façade pointing strategy, a canny edge detector was run on the map of the building (Figure 7A) and a vector was generated from each viewpoint to the nearest pixel on the edge of the building (Figure 7B). The vectors located on obstacles were then eliminated by examining whether any of their pixels were located on the map pixels with grey value equal to zero (Figure 7C). The invisible points on the façade building were then identified by (1) running the Harris corner detector for points located on the corners of the building, and (2) finding the edge points that their corresponding viewpoints vectors were eliminated due to obstacles (Figure 7D). Finally, in the façade pointing strategy, the direction of the six nearest viewpoints to each of the invisible points was modified towards the invisible points (Figure 7E).

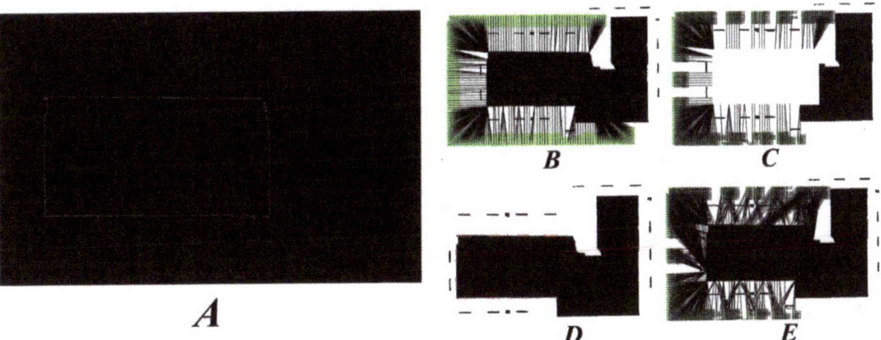

B   **C**

A

D   **E**

Figure 7. The points on the façade of the refectory building extracted using the edge detection algorithm (**A**); the direction of viewpoints towards the façade points without considering the obstacles (**B**); the direction of viewpoints towards the façade with considering the obstacles (**C**); the invisible points on the façade obstacles were eliminated similarly to the technique used in Figure 7C. The viewpoint orientation located in front of obstacles (**D**); the direction of viewpoints towards the faced points modified in order to see invisible points (**E**).

As shown in Figure 8, in the centre pointing strategy, the directions of viewpoints were shown by drawing vectors between the viewpoints and the centre of building. The vectors located on the parameters were then computed using Equations (2)–(5).

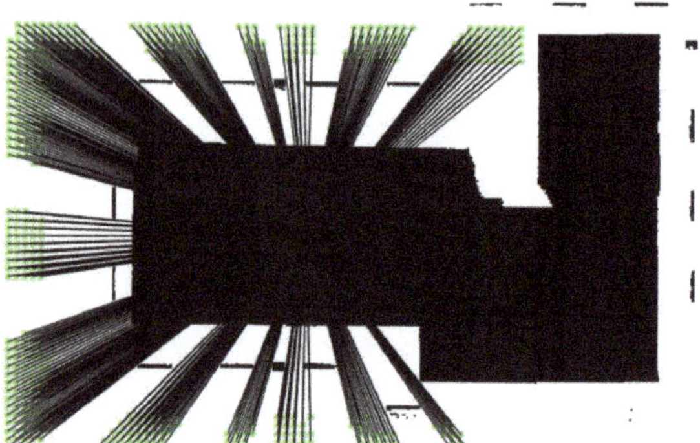

Figure 8. The direction of the viewpoints towards the centre of the building.

3.1.4. Clustering and Selecting Vantage Viewpoints

To have a more complete point cloud of the building in the vertical (z) direction, the number of viewpoints was doubled in order to have two viewpoints with the same location in the x and y coordinates, but with two different values in the z direction (0.4 and 1.6 m from the ground). Figure 9 illustrates the initial candidate viewpoints for the façade pointing approach and the four-zone cone (see Section 3.3) of two points in the CAD model. As can be concluded from this figure, by increasing the incidence angles, the aperture of the cone will decrease, and thus the points considered visible in the viewpoints will be closer to each other. This assumption results in more viewpoints by increasing the incidence angle.

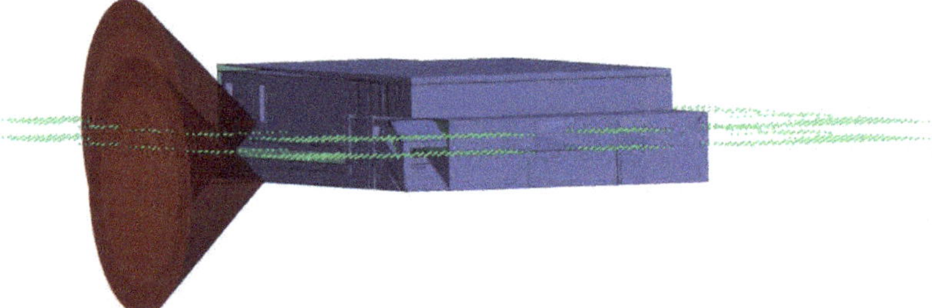

Figure 9. The candidate viewpoints and the cone of two points of the initial mesh for façade pointing. For better visualization, only the cone with the two initial points is presented.

This fact is proved by setting different values for the incidence angle and running the algorithm for the clustering and selection of the vantage image. The results are presented in Figure 10. The number of viewpoints for the four different incidence angles is provided in Figure 6. Although the minimum number of viewpoints was obtained with an incidence angle of 20 degrees, a low number of images could increase the probability of failure of matching procedures in SfM due to the wide angle between the optical axes of adjacent cameras. This issue can be seen in Figure 11, which shows the gap in the positions of viewpoints in the corner of the building (the red box in the Figure) as well as the failure in image alignment in SfM for the dataset with incidence angles set below 60 degrees (the bottom of Figure 11A,B). In the experiments, with a trial-and-error approach, it was found that any value below 80 degrees for this parameter could result in a failure in image alignment. This happened when running hybrid approach in the simulation.

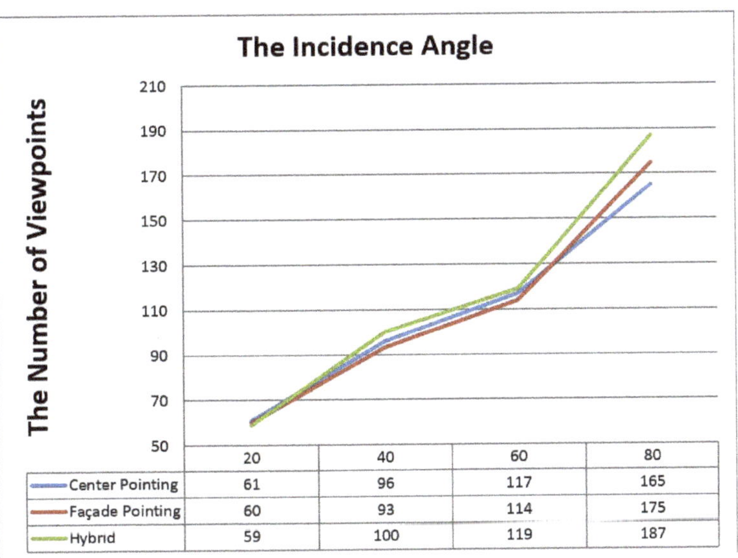

	20	40	60	80
Center Pointing	61	96	117	165
Façade Pointing	60	93	114	175
Hybrid	59	100	119	187

Figure 10. The number of viewpoints for the centre, façade and hybrid pointing imaging network design in different setting of incidence angles (20, 40, 60 and 80 degrees).

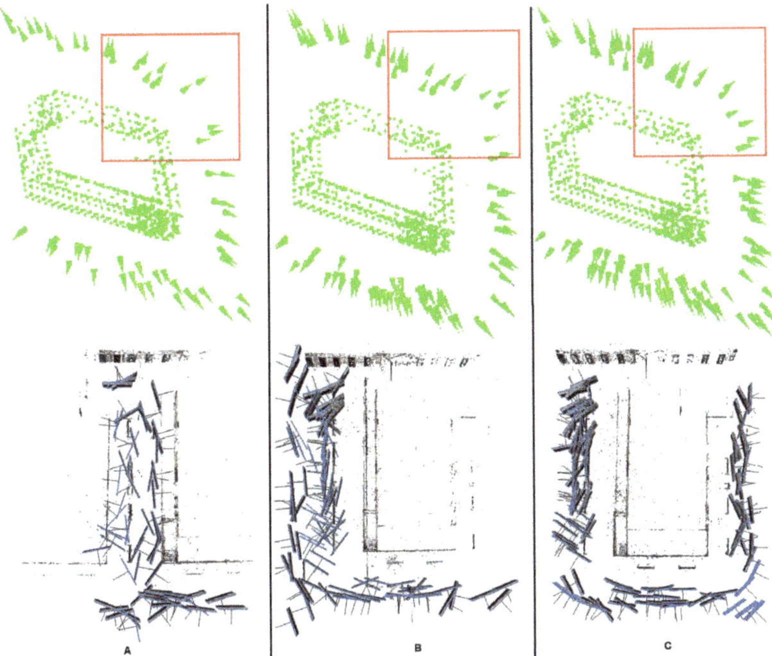

Figure 11. The selected viewpoints of the centre pointing imaging network for different incidence angles ((**A**): 20, (**B**): 40 and (**C**): 60 degrees) in viewpoint selection step (**top**) and SfM step (**bottom**).

Given the candidate viewpoints in the centre, façade and hybrid approaches, clustering and selecting procedures were applied, while 60 degrees was set as incidence angle. The clustering and selecting algorithms (Section 3.3) were used to select (Figure 12) a set of viewpoints, which were selected at heights of 0.4 and 1.6 m, as follows:

- Centre pointing approach: 96 viewpoints were selected out of 1020 initial candidate viewpoints;
- Façade pointing: 107 viewpoints were selected out of 5218 initial viewpoints;
- Hybrid approach: 119 viewpoints were selected out of 6238 initial candidates.
- Centre & façade pointing: 213 viewpoints were chosen as a dataset including the output of both of the first two approaches.

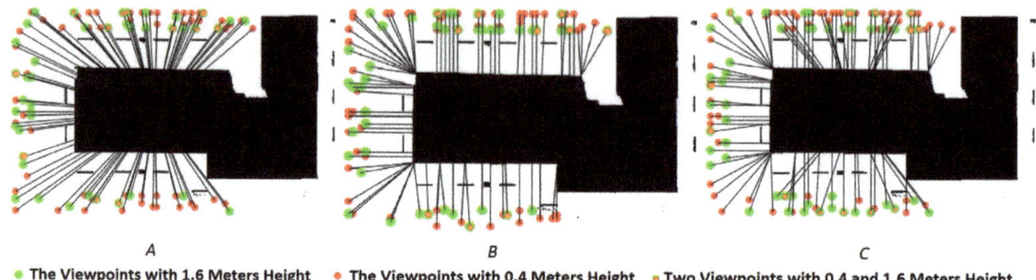

Figure 12. The final vantage viewpoints selected from the candidate viewpoints for the façade pointing (**A**), centre pointing (**B**), and hybrid (**C**) imaging networks.

3.1.5. Image Acquisition and Dense Point Cloud Generation

Given the candidate viewpoints, the robot was moved around the scene to capture the images in the designed viewpoints for all four approaches. The captured images were then processed to derive dense point clouds. Figure 13 shows a top view of the camera poses and point clouds for all four imaging network designs. Three regions (R1, R2 and R3) were considered to evaluate the quality of the derived point clouds.

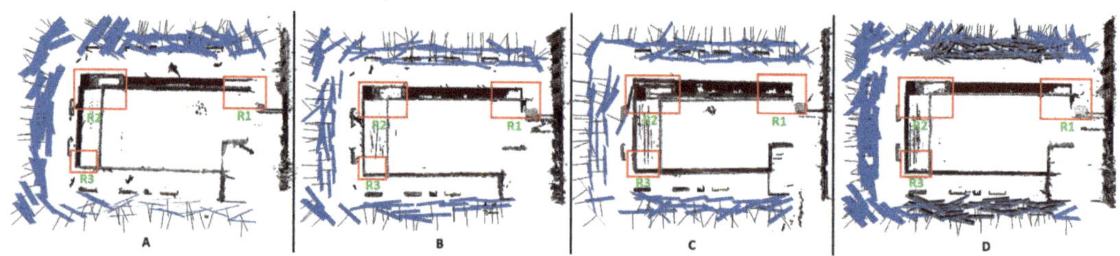

Figure 13. The captured images and the point clouds of the simulated building in centre pointing (**A**), façade pointing (**B**), hybrid (**C**), and combined centre & façade (**D**) imaging network design. Three areas (R1, R2, R3) are identified where a quality check was performed.

Figure 14 illustrates the point clouds of the building generated with the four proposed approaches. To compare the point clouds, three areas (shown as R1, R2 and R3 in Figure 13) were taken into account. Clearly, the best point cloud was generated with the images of centre & façade approach. The point cloud generated using images captured with the hybrid approach shows errors, noise and incompleteness in R2 (the red box for R2 in Figure 13C). This was due to the low number of viewpoints selected in the corners of the building with respect with other three image acquisition approaches. This issue resulted in the failure of image alignment in these regions. These results clarified the importance of having nearby viewpoints with smooth orientation changes in the corners of buildings.

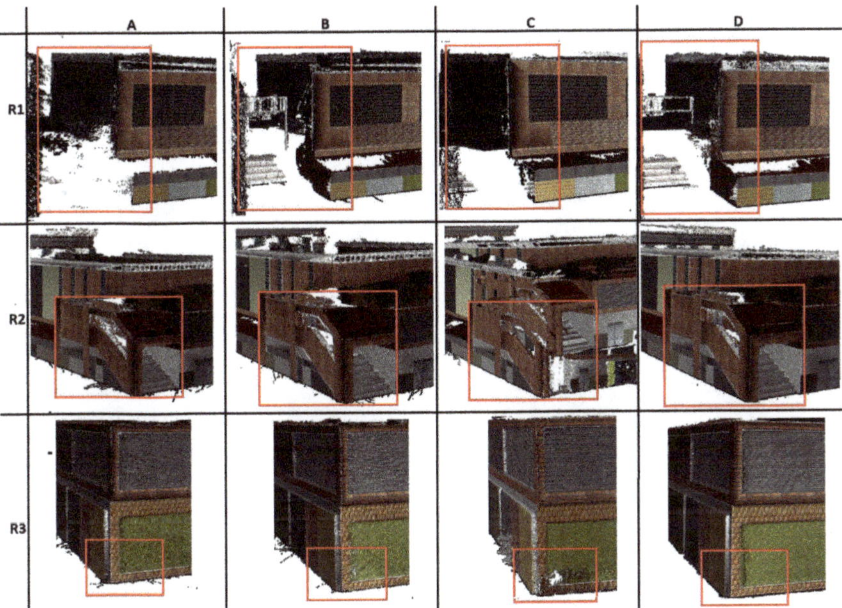

Figure 14. The details of the results in the three selected areas (R1, R2, R3) for all of the image acquisition approaches: centre pointing (**A**), façade pointing (**B**), hybrid (**C**) and combined (**D**). The hybrid strategy (**C**) showed incomplete results.

3.2. Simulation Experiments on a Building with Complex Shape

To evaluate the performance of the method for a building with a complex footprint shape, a building was designed using SketchUp software in such a way that it included different curved walls and corners with several obstacles in front of each walls. As illustrated in Figure 15, the model was also decorated with different colourful patterns to overcome the problem of textureless surfaces in the SfM and MVS algorithms. The model was then imported into the ROS Gazebo environment to be employed in the 3D reconstruction procedure presented in this work. To make the evaluation procedure more challenging, a part of the building was considered for 3D reconstruction (the area painted orange in Figure 16), and another part played a role as a self-occlusion area.

Given the building model in ROS Gazebo, the robot was used to generate a map of the building environment with GMapping algorithm. By setting camera parameters and expected accuracy at levels similar to those in the previous project, the minimum and maximum distances for the camera placement were computed (15,390 mm and 5130 mm, respectively) and converted into map units. As can be seen in Figure 16, the map was used in the present method to generate sample viewpoints (Figure 16A), as well as initial candidate viewpoints for both the centre and façade pointing approaches (Figure 16B,C). The generated candidate viewpoints were then imported into the clustering and selection approaches in order to produce four different outputs, including centre pointing, façade pointing, hybrid approach (Figure 16D–F), and centre & façade pointing. As the only difference between this project and the previous one with respect to setting the parameters of the clustering and selection approach, the incidence angle parameter was set at 80 degrees. This was done in order to prevent any failures in the photo alignment procedure in SfM.

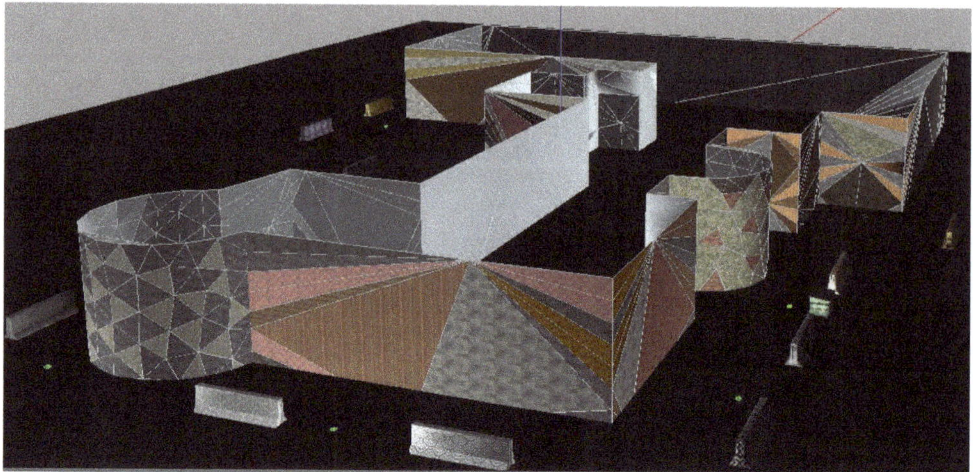

Figure 15. The complex building used to evaluate the performance of the algorithm.

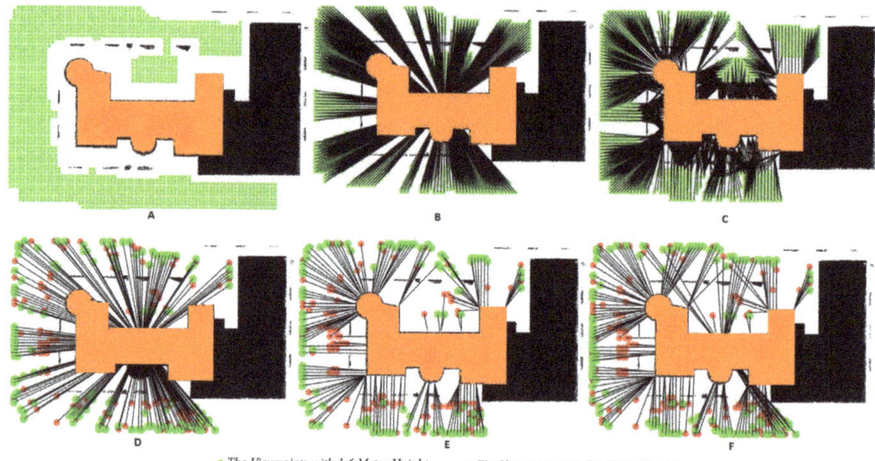

Figure 16. The steps of centre, façade, and hybrid pointing approaches for the imaging network design of complex buildings. The buffer of optimum camera positions (**A**); the viewpoint directions toward the centre of the building (**B**) and toward the façade (**C**); the outputs of the clustering and selection approach on centre pointing viewpoints (**D**); façade pointing viewpoints (**E**) and both of them (**F**).

Having generated the viewpoints for all of the approaches, they were used in the next step to navigate the robot around the building, and to capture images in the designed poses. The captured images were then imported into the SfM and MVS approaches in order to generate a dense point cloud of the building. The point clouds of one side of the building, which have a greater complexity than the output of each of the approaches, are displayed in Figure 17. At first glance, the best results were achieved when using the centre & façade pointing approach.

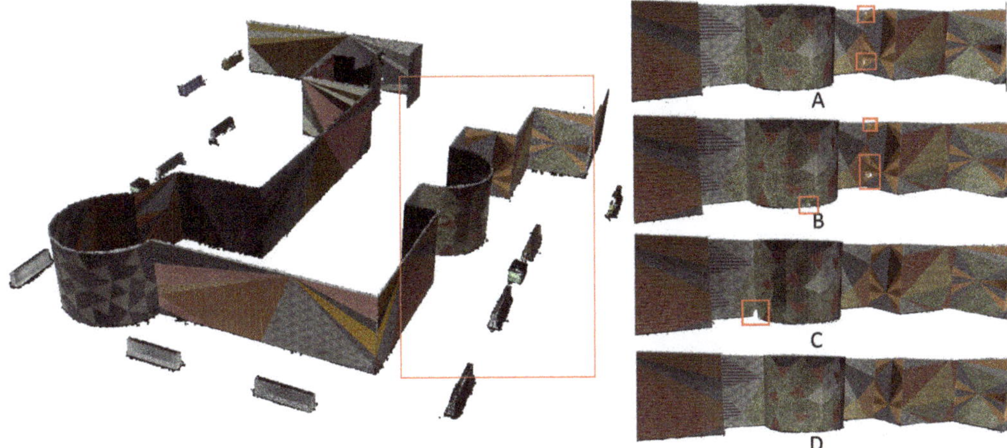

Figure 17. The final point cloud of the complex building (**left**) and the point clouds of the selected area in the red box for the centre pointing (**A**), façade pointing (**B**), hybrid (**C**) and centre & façade pointing (**D**) approaches. The gaps in the point cloud are shown using red boxes at the right of the figures.

Table 3 shows the number of initial viewpoints and the selected viewpoints, and the number of points in the final point cloud for each of the implemented approaches for a complex building. Similar to the previous project, the centre & façade pointing approach resulted in more complete point cloud when using 570 images. If the computation expenses are important for this comparison, the best approach is the hybrid. It performed better in this project with respect to the previous one due to the incidence angle being increased from 70 to 80 degrees, leading to a denser imaging network. Although the number of selected viewpoints for centre pointing (292) was close to this number for the hybrid approach (301), the worst results were achieved when running this approach due to the lack of flexibility of this approach with respect to overcoming the occluded area. Façade pointing also had limitations with respect to the 3D reconstruction of walls located in front of other walls (Figure 17A,B), but this approach resulted in more points than the centre pointing and hybrid approaches, with an even lower number of images (278).

Table 3. The results of running the four approaches on the complex building.

	Centre Pointing	Façade Pointing	Hybrid	Centre & Façade Pointing
The Number of Initial Viewpoints	2386	10,240	12,626	——
The Number of Selected Viewpoints	292	278	301	570
The Number of Points in the Point Cloud	9,175,444	11,156,211	10,648,205	11,630,850

3.3. Real-World Experiments

To evaluate the proposed algorithm in a real-world scenario, the refectory building of the civil department of K. N. Toosi University of Technology was considered as a case study (Figure 18, left). A map of the building and its surrounding environment was generated using classic surveying and geo-referencing procedures (Figure 18, right).

Moreover, in order to compare the results of the presented approaches with a standard method, known as continuous image capturing, for the image-based 3D reconstruction of a building a DSLR Nikon D5500 was used to capture images of the building from a suitable distance, where the whole height of each wall of the building can be seen in the images. In this camera, there is an option to capture high-resolution still images continuously every fifth of a second. The images were captured from the building in two complete rings at two different heights by rotating around the building twice continuously (Figure 19, left). Having captured the images, due to the huge number of images (1489 images) they were imported into a server computer with 24 CPU cores and 113 GiB RAM, as well as a GeForce RTX 2080 NVIDIA graphics card for running SfM and MVS procedures to generate a dense point cloud of the building. It took 200 min to complete the MVS procedure. As another common method for 3D reconstruction of the building in a process called continuous image capturing & clustering and selection, the captured images in the first method were used in the clustering and selection approach presented in Section 2.3 of this article to reduce the number of the images. In this procedure, the incidence angle was set to 80 degrees, and 236 images were selected as optimum images for 3D reconstruction (Figure 19, right). Running MVS on the selected images in the server computer took 14 min to generate the dense point cloud.

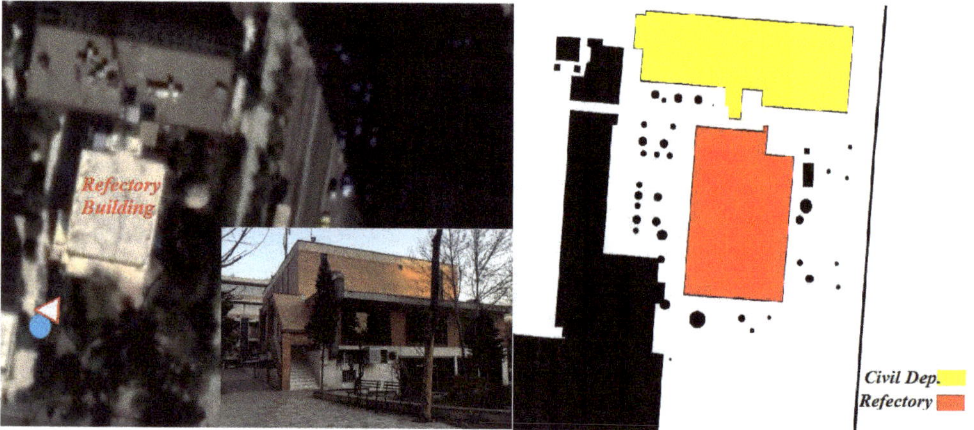

Figure 18. A cropped satellite view of the civil department and the refectory buildings augmented with a terrestrial image captured from the refectory building (**left**). The available surveying map of the building and their surrounding objects (**right**).

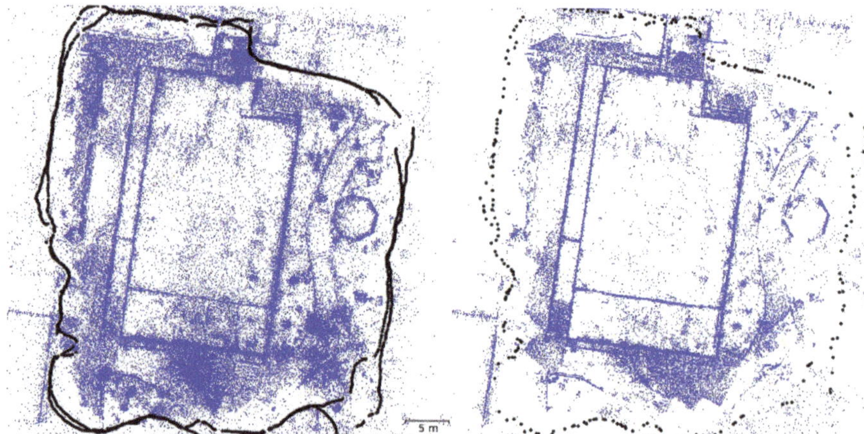

Figure 19. The outputs of running the SfM procedure on the images of continuous image capturing (**left**) and continuous image capturing & clustering and selection (**right**) modes. The black dots in the figures show the position of the camera, and the blue dots represent the sparse point cloud of the building and its environment.

Starting from the available map, similar to the simulation section, the steps of the algorithm (Section 3) were followed (Figure 20) to generate viewpoints for all four approaches. The clustering and selecting procedure finally chose 176 viewpoints in centre pointing, 177 viewpoints in façade pointing and 178 viewpoints in hybrid, out of 232, 572 and 804 candidate viewpoints, respectively. All of the viewpoints selected in the first two approaches (355 viewpoints) were chosen as the output for the centre & façade pointing (Figure 21).

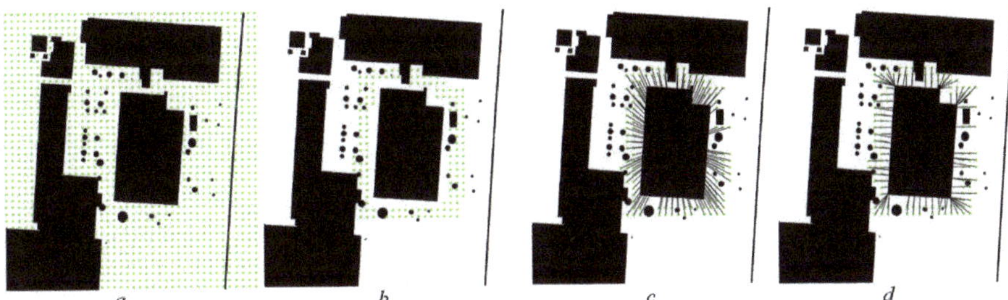

Figure 20. The initial sample viewpoints (**a**), the sample viewpoints located at the optimal range from the building (**b**), and the direction of each viewpoint for centre pointing (**c**) and façade pointing (**d**).

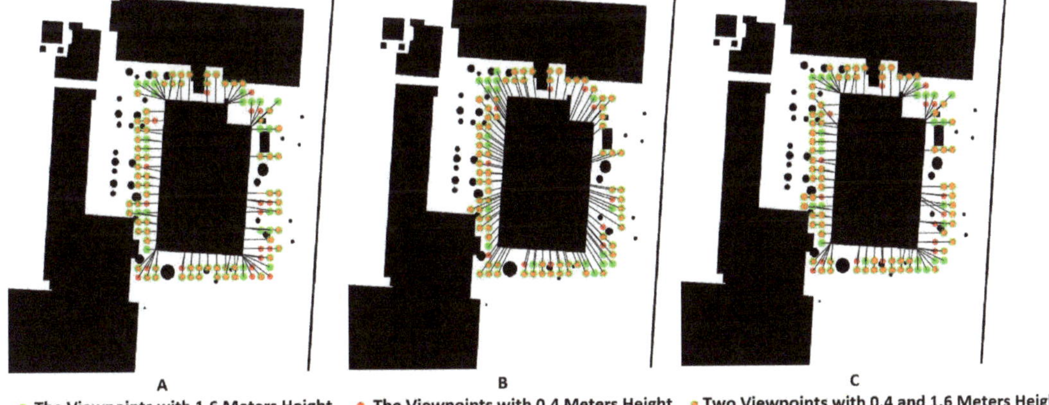

Figure 21. The final vantage viewpoints selected from the candidate viewpoints of the real-world project for the façade pointing (**A**), centre pointing (**B**), and hybrid (**C**) imaging networks.

Having designed four imaging networks, a DSLR Nikon Camera (D5500) was implemented on a tripod to capture images of the building at the designated viewpoints. A focal length of 18 mm and a F-Stop of 6.3 were set for the camera. These values were estimated using a trial-and-error approach during the clustering and selection step (Section 2.3) by setting different values for these parameters and checking the final accuracy of the intersecting points. All of the captured images in all of the approaches (façade pointing, centre pointing, hybrid and centre & façade pointing) were then processed in order to derive camera poses and dense point clouds (Figure 22).

The 3D coordinates of the 30 Ground Control Points (GCPs) placed on the building façades were measured using a total station and were manually identified in the images. Fifteen points were then used to constrain the SfM bundle adjustment solution as a ground control (the odd numbers in Figure 22), and the other 15 (the even numbers Figure 22) were used as check points. Figure 22 displays the error ellipsoids of the GCPs for the presented approaches including the centre (Figure 22A), façade (Figure 22B), hybrid (Figure 22C) and centre & façade (Figure 22D) pointing datasets. The size of the error ellipsoids for the façade pointing dataset was almost twice as large as the size of the error ellipsoids for centre pointing. This could be due to the better configuration of rays coming from the cameras to each point in the centre pointing dataset which leads to better ray intersection angles. These angles in the façade pointing dataset are small, resulting in less accurate coordinates, but a more favourable geometry for dense matching and dense point cloud generation.

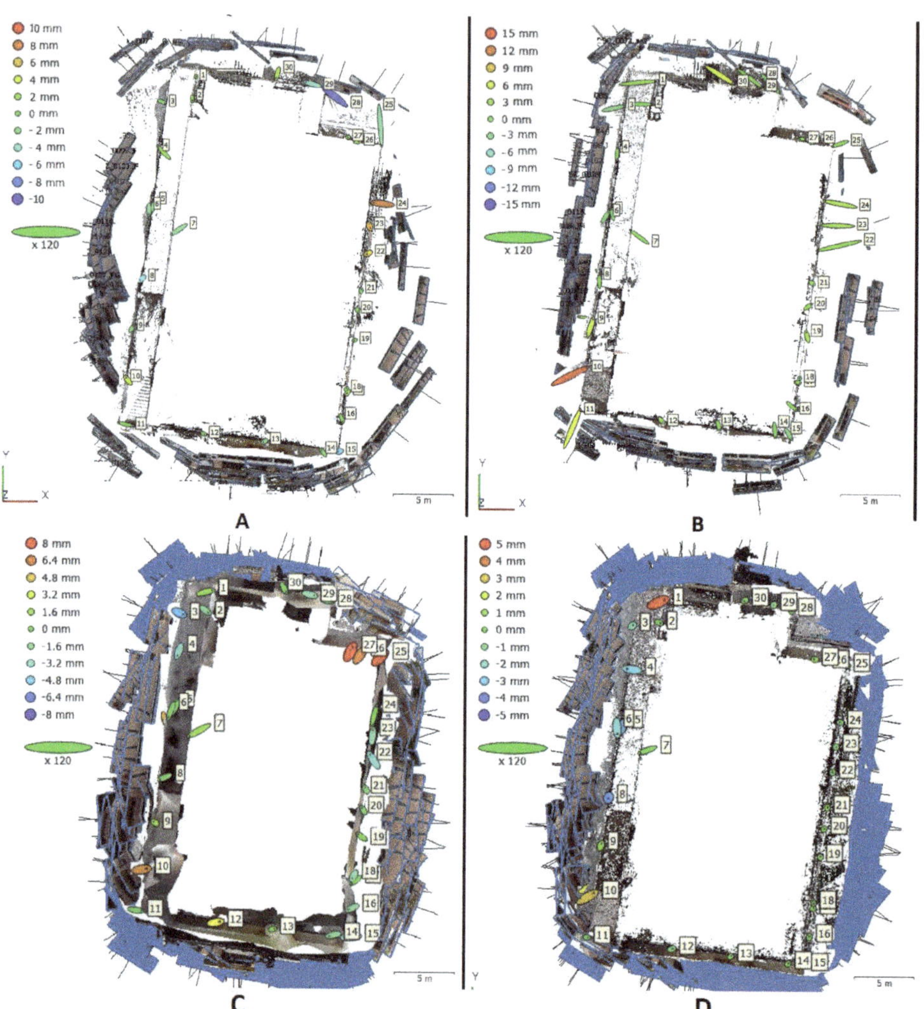

Figure 22. The four recovered image networks for centre pointing (**A**), façade pointing (**B**), hybrid (**C**) and centre & façade datasets (**D**). The error ellipsoids of GCPs, also demonstrated through the colourful ellipses to the left side of the building. For better visualization, the scale of the ellipses is multiplied by 120.

The GCPs were also used when evaluating the accuracy of the point clouds generated using the two common methods. As shown in Figure 23, in the bottom left corner of the building map, the distance from the camera to the building was reduced due to workspace limitations. This resulted in a reduction of the accuracy on the GCPs at this corner in comparison with other corners of the building. The results also indicate that having more images does not always lead to a better accuracy for GCPs, and more images produce more noise in the observations, with this noise at some point leading to a loss of accuracy.

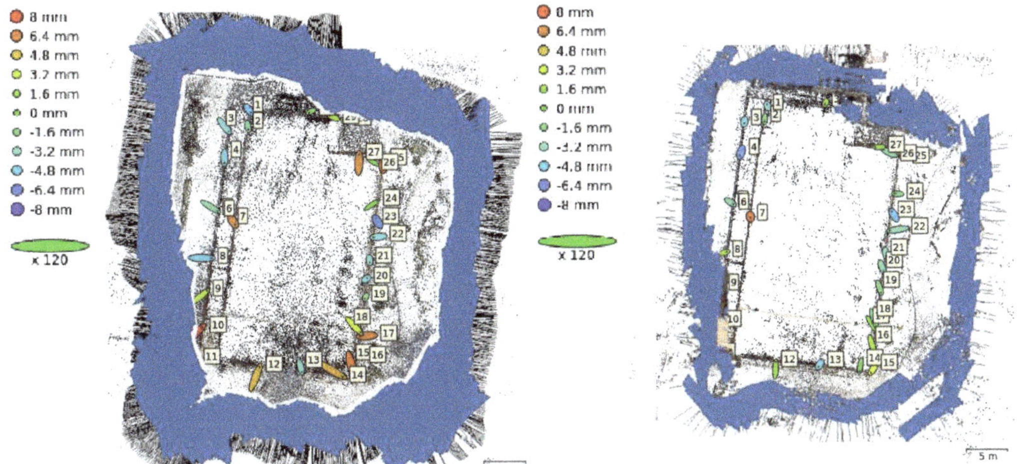

Figure 23. The error ellipsoids on GCPs for the continuous image capturing and the continuous image capturing & clustering and selecting approaches.

Figure 24 illustrates the total error of the GCPs for all datasets. It can be observed from this figure that the points located around the middle of the building have less error than the points located at the corners of the building for all approaches. Moreover, the mean of GCP error for the façade pointing datasets is almost two times bigger than this value for the centre pointing datasets. The maximum error of GCP for all of the presented approaches, with the exception of centre pointing, is related to the error of estimating the X coordinates. As mentioned in the paragraphs above, this is due the stronger configuration of images in centre pointing datasets with triangle intersections that are closer to equilateral triangles.

To evaluate the proposed approaches in comparison with the two standard approaches, two criteria based on completeness and accuracy of the final dense point cloud were taken into account. Firstly, the quality of the point clouds was visually evaluated in three corners of the building, similar to the simulation project (Sections 3.1 and 3.2). Figure 25 shows the quality of the point clouds in the mentioned regions. The worst point clouds were generated when using the centre pointing dataset (Figure 25A), and the most complete point cloud with the fewest gaps was generated using the continuous capturing images dataset. Following this method, the continuous capture of images & clustering and selection approach, and the centre & façade approach obtained the second and third ranks for the generation of complete point clouds (Figure 25D,E). The hybrid dataset resulted in a more complete point cloud than the façade pointing dataset. Although the common methods were able to generate dense point clouds, the point clouds of these approaches included more noise and outliers due to the blurred images in the dataset.

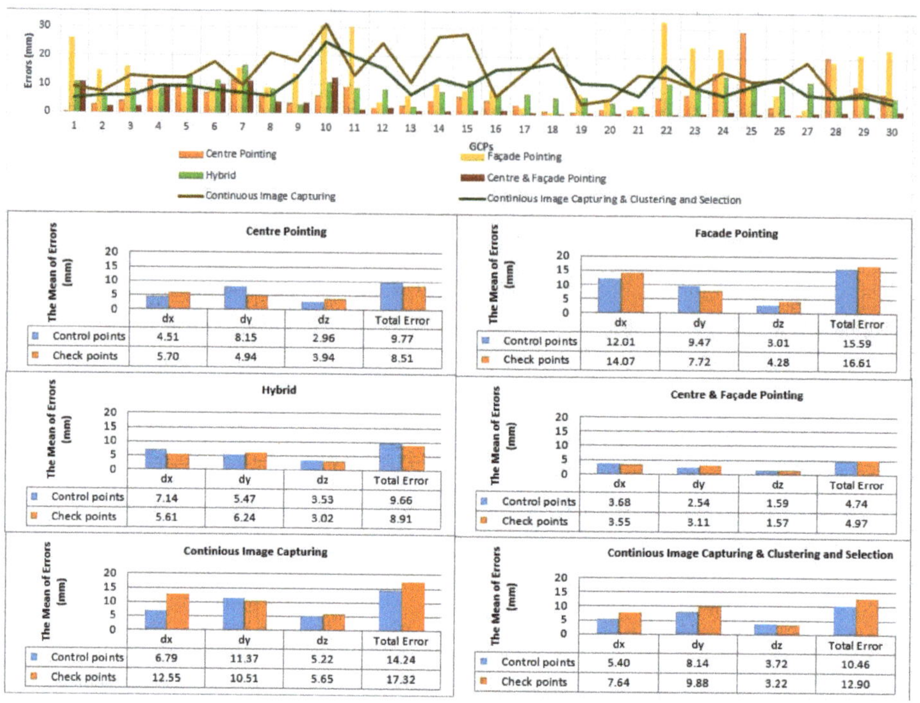

Centre Pointing

The Mean of Errors (mm)

	dx	dy	dz	Total Error
Control points	4.51	8.15	2.96	9.77
Check points	5.70	4.94	3.94	8.51

Façade Pointing

The Mean of Errors (mm)

	dx	dy	dz	Total Error
Control points	12.01	9.47	3.01	15.59
Check points	14.07	7.72	4.28	16.61

Hybrid

The Mean of Errors (mm)

	dx	dy	dz	Total Error
Control points	7.14	5.47	3.53	9.66
Check points	5.61	6.24	3.02	8.91

Centre & Façade Pointing

The Mean of Errors (mm)

	dx	dy	dz	Total Error
Control points	3.68	2.54	1.59	4.74
Check points	3.55	3.11	1.57	4.97

Continious Image Capturing

The Mean of Errors (mm)

	dx	dy	dz	Total Error
Control points	6.79	11.37	5.22	14.24
Check points	12.55	10.51	5.65	17.32

Continious Image Capturing & Clustering and Selection

The Mean of Errors (mm)

	dx	dy	dz	Total Error
Control points	5.40	8.14	3.72	10.46
Check points	7.64	9.88	3.22	12.90

Figure 24. The errors of GCP coordinates for all four approaches (**top**). The mean of errors of the control and check points in X, Y and Z directions and the total errors for all approaches (**bottom**).

Then, as no ground truth data were available, the point cloud completeness was evaluated by counting the number of points on five-yard mosaics (Figure 26) as well as on the whole building. As shown in Figure 27A, all the presented approaches except the centre pointing dataset were able to provide more points on the mosaics than the standard approaches. Moreover, in the case of the number of points on the whole building (Figure 27B), the centre & façade pointing and hybrid datasets resulted in point clouds with more points (33 and 30 million points, respectively). Façade pointing led to more points for the whole building compared to the centre pointing dataset.

The noise level of the point clouds was evaluated by estimating the average standard deviations of a fitted plane on the mosaics. To evaluate the flatness of the mosaic surfaces, accurate 3D point clouds were separately generated for them in the lab by capturing many convergent images at close range (0.6 m), and a plan was fitted to each of the point clouds. The results showed that the surface of the mosaics fit on a plane with a standard deviation of around 0.2 mm. As illustrated in Figure 27C, the average standard deviations of fitting a plane on the mosaics point clouds generated with the hybrid and centre pointing approaches were almost identical (2.9 mm). While the best results were achieved by using the centre & façade dataset (1.4 mm), the noisiest point cloud was generated by the dataset of the continuous image capturing approach as the common method with the average standard deviation of 18 mm. Exploiting the clustering and selection approach on the continuous image capturing dataset led to a reduction of noise to one-sixth of its value (2.8 mm).

Considering both the number of points and the standard deviation of the fitted plane, it can be concluded that although the number of images in the centre & façade dataset is almost twice that of the other presented approaches, it is the best approach in terms of completeness and accuracy criteria. If the number of images is crucial in terms of the processing time and computer memory required, then hybrid and façade pointing can be considered as the best methods, respectively.

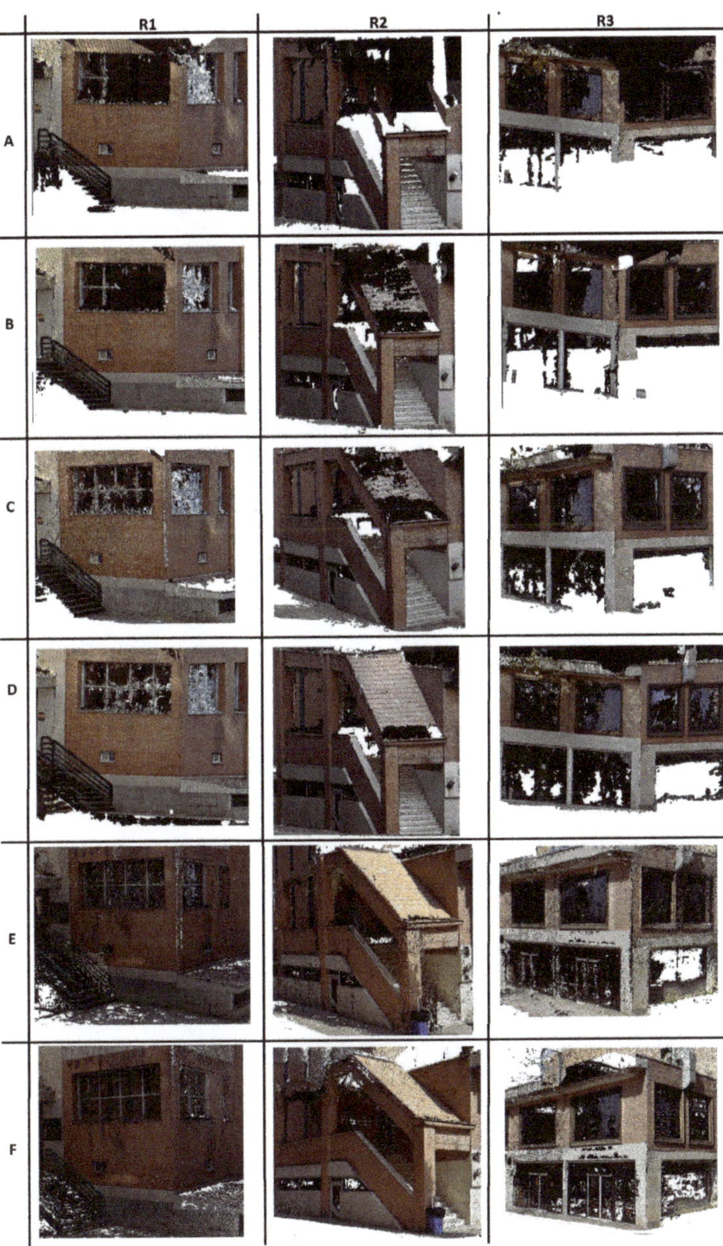

Figure 25. The dense point clouds generated with centre pointing (**A**), façade pointing (**B**), hybrid (**C**), centre & façade pointing (**D**), continuous image capturing (**E**), and continuous image capturing & clustering and selection (**F**) approaches.

Figure 26. The locations of the mosaics placed on the building façades.

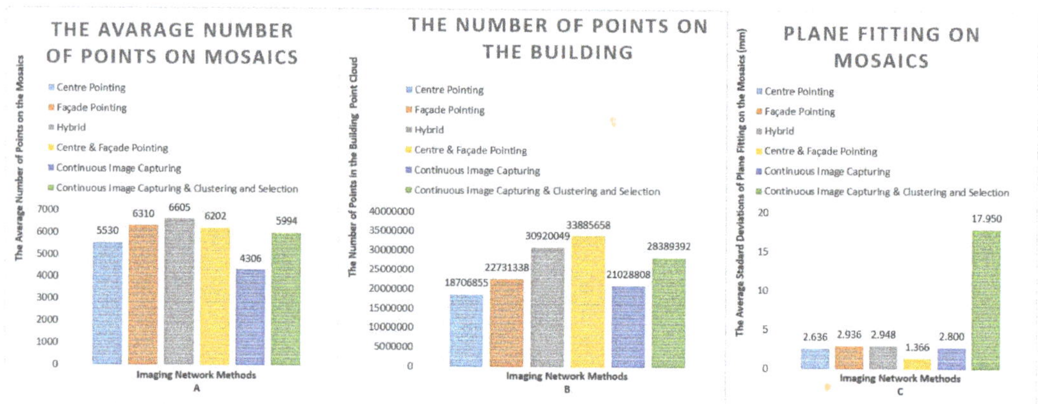

Figure 27. the average number of points on the five mosaics (**A**) and the whole building (**B**); the average standard deviations of plane fitting on the mosaics point clouds (**C**).

4. Discussion

This work presented an image-based 3D pipeline for the reconstruction of a building using an unmanned ground vehicle (UGV) or a human agent coupled with an imaging network design (view planning) algorithm. Four different approaches, including façade, centre, hybrid, and centre & façade pointing, were designed, developed and compared with each other in both simulated and real-world environments. Moreover, two other methods—continuous image capturing, and continuous image capturing & clustering and selection approaches—were considered as standard methods in real-world experiments for evaluating the performance of the presented methods. The results showed that the first standard method requires a fast computer, and even when using a server computer, a noisy point cloud is generated using this approach. Although clustering and selecting vantage images on this dataset reduced the noise considerably, the number of points on the building and the density of the points were dramatically reduced. Although the façade pointing approach could lead to more complete point clouds due to images with parallel optical axes more suitable for MVS algorithms, the accuracy of individual points in the centre pointing scenario was better, due to stronger intersection angles. Using all of the images of both of the previous approaches (centre & façade pointing) led to a more complete and more accurate point cloud than in the two first approaches (façade pointing and centre pointing).

Clustering and selecting vantage viewpoints of the candidate viewpoints using both centre and façade pointing directions (hybrid approach) may result in a failure of alignment in SfM if the incidence angle is set below 80 degrees. This happened for the first simulation dataset. Obviously, more complete and accurate point clouds can be achieved by using the centre & façade pointing approach, with the disadvantages of greater processing time and greater requirement of computer power.

5. Conclusions

This paper proposes a novel imaging network design algorithm for façade 3D reconstruction using a UGV. In comparison with other state-of-the-art algorithms in this field, such as that presented in [21], the presented method takes into account range-related constraints when defining the suitable range from the building, and the clustering and selecting approach is performed using a visibility matrix defined based on a four-zone cone instead of filtering for coverage and filtering for accuracy. Moreover, instead of defining the viewpoint orientation towards the façade in [21], four different viewpoint directions were defined and compared with one another.

In this work, in order to generate the input dataset, 2D maps were obtained usingh SLAM and surveying techniques. In the case of using the presented method for any other building, the 2D maps can also be obtained by using Google Maps or a rapid imagery flight with a mini-UAV. For a rough 3D model of the building, the definition of a thickness of the building's footprint was used in this work. In future work, rapid 3D modelling software such as SketchUp or video photogrammetry with the ability to capture image sequences could also be used.

In terms of capturing images, in the simulation experiments in this work, a navigation system was used to capture images in the designed poses. The navigation system was explained in another article [61]. Although the images of the real building were captured by an operator carrying a DSLR camera, this could also be performed with a real UGV or UAV.

Starting from the proposed imaging network methods, several research topics can be defined as a follow-up:

- Develop another imaging network for a UGV equipped with a digital camera mounted on a pan-tilt unit; so far, it was assumed that the robot is equipped with a camera fixed to the body of the robot, and with no rotations allowed.
- Deploy the proposed imaging network on mini-UAV; in this work, the top parts of the building were ignored (not seen) for 3D reconstruction purposes due to onboard camera limitations, whereas the fusion with UAV images would allow a complete survey of a building.
- Use the clustering and selection approach for key frame selection of video sequences for 3D reconstruction purposes.

Author Contributions: Data curation, A.H.; Investigation, A.H. and F.R.; Software, A.H.; Supervision, F.R.; Writing—original draft, A.H.; Writing—review & editing, F.R. All authors have read and agreed to the published version of the manuscript.

Funding: This research received no external funding.

Data Availability Statement: Not applicable.

Acknowledgments: This work was part of a big project in close-range photogrammetry and robotics lab funded by K. N. Toosi University, Iran. The authors are thankful to Masoud Varshosaz and Hamid Ebadi for cooperating in defining the proposal of the project.

Conflicts of Interest: The authors declare no conflict of interest.

References

1. Adán, A.; Quintana, B.; Prieto, S.; Bosché, F. An autonomous robotic platform for automatic extraction of detailed semantic models of buildings. *Autom. Constr.* **2020**, *109*, 102963. [CrossRef]
2. Al-Kheder, S.; Al-Shawabkeh, Y.; Haala, N. Developing a documentation system for desert palaces in Jordan using 3D laser scanning and digital photogrammetry. *J. Archaeol. Sci.* **2009**, *36*, 537–546. [CrossRef]
3. Valero, E.; Bosché, F.; Forster, A. Automatic Segmentation of 3D Point Clouds of Rubble Masonry Walls, and Its Ap-plication To Building Surveying, Repair and Maintenance. *Autom. Constr.* **2018**, *96*, 29–39. [CrossRef]
4. Macdonald, L.; Ahmadabadian, A.H.; Robson, S.; Gibb, I. High Art Revisited: A Photogrammetric Approach. In *Electronic Visualisation and the Arts*; BCS Learning and Development Limited: Swindon, UK, 2014; pp. 192–199.
5. Noh, Z.; Sunar, M.S.; Pan, Z. A Review on Augmented Reality for Virtual Heritage System. In *Transactions on Petri Nets and Other Models of Concurrency XV*; von Koutny, M., Pomello, L., Kordon, F., Eds.; Springer Science and Business Media LLC: Berlin/Heidelberg, Germany, 2009; pp. 50–61.
6. Nocerino, E.; Menna, F.; Remondino, F. Accuracy of typical photogrammetric networks in cultural heritage 3D modeling projects. *ISPRS Int. Arch. Photogramm. Remote. Sens. Spat. Inf. Sci.* **2014**, *45*, 465–472. [CrossRef]
7. Logothetis, S.; Delinasiou, A.; Stylianidis, E. Building Information Modelling for Cultural Heritage: A review. *ISPRS Ann. Photogramm. Remote Sens. Spat. Inf. Sci.* **2015**, *2*, 177–183. [CrossRef]
8. Remondino, F.; Nocerino, E.; Toschi, I.; Menna, F. A Critical Review of Automated Photogrammetric Processing Of Large Datasets. *ISPRS Int. Arch. Photogramm. Remote Sens. Spat. Inf. Sci.* **2017**, *42*, 591–599. [CrossRef]
9. Amini, A.S.; Varshosaz, M.; Saadatseresht, M. Development of a New Stereo-Panorama System Based on off-The-Shelf Stereo Cameras. *Photogramm. Rec.* **2014**, *29*, 206–223. [CrossRef]
10. Watkins, S.; Burry, J.; Mohamed, A.; Marino, M.; Prudden, S.; Fisher, A.; Kloet, N.; Jakobi, T.; Clothier, R. Ten questions concerning the use of drones in urban environments. *Build. Environ.* **2020**, *167*, 106458. [CrossRef]
11. Alsadik, B.; Remondino, F. Flight Planning for LiDAR-Based UAS Mapping Applications. *ISPRS Int. J. Geo Inf.* **2020**, *9*, 378. [CrossRef]
12. Koch, T.; Körner, M.; Fraundorfer, F. Automatic and Semantically-Aware 3D UAV Flight Planning for Image-Based 3D Recon-struction. *Remote Sens.* **2019**, *11*, 1550. [CrossRef]
13. Hosseininaveh, A.; Robson, S.; Boehm, J.; Shortis, M. Stereo-Imaging Network Design for Precise and Dense 3d Re-construction. *Photogramm. Rec.* **2014**, *29*, 317–336.
14. Hosseininaveh, A.; Robson, S.; Boehm, J.; Shortis, M. Image selection in photogrammetric multi-view stereo methods for metric and complete 3D reconstruction. In Proceedings of the SPIE-The International Society for Optical Engineering, Munich, Germany, 23 May 2013; Volume 8791.
15. Hosseininaveh, A.; Serpico, S.; Robson, M.; Hess, J.; Boehm, I.; Pridden, I.; Amati, G. Automatic Image Selection in Photogrammet-ric Multi-View Stereo Methods. In Proceedings of the International Symposium on Virtual Reality, Archaeology and Intelligent Cultural Heritage, Brighton, UK, 19–21 November 2012.
16. Hosseininaveh, A.; Yazdan, R.; Karami, A.; Moradi, M.; Ghorbani, F. Clustering and selecting vantage images in a low-cost system for 3D reconstruction of texture-less objects. *Measurement* **2017**, *99*, 185–191. [CrossRef]
17. Vasquez-Gomez, J.I.; Sucar, L.E.; Murrieta-Cid, R.; Lopez-Damian, E. Volumetric Next-best-view Planning for 3D Object Reconstruction with Positioning Error. *Int. J. Adv. Robot. Syst.* **2014**, *11*, 159. [CrossRef]
18. Alsadik, B.; Gerke, M.; Vosselman, G. Automated Camera Network Design for 3D Modeling of Cultural Heritage Objects. *J. Cult. Herit.* **2013**, *14*, 515–526. [CrossRef]
19. Mahami, H.; Nasirzadeh, F.; Ahmadabadian, A.H.; Nahavandi, S. Automated Progress Controlling and Monitoring Using Daily Site Images and Building Information Modelling. *Buildings* **2019**, *9*, 70. [CrossRef]
20. Mahami, H.; Nasirzadeh, F.; Ahmadabadian, A.H.; Esmaeili, F.; Nahavandi, S. Imaging network design to improve the automated construction progress monitoring process. *Constr. Innov.* **2019**, *19*, 386–404. [CrossRef]
21. Palanirajan, H.K.; Alsadik, B.; Nex, F.; Elberink, S.O. Efficient Flight Planning for Building Façade 3d Reconstruction. *ISPRS Int. Arch. Photogramm. Remote. Sens. Spat. Inf. Sci.* **2019**, *XLII-2/W13*, 495–502. [CrossRef]
22. Kriegel, S.; Bodenmüller, T.; Suppa, M.; Hirzinger, G. A surface-based Next-Best-View approach for automated 3D model completion of unknown objects. In Proceedings of the 2011 IEEE International Conference on Robotics and Automation, Shanghai, China, 9–13 May 2011; pp. 4869–4874.
23. Isler, S.; Sabzevari, R.; Delmerico, J.; Scaramuzza, D. An Information Gain Formulation for Active Volumetric 3D Reconstruction. In Proceedings of the 2016 IEEE International Conference on Robotics and Automation (ICRA), Stockholm, Sweden, 16–21 May 2016; pp. 3477–3484.
24. Monica, R.; Aleotti, J. Surfel-Based Next Best View Planning. *IEEE Robot. Autom. Lett.* **2018**, *3*, 3324–3331. [CrossRef]
25. Furukawa, Y. Clustering Views for Multi-View Stereo (CMVS). 2010. Available online: https://www.di.ens.fr/cmvs/ (accessed on 13 May 2021).
26. Furukawa, Y.; Curless, B.; Seitz, S.M.; Szeliski, R. Towards Internet-scale multi-view stereo. In Proceedings of the 2010 IEEE Computer Society Conference on Computer Vision and Pattern Recognition, San Francisco, CA, USA, 13–18 June 2010; pp. 1434–1441.

27. Agarwal, S.; Furukawa, Y.; Snavely, N.; Simon, I.; Curless, B.; Seitz, S.M.; Szeliski, R. Building rome in a day. *Commun. ACM* **2011**, *10*, 105–112. [CrossRef]
28. Scott, W.R.; Roth, G.; Rivest, J.-F. View Planning for Automated Three-Dimensional Object Reconstruction and Inspection. *ACM Comput. Surv.* **2003**, *35*, 64–96. [CrossRef]
29. Hosseininaveh, A.A.; Sargeant, B.; Erfani, T.; Robson, S.; Shortis, M.; Hess, M.; Boehm, J. Towards Fully Automatic Reliable 3D Ac-quisition: From Designing Imaging Network to a Complete and Accurate Point Cloud. *Robot. Auton. Syst.* **2014**, *62*, 1197–1207. [CrossRef]
30. Vasquez-Gomez, J.I.; Sucar, L.E.; Murrieta-Cid, R. View Planning for 3D Object Reconstruction with a Mobile Manipulator Robot. In Proceedings of the 2014 IEEE/RSJ International Conference on Intelligent Robots and Systems, Chicago, IL, USA, 14–18 September 2014; pp. 4227–4233.
31. Fraser, S. Network Design Considerations for Non-Topographic Photogrammetry. *Photogramm. Eng. Remote Sens.* **1984**, *50*, 115–1126.
32. Tarbox, G.H.; Gottschlich, S.N. Planning for Complete Sensor Coverage in Inspection. *Comput. Vis. Image Underst.* **1995**, *61*, 84–111. [CrossRef]
33. Scott, W.R. Model-based view planning. *Mach. Vis. Appl.* **2007**, *20*, 47–69. [CrossRef]
34. Chen, S.; Li, Y. Automatic Sensor Placement for Model-Based Robot Vision. *IEEE Trans. Syst. Man, Cybern. Part B* **2004**, *34*, 393–408. [CrossRef]
35. Karaszewski, M.; Adamczyk, M.; Sitnik, R. Assessment of next-best-view algorithms performance with various 3D scanners and manipulator. *ISPRS J. Photogramm. Remote Sens.* **2016**, *119*, 320–333. [CrossRef]
36. Zhou, X.; Yi, Z.; Liu, Y.; Huang, K.; Huang, H. Survey on path and view planning for UAVs. *Virtual Real. Intell. Hardw.* **2020**, *2*, 56–69. [CrossRef]
37. Nocerino, E.; Menna, F.; Remondino, F.; Saleri, R. Accuracy and Block Deformation Analysis in Automatic UAV and Terrestrial Photogrammetry–Lesson Learnt. *ISPRS Ann. Photogramm. Remote Sens. Spat. Inf. Sci.* **2013**, *2*, 203–208. [CrossRef]
38. Jing, W.; Polden, J.; Tao, P.Y.; Lin, W.; Shimada, K. View planning for 3D shape reconstruction of buildings with unmanned aerial vehicles. In Proceedings of the 2016 14th International Conference on Control, Automation, Robotics and Vision (ICARCV), Phuket, Thailand, 13–15 November 2016; pp. 1–6.
39. Zheng, X.; Wang, F.; Li, Z. A multi-UAV cooperative route planning methodology for 3D fine-resolution building model reconstruction. *ISPRS J. Photogramm. Remote Sens.* **2018**, *146*, 483–494. [CrossRef]
40. Almadhoun, R.; Abduldayem, A.; Taha, T.; Seneviratne, L.; Zweiri, Y. Guided Next Best View for 3D Reconstruction of Large Complex Structures. *Remote Sens.* **2019**, *11*, 2440. [CrossRef]
41. Mendoza, M.; Vasquez-Gomez, J.I.; Taud, H.; Sucar, L.E.; Reta, C. Supervised learning of the next-best-view for 3d object reconstruction. *Pattern Recognit. Lett.* **2020**, *133*, 224–231. [CrossRef]
42. Huang, R.; Zou, D.; Vaughan, R.; Tan, P. Active Image-Based Modeling with a Toy Drone. In Proceedings of the 2018 IEEE International Conference on Robotics and Automation (ICRA), Brisbane, Australia, 21–25 May 2018; pp. 1–8.
43. Hepp, B.; Nießner, M.; Hilliges, O. Plan3d: Viewpoint and Trajectory Optimization for Aerial Multi-View Stereo Recon-struction. *ACM Trans. Graph.* **2018**, *38*, 1–17. [CrossRef]
44. Krause, A.; Golovin, D. Submodular Function Maximization. *Tractability* **2014**, *3*, 71–104.
45. Roberts, M.; Shah, S.; Dey, D.; Truong, A.; Sinha, S.; Kapoor, A.; Hanrahan, P.; Joshi, N. Submodular Trajectory Optimization for Aerial 3D Scanning. In Proceedings of the 2017 IEEE International Conference on Computer Vision (ICCV), Venice, Italy, 22–29 October 2017; pp. 5334–5343.
46. Smith, N.; Moehrle, N.; Goesele, M.; Heidrich, W. Aerial Path Planning for Urban Scene Reconstruction: A Continuous Optimization Method and Benchmark. *ACM Trans. Graph.* **2019**, *37*, 183. [CrossRef]
47. Arce, S.; Vernon, C.A.; Hammond, J.; Newell, V.; Janson, J.; Franke, K.W.; Hedengren, J.D. Automated 3D Reconstruction Using Op-timized View-Planning Algorithms for Iterative Development of Structure-from-Motion Models. *Remote Sens.* **2020**, *12*, 2169. [CrossRef]
48. Yuhong. Robot Operating System (ROS) Tutorials (Indigo Ed.). 2018. Available online: http://wiki.ros.org/ROS/Tutorials (accessed on 26 June 2018).
49. Gazebo. Gazebo Tutorials. 2014. Available online: http://gazebosim.org/tutorials (accessed on 9 March 2020).
50. Gazebo. Tutorial: ROS Integration Overview. 2014. Available online: http://gazebosim.org/tutorials?tut=ros_overview (accessed on 9 March 2020).
51. Husky. Available online: http://wiki.ros.org/husky_navigation/Tutorials (accessed on 27 June 2018).
52. Grisetti, G.; Stachniss, C.; Burgard, W. Improved Techniques for Grid Mapping with Rao-Blackwellized Particle Filters. *IEEE Trans. Robot.* **2007**, *23*, 34–46. [CrossRef]
53. Wu, C. Visualsfm: A Visual Structure from Motion System. 2011. Available online: http://ccwu.me/vsfm/doc.html (accessed on 13 May 2021).
54. Trimble Inc. Sketchup Pro 2016. 2016. Available online: https://www.sketchup.com/ (accessed on 3 November 2016).
55. Otsu, N. A threshold selection method from gray-level histograms. *IEEE Trans. Syst. Man Cybern.* **1979**, *9*, 62–66. [CrossRef]
56. Hosseininaveh, A. *Photogrammetric Multi-View Stereo and Imaging Network Design*; University College London: London, UK, 2014.

57. Ahmadabadian, H.; Robson, S.; Boehm, J.; Shortis, M.; Wenzel, K.; Fritsch, D. A Comparison of Dense Matching Algorithms for Scaled Surface Reconstruction Using Stereo Camera Rigs. *ISPRS J. Photogramm. Remote Sens.* **2013**, *78*, 157–167. [CrossRef]
58. Mousavi, V.; Khosravi, M.; Ahmadi, M.; Noori, N.; Haghshenas, S.; Hosseininaveh, A.; Varshosaz, M. The performance evaluation of multi-image 3D reconstruction software with different sensors. *Measurement* **2018**, *120*, 1–10. [CrossRef]
59. Agisoft PhotoScan Software. Agisoft Metashape. Available online: https://www.agisoft.com/ (accessed on 30 January 2020).
60. Grisetti, G.; Stachniss, C.; Burgard, W. Improving Grid-based SLAM with Rao-Blackwellized Particle Filters by Adaptive Proposals and Selective Resampling. In Proceedings of the 2005 IEEE International Conference on Robotics and Automation, Barcelona, Spain, 18–22 April 2005.
61. Hosseininaveh, A.; Remondino, F. An Autonomous Navigation System for Image-Based 3D Reconstruction of Façade Using a Ground Vehicle Robot. *Autom. Constr.* **2021**. under revision.

 *remote sensing*

Article

Outdoor Mobile Mapping and AI-Based 3D Object Detection with Low-Cost RGB-D Cameras: The Use Case of On-Street Parking Statistics

Stephan Nebiker *, Jonas Meyer, Stefan Blaser, Manuela Ammann and Severin Rhyner

Institute of Geomatics, FHNW University of Applied Sciences and Arts Northwestern Switzerland, Hofackerstrasse 30, 4132 Muttenz, Switzerland; jonas.meyer@fhnw.ch (J.M.); stefan.blaser@fhnw.ch (S.B.); manuela.ammann@fhnw.ch (M.A.); severin.rhyner@fhnw.ch (S.R.)
* Correspondence: stephan.nebiker@fhnw.ch

Abstract: A successful application of low-cost 3D cameras in combination with artificial intelligence (AI)-based 3D object detection algorithms to outdoor mobile mapping would offer great potential for numerous mapping, asset inventory, and change detection tasks in the context of smart cities. This paper presents a mobile mapping system mounted on an electric tricycle and a procedure for creating on-street parking statistics, which allow government agencies and policy makers to verify and adjust parking policies in different city districts. Our method combines georeferenced red-green-blue-depth (RGB-D) imagery from two low-cost 3D cameras with state-of-the-art 3D object detection algorithms for extracting and mapping parked vehicles. Our investigations demonstrate the suitability of the latest generation of low-cost 3D cameras for real-world outdoor applications with respect to supported ranges, depth measurement accuracy, and robustness under varying lighting conditions. In an evaluation of suitable algorithms for detecting vehicles in the noisy and often incomplete 3D point clouds from RGB-D cameras, the 3D object detection network PointRCNN, which extends region-based convolutional neural networks (R-CNNs) to 3D point clouds, clearly outperformed all other candidates. The results of a mapping mission with 313 parking spaces show that our method is capable of reliably detecting parked cars with a precision of 100% and a recall of 97%. It can be applied to unslotted and slotted parking and different parking types including parallel, perpendicular, and angle parking.

Keywords: parking statistics; vehicle detection; mobile mapping; robot operating system; 3D camera; RGB-D; performance evaluation; convolutional neural networks; smart city

Citation: Nebiker, S.; Meyer, J.; Blaser, S.; Ammann, M.; Rhyner, S. Outdoor Mobile Mapping and AI-Based 3D Object Detection with Low-Cost RGB-D Cameras: The Use Case of On-Street Parking Statistics. *Remote Sens.* **2021**, *13*, 3099. https://doi.org/10.3390/rs13163099

Academic Editors: Ville Lehtola, Andreas Nüchter and François Goulette

Received: 30 June 2021
Accepted: 31 July 2021
Published: 5 August 2021

1. Introduction

We are currently witnessing a transformation of urban mobility from motorized individual transport towards an increasing variety of multimodal mobility offerings, including public transport, dedicated bike paths, and various ridesharing services for cars, bikes, e-scooters, and the like. These offerings, on the one hand, are expected to decrease the need for on-street parking spaces and the undesirable traffic associated with searching for available parking spots, which has been shown to account for an average of 30% of the total traffic in major cities [1]. On the other hand, government agencies are interested in freeing street space—for example, that which is currently occupied by on-street parking—to accommodate new lanes for bikes etc. to support and promote more sustainable traffic modes.

On-street parking statistics support government agencies and policy makers in reviewing and adjusting parking space availability, parking rules and pricing, and parking policies in general. However, creating parking statistics for city districts or even entire cities is a very labor-intensive process. For example, parking statistics for the city of Basel, Switzerland were obtained in 2016 and 2019 using low-cost GoPro videos captured from an e-bike in combination with manual interpretation by human operators [2]. This interpretation is time-consuming and costly and thus limits the number of observation epochs, the time spans, and the repeatability of current on-street parking statistics. However, the human interpretation also has a number of advantages: firstly, it can cope with all types of on-street parking such as parallel parking, angle parking, and perpendicular parking; secondly, it can be utilized with low georeferencing accuracies, since the assignment of cars to individual parking slots or areas is part of the manual interpretation process.

A future solution for creating parking statistics at a city-wide scale should (a) support low-cost platforms and sensors to ensure the scalability of the data acquisition, (b) be capable of handling all relevant parking types, (c) provide an accurate detection of vehicles, and (d) support a fully automated and robust assignment to individual parking slots or unslotted parking areas of the respective GIS database. The feasibility of reliable roadside parking statistics using observations from mobile mapping systems has been successfully demonstrated by Mathur et al. [3], Bock et al. [4], and Fetscher [5]. However, the first two solutions are limited to parallel roadside parking, the second relies on an expensive mobile mapping system with two high-end LiDAR sensors, and the third utilizes 3D street-level imagery that does not provide the required revisit frequencies for time-of-the-day occupancy statistics.

In the time since the above-cited studies, two major developments relevant to this project have occurred: (a) the development of increasingly powerful low-cost (3D) mapping sensors and (b) the development of AI-based object and in particular vehicle detection algorithms. Both developments are largely driven by autonomous driving and mobile robotics. Low-cost 3D sensors or RGB-D cameras with depth sensors either based on active or passive stereo or on solid-state LiDAR could also play an important role in 3D mobile mapping and automated object detection and localization.

Low-cost RGB-D cameras have found widespread use in indoor applications such as gaming and robotics. Outdoor use has been limited due to several reasons, e.g., demanding lighting conditions or the requirement for longer measurement ranges. However, recent progress in 3D sensor development, including new stereo depth estimation technologies, increasing measurement ranges, and advancements in solid-state LiDAR, could soon make outdoor applications a reality.

Our paper investigates the use of low-cost RGB-D cameras in a demanding outdoor mobile mapping use case and features the following main contributions:

- A mobile mapping payload based on the Robot Operating System (ROS) with a low-cost global navigation satellite system/inertial measurement unit (GNSS/IMU) positioning unit and two low-cost Intel RealSense D455 3D cameras
- Integration of the above on an electric tricycle as a versatile mobile mapping research platform (capable of carrying multiple sensor payloads)
- A performance evaluation of different low-cost 3D cameras under real-world outdoor conditions
- A neural network-based approach for 3D vehicle detection and localization from RGB-D imagery yielding position, dimension, and orientation of the detected vehicles
- A GIS-based approach for clustering of vehicle detections and to increase the robustness of the detections
- Test campaigns to evaluate the performance and limitations of our method.

The paper commences with a literature review on the following main aspects: (a) smart parking and on-street parking statistics, (b) low-cost 3D sensors and applications, and (c) vehicle detection algorithms. In Section 3, we provide an overview of the system and workflow, introduce our data capturing system, and discuss the data anonymization and 3D vehicle detection approaches. In Section 4, we introduce the study area and the measurement campaigns used for our study. In Section 5, we discuss experiments and results for the following key issues: georeferencing, low-cost 3D camera performance in indoor and outdoor environments, and AI-based 3D vehicle detection.

2. Related Work

2.1. Smart Parking and On-Street Parking Statistics

Numerous works deal with the optimal utilization of parking space, on the one hand, and with approaches to limit undesired traffic in search of free space, on the other. Polycarpou et al. [6], Paidi et al. [7], and Barriga et al. [8] provide comprehensive overviews of smart parking solutions. Most of these approaches rely on ground-based infrastructure. Therefore, they are typically limited to indoor parking or to off-street parking lots. Several studies are aimed at supporting drivers in the actual search for free parking spaces [9,10]. However, these approaches are limited to the vicinity of the current vehicle position and are not intended for global-scale mapping. In contrast to the large number of studies on smart parking, only a few works address the acquisition of on-street parking statistics for city districts or even entire cities. These works can be distinguished by the sensing technology (ultrasound, LiDAR, 2D and 3D imagery), the detection algorithm and type (e.g., gap or vehicle), and the supported parking types.

In one of the earlier studies called ParkNet, Mathur et al. [3] equipped probe vehicles with GPS and side-looking ultrasonic range finders mounted to the passenger door to determine parking spot occupancy. In ParkNet, the geotagged range profile data are sent to a central server, which creates a real-time map of the parking availability. The authors further propose an environmental finger printing approach to address the challenges of GPS positioning uncertainty with position errors in the range of 5–10 m. In their paper, the authors claim a 95% accuracy in terms of parking spot counts and a 90% accuracy of the parking occupancy maps. The main limitations of their approach are that the ultrasound range finders are unable to distinguish between actual cars and other objects with a similar sensory response (e.g., cyclists and flowerpots) and that the approach is limited to parallel curbside parking.

In a more recent study, Bock et al. [4] describe a procedure for extracting on-street parking statistics from 3D point clouds that have been recorded with two 2D LiDAR sensors mounted on a mobile mapping vehicle. Parked vehicles are detected in a two-step approach: an object segmentation followed by an object classification using a random forest classifier. With their processing chain, the authors present results with a precision of 98.4% and a recall of 95.8% and demonstrate its suitability for time-of-the-day parking statistics. The solution supports parallel and perpendicular parking, but its practical use is limited due to the expensive high-end dual LiDAR mobile mapping system.

More recently, there have been several studies investigating image-based methods for detecting parked vehicles or vacant parking spaces. Grassi et al. [11], for example, describe ParkMaster, an in-vehicle, edge-based video analytics service for detecting open parking spaces in urban environments. The system uses video from dash-mounted smartphones to estimate each parked car's approximate location—assuming parallel parking only. In their experiments in three different cities, they achieved an average accuracy of parking estimates close to 90%. In the latest study, Fetscher [5] uses 3D street-level imagery [12] to derive on-street parking statistics. The author first employs Facebook's Detectron2 [13] object detection algorithms to detect and segment cars in 2D imagery. These segments are subsequently used to mask the depth maps of 3D street-level imagery and to derive 3D point clouds of the candidate objects. The author then presents two methods for localizing the vehicle positions in the point clouds: a corner detection approach and a clustering approach, which yield detection accuracies of 97% and 98.3%, respectively, and support parallel, angle, and perpendicular parking types.

2.2. Low-Cost 3D Sensors and Applications

Gaming, mobile robotics, and autonomous driving are the main driving forces in the development of low-cost 3D sensors. 3D cameras integrating depth sensors with imaging sensors have the potential advantages of improved scene understanding through combinations or fusion of image-based and point cloud-based object recognition and of direct 3D object localization with respect to the camera pose. 3D or RGB-D cameras provide two types of co-registered data covering a similar field of view: imagery (RGB) and range or depth (D) data. Ulrich et al. [14] provide a good overview of the different RGB-D camera and depth estimation technologies, including passive stereoscopy, structed light, time-of-flight (ToF), and active stereoscopy [14]. Low-cost depth and RGB-D cameras have been widely researched for various close-range applications in indoor environments. These include applications such as gesture recognition, sign language recognition [15], body pose estimation [16], and 3D scene reconstruction [17].

2.2.1. RGB-D Cameras in Mapping Applications

There are numerous works investigating the use of low-cost action and smartphone cameras for mobile mapping purposes [18–20] and an increasing number investigating the use of RGB-D cameras for indoor mapping [21,22]. By contrast, there are only a few published studies on outdoor applications of RGB-D cameras [23]. Brahmanage et al. [23], for example, investigate simultaneous localization and mapping (SLAM) in outdoor environments using an Intel RealSense D435 RGB-D camera. They discuss the challenges of noisy or missing depth information in outdoor scenes due to glare spots and high illumination regions. Iwaszczuk et al. [24] discuss the inclusion of an RGB-D sensor on their mobile mapping backpack and stress the difficulties with measurements under daylight conditions.

2.2.2. Performance Evaluation of RGB-D Cameras

If RGB-D cameras are to be used for measuring purposes and specifically for mapping applications, knowing their accuracy and precision is a key issue. There are a number of studies evaluating the performance of RGB-D sensors in indoor environments [14,25,26] and a study by Vit and Shani in a close-range outdoor scenario [27]. Halmetschlager-Funek et al. [25] evaluated 10 depth cameras for bias, precision, lateral noise, different light conditions and materials, and multiple sensor setups in an indoor environment with ranges up to 2 m. Ulrich et al. 2020 [14] tested different 3D camera technologies in their research on face analysis and ranked different technologies with respect to their application to recognition, identification, and other use cases. In their study, active and passive stereoscopy emerged as the best technology. Lourenço and Araujo [26] performed an experimental analysis and comparison of the depth estimation by the RGB-D cameras SR305, D415, and L515 from the Intel RealSense product family. These three cameras use three different depth sensing technologies: structured light projection, active stereoscopy, and ToF. The authors tested the performance, accuracy, and precision of the cameras in an indoor environment with controlled and stable lighting. In their experimental setup, the L515 using solid-state LiDAR ToF technology provided more accurate and precise results than the other two. Finally, Vit and Shani [27] investigated four RGB-D sensors, namely, Astra S, Microsoft Kinect II, Intel RealSense SR300, and Intel RealSense D435, for their agronomical use case of field phenotyping. In their close-range outdoor experiments with measuring ranges between 0.2 and 1.5 m, Intel's RealSense D435 produced the best results in terms of accuracy and exposure control.

2.3. Vehicle Detection

Vehicle detection is a subtask of object detection, which focuses on detecting instances of semantic objects. The object detection task can be defined as the fusion of object recognition and localization [28]. For object detection in 2D space within an image plane, different types of traditional machine learning (ML) algorithms can be applied. Such algorithms are usually based on various kinds of feature descriptors combined with appropriate classifiers such as support vector machine (SVM) or random forests. In recent years, traditional approaches have been replaced by neural networks with increasingly deep network architectures. This allows the use of high dimensional input data and automatic recognition of structures and representations needed for detection tasks [29].

The localization of detected objects within the image plane is insufficient for many tasks such as path planning or collision avoidance in the field of autonomous driving.

To estimate the exact position, size, and orientation of an object in a geodetic reference frame (subsequently referred to as world coordinates), the third dimension is required [30]. Arnold et al. [30] divide 3D object detection (3DOD) into three main categories based on different sensor modality.

2.3.1. Monocular

3D object detection methods using exclusively monocular RGB images are usually based on a two-step approach since no depth information is available. First, 2D candidates are detected within the image, before in the second step 3D bounding boxes representing the object are computed based on the candidates. Either neural networks, geometric constraints, or 3D model matching are used to predict the 3D bounding boxes [29].

2.3.2. Point Cloud

Point clouds can be obtained by different sensors such as stereo cameras, LiDAR, or solid-state LiDAR. The 3D object detection methods based on point clouds can be subdivided into projection, volumetric representations, and point-nets methods [30]. To use the well-researched and tested network architectures from the field of 2D object detection, some projection-based methods convert the raw point clouds into images. Other projection-based approaches transform the point clouds into depth maps or project them onto the ground plane, leveraging bird's eye-projection techniques. The reconstruction of the 3D bounding box can be performed by position and dimension regression [30]. Volumetric approaches transform the point cloud in a pre-processing step into a 3D grid or a voxel structure. The prediction of the 3DOD is done by fully convolutional networks (FCNs) [30]. Methods leveraging PointNet architectures such as PointRCNN [31] or PV-RCNN [32] do not require a pre-processing step such as projection or voxelization. They can process raw point clouds directly and return the 3D bounding boxes of objects of interest [30]. The leaderboard of the KITTI 3D object detection benchmark [33] shows that most of the currently top ranked methods for 3D object detection [34–36] use point clouds as input data.

2.3.3. Fusion

Fusion-based approaches combine both RGB images and point clouds. Since images provide texture information and point clouds supply depth information, fusion-based approaches use both information to improve the performance and reliability of 3DOD. These methods usually rely on region proposal networks (RPNs) from RGB images such as Frustum PointNets [37] or Frustum ConvNet [38].

3. Materials and Methods

3.1. Overview of System and Workflow

In the following sections, we introduce our mobile mapping platform and payload incorporating two low-cost 3D cameras and a low- to mid-range GNSS/INS system. This is followed by the description of the workflow for deriving on-street parking statistics from georeferenced 3D imagery. The main components of this workflow are illustrated in Figure 1.

3.2. Data Capturing System

For our investigations, we developed a prototypic RGB-D image-based mobile mapping (MM) sensor payload using low-cost components. This enables easy industrialization and scaling of the system in the future. At the present stage of development, we used an electric tricycle as an MM platform. It includes two racks in the front and rear (Figure 2a) where our sensor payloads can be easily attached.

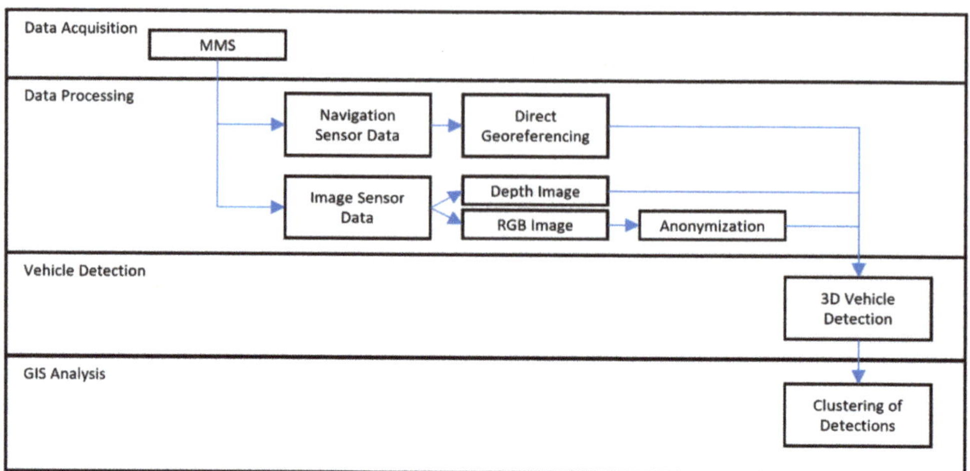

Figure 1. Overview of the workflow for acquiring, processing, and analyzing RGB-D imagery to detect parked vehicles and generate on-street parking statistics.

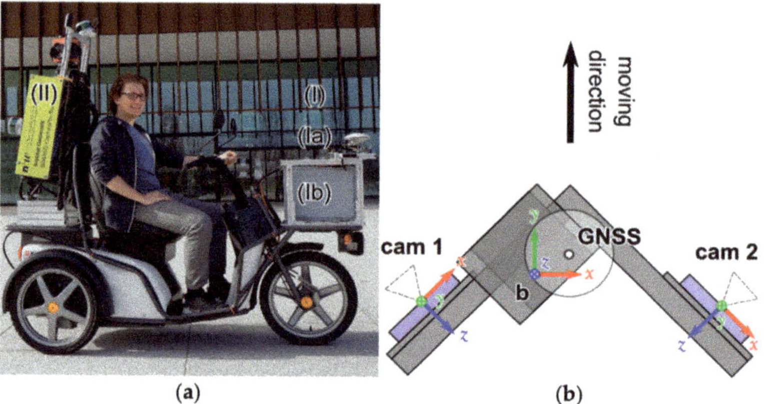

Figure 2. (a) Electrical tricycle mobile mapping platform with the low-cost sensor setup, which is mounted on the front luggage carrier (I) with (Ia) multi-sensor frame and (Ib) computer for data registration. Our backpack MMS that is fixed on the back luggage carrier (II) was used as a reference system for our investigations. (b) Outline of our low-cost multi-sensor frame showing sensor coordinate frames as the body frame b, both RealSense coordinate frames cam 1 and 2, and the GNSS L1 phase center GNSS. View frusta are indicated with dashed lines.

3.2.1. System Components

Our developed MM payload includes both navigation and mapping sensors as well as a computer for data pre-processing and data storage. The GNSS and IMU-based navigation unit SwiftNav Piksi Multi consists of a multi-band and multi-constellation GNSS RTK receiver board and a geodetic GNSS antenna. The GNSS receiver board also includes the consumer-grade IMU Bosch BMI160. Furthermore, the navigation unit provides numerous interfaces, e.g., for external precise hardware-based timestamp creation [39].

For mapping, we used the RGB-D camera Intel RealSense D455. The manufacturer specifies an active stereo depth resolution up to 1280×720 pixels and a depth diagonal field of view over $90°$ (see Table 1). Both depth and RGB cameras use a global shutter. The specified depth sensor range is from 0.4 m to over 10 m, but the range can vary depending on the lighting conditions [40]. In addition, the RGB-D camera supports precise hardware-based triggering using electric pulses, which is crucial for kinematic applications. However, external hardware-based triggering is currently only provided for depth images.

Table 1. Sensor specifications of the Intel RealSense D455 [40].

	RGB Sensors	**Depth Sensors**
Shutter Type	Global Shutter	Global Shutter
Image Sensor	OV9782	OV9282
Max Framerate	90 fps (with max resolution 30 fps)	90 fps (with max resolution 30 fps)
Resolution	1 MP (1280 × 800 px/3 μm)	1 MP (1280 × 720 px/3 μm)
Field of View	H:87° ± 3/V:58° ± 1/D:95° ± 3	H:87° ± 3/V:58° ± 1/D:95° ± 3

Finally, our MM payload includes the embedded single-board computer nVidia Jetson TX2, which includes a powerful nVidia Pascal-family GPU with 256 cuda cores that enables AI-based edge computing with low energy consumption [41].

3.2.2. System Configuration

Our MM sensor payload consists of a robust aluminum frame to which we stably attached our sensor components. We mounted our sensor frame on the front rack of the electrical tricycle. Both navigation and mapping sensors sit on top of the sensor frame (Figure 2a(Ia)), while the computer and the electronics for power supply and sensor synchronization are in the gray box below the sensors (Figure 2a(Ib)).

The aluminum profiles for the sensor configuration on top of the sensor frame are each angled 45° to the corresponding side. We fixed two RGB-D Intel RealSense cameras on it, so that the first camera, cam 1, points to the front left and the second camera, cam 2, points to the front right (Figure 2b). The second camera thus will detect parking spaces and vehicles that are often located on the right-hand side of the road in urban areas. In the case of one-way streets, the first camera will also detect parking spaces on the left side of the road. Furthermore, the oblique mounting ensures common image features in successive image epochs in moving direction.

We mounted the GNSS and IMU-based navigation unit SwiftNav Piksi Multi on top of the sensor configuration, so that the GNSS antenna is as far up and as far forward as possible and does not obscure the field of view of the cameras or the driver's field of view. At the same time, the GNSS signal should be obscured as little as possible by the driver or by other objects.

In addition, we mounted our self-developed BIMAGE backpack mobile mapping system (MMS) [42,43] on the rear rack of the electrical tricycle. The BIMAGE backpack is a portable high-performance mobile mapping system, which we used in this project as a reference system for our performance investigations.

3.2.3. System Software

For this stage of development, we designed the system software for data capturing as well as for raw data registration. However, our software has a modular and flexible design and is based on the graph-based robotic framework Robot Operating System (ROS) [44]. This forms an ideal basis for further development steps towards edge computing and on-board AI detection. Furthermore, the ROS framework is easily adoptable and expandable in terms of how to integrate new sensors.

We use the ROS Wrapper for Intel RealSense Devices [45], which is provided and maintained by Intel for the RealSense camera control. For hardware-based device triggering, we use our self-developed ROS trigger node.

3.2.4. Data Acquisition and Georeferencing

Generally, an MM campaign using GNSS/INS-based direct georeferencing starts and ends with the system initialization. The initialization process determines initial position, speed, and rotation values for the Kalman filter used for the state estimation, whereby the azimuth component between the local body frame and the global navigation coordinate frame is the most critical component. The initialization requires a dynamic phase, which from experience requires about three minutes with good GNSS coverage. The initialization procedure at the beginning and at the end of a campaign enables combined forward as well as backward trajectory post-processing, thus ensuring an optimal trajectory estimation.

During the campaign, the computer triggers both RealSense RGB-D cameras as well as the navigation unit with 5 fps. The navigation unit generates precise timestamps of the camera triggering and continuously registers GNSS and IMU raw data on a local SD card. At the same time, the computer receives both RGB and depth image raw data, which are stored in a so-called ROS bag file on an external solid-state drive (SSD).

A first post-processing workflow converts the image raw data into RGB-D images and performs a combined tightly coupled forward and backward GNSS and IMU sensor data fusion and trajectory processing in Waypoint Inertial Explorer [46]. By interpolating the precise trigger timestamps, each RGB-D image finally receives an associated directly georeferenced pose.

3.3. Data Pre-Processing

3.3.1. Data Anonymization

Anonymization of the image data was a critical issue for the City of Basel as a project partner. Therefore, the anonymization workflow was developed in close cooperation with the state data protection officer and verified by the same at the end. The open-source software "Anonymizer" [47] is used to anonymize personal image data in the street environment. The anonymization process of images is divided into two steps. First, faces and vehicle license plates are detected in the input images by a neural network pre-trained on a non-open dataset [47]. In the second step, the detected objects are blurred by a Gaussian filter and an anonymized version of the input image is saved. In "Anonymizer", both the probability scores of the detected objects and the intensity of the blurring can be chosen. The parameters used were selected empirically, whereby a good compromise had to be found between reliable anonymization and as few unnecessarily obscured areas in the images as possible.

3.3.2. Conversion of Depth Maps to Point Clouds

All vehicle detection algorithms tested and used in this project require point clouds as input data. As point clouds are not directly stored in our system configuration, an additional pre-processing step is necessary. In this step, the point clouds are computed using the geometric relationships between camera geometry and depth maps. The resulting point clouds are cropped to a range of 0.4–8 m, because noise increases strongly beyond this range (see experiments and results in Section 4.3) and are saved as binary files. Figure 3 shows three typical urban scenes with parked cars in the top row, the corresponding depth maps in the middle row, and perspective views of the resulting point clouds in the bottom row. The depth maps in the middle row show some significant data gaps, especially in areas with very high reflectance, which are typical for shiny car bodies (in particular, Figure 3e). The bottom row illustrates the relatively high noise level in the point clouds, which results from the low-cost 3D cameras.

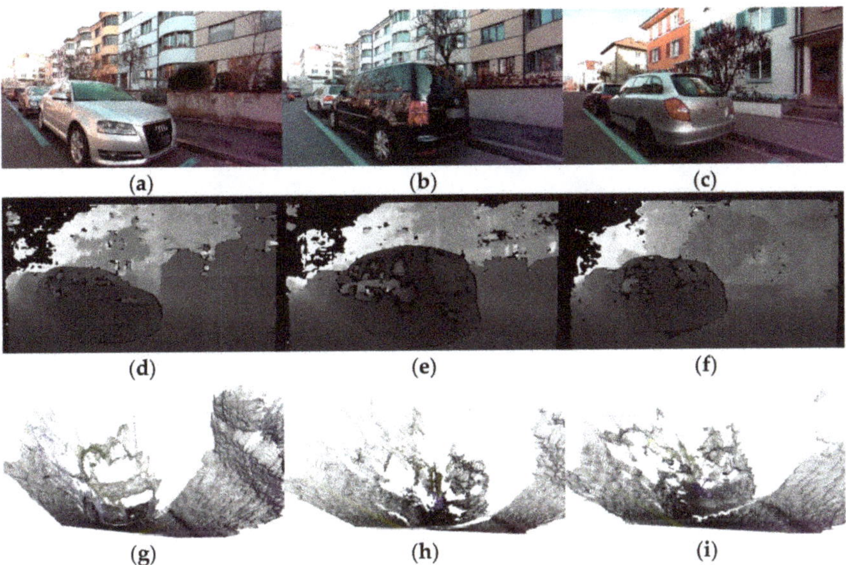

Figure 3. Three different parked cars: (**a**–**c**) show the RGB images, (**d**–**f**) depict the associated depth maps, and (**g**–**i**) represent the processed point clouds. The point clouds were colored in this visualization for better understanding of the scene. However, the data pre-processing does not include coloring of the point clouds.

3.4. 3D Vehicle Detection and Mapping

3.4.1. AI-Based 3D Detection

The detection of other road users and vehicles in traffic is an important aspect in the field of autonomous driving. To solve the problem of accurate and reliable 3D vehicle detections, there are a variety of approaches. A good overview of the available approaches and their performance is provided by the leaderboard of the KITTI 3D object detection benchmark [33]. Because the best approaches on the leaderboard are all based on point clouds, only those were considered in this project. At the time of the investigations, the best approach was PV-RCNN [32]. In the freely available open-source project OpenPCDet [48], besides the official implementation of PV-RCNN, other point cloud-based approaches for 3D object detection are provided. These are:

- PointPillars [49]
- SECOND [50]
- PointRCNN [31]
- Part-A2 net [51]
- PV-RCNN [32].

All the approaches listed above were trained using the point clouds from the KITTI 3D object detection benchmark [52]. It should be noted that only point clouds of the classes car, pedestrian, and cyclist were used for the training. The KITTI point clouds were acquired with a high-end Velodyne HDL-64E LiDAR scanner mounted on a car about 1.8 m above the ground [52]. Figure 4 shows a comparison between the point clouds provided in the KITTI benchmark and our own point clouds acquired with the RealSense D455. The KITTI point clouds are sparse, and the edges of the vehicles are clearly visible (Figure 4a). By contrast, our point clouds are much denser, but edges cannot really be detected (Figure 4b). In addition, our point clouds also have significantly higher noise. This can be seen very well on surfaces that should be even, such as the road surface or the sides of the vehicle.

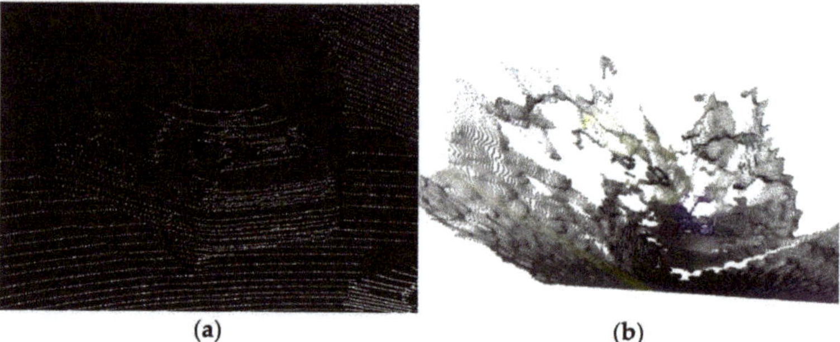

(a)   (b)

Figure 4. Comparison of point clouds. (**a**) Point cloud acquired with Velodyne HDL-64E LiDAR scanner provided in KITTI benchmark [52]. (**b**) Point cloud obtained from RealSense D455 (own data).

Due to the higher noise in our own point clouds and the lower mounting of the sensor by about half a meter compared with the KITTI benchmark, the question arose as to whether the models pre-trained with the KITTI datasets could be successfully applied to our data. For this purpose, all approaches provided in OpenPCDet were evaluated in a small test area with 32 parked cars and 2 parked vans (Figure 5). PointRCNN [31] detected 32 out of 34 vehicles correctly and did not provide any false detections. In contrast, all other approaches could only detect very few vehicles correctly (13 or less out of 34). Based on these results, we decided to use the method PointRCNN for 3D vehicle detection, which consists of two stages. In stage 1, 3D proposals are generated directly from the raw point cloud in a bottom-up manner via segmenting the point cloud of the whole scene into foreground points and background. In the second stage, the proposals are refined in the canonical coordinates to obtain the final detection results [31].

(a)   (b)

Figure 5. Test site to evaluate the approaches provided in OpenPCDet on our own data. The test road leads through a residential neighborhood in the city of Basel with parking spaces on the left and right side. (**a**) Aerial image of the test road; (**b**) map with the parking spaces (blue) (data source: Geodaten Kanton Basel-Stadt).

Parking space management in geographic information systems (GIS) relies on two-dimensional data in a geodetic reference frame, further referred to as world coordinates. Therefore, the detected 3D bounding boxes must be transformed from local sensor coordinates to world coordinates, converted into 2D bounding boxes, and exported in an appropriate format. For this purpose, the OpenPCDet software has been extended to include these aspects. The transformation to world coordinates is performed by applying the sensor poses from direct georeferencing as transformation. The conversion from 3D to 2D bounding boxes is done by projecting the base plane of the 3D bounding box onto the xy-plane by omitting the z coordinates. The resulting 2D bounding boxes as well as the associated class labels and the probability scores of the detected vehicles are exported in GeoJSON format [53] for further processing. The availability of co-registered RGB imagery and depth data resulting from the use of RGB-D sensors has several advantages: it allows the visual verification of the detection results in the imagery, e.g., as part of the feedback loop of a future production system. The co-registered imagery could also be used to facilitate future retraining of existing vehicle classes or for the training of new, currently unsupported vehicle classes.

3.4.2. GIS Analysis

Following the object detection, we currently employ a GIS analysis in QGIS [54] to obtain the number of parked vehicles from the detection results. Since the vehicle detection algorithm returns all detection results, the highly redundant 2D bounding boxes must first be filtered based on their probability scores. For the parking statistics, we only use detected vehicles with a score greater than 0.9 (90% probability). For each vehicle, several 2D-bounding boxes remain after filtering (Figure 6); hence, we perform clustering.

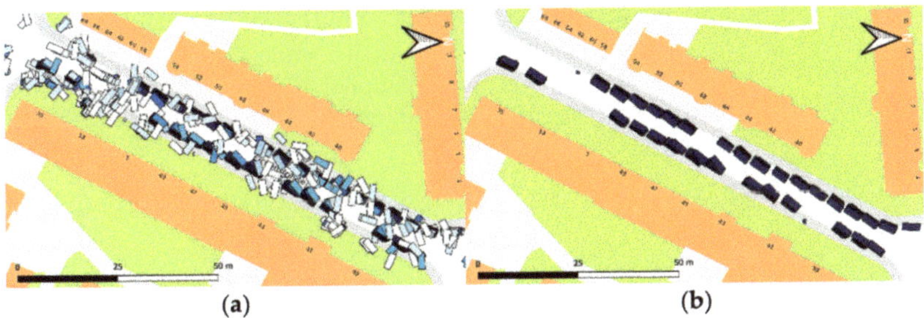

Figure 6. Result of the vehicle detection algorithm (**a**) and filtered result according to the probability score (**b**) (data source for background map: Geodaten Kanton Basel-Stadt).

First, each of the filtered 2D bounding boxes is assigned a unique ID and its centroid is calculated. Then the centroids are clustered using the density-based DBSCAN algorithm [55]. Each centroid point is checked as to whether it has a minimum number of neighboring points (minPts) within a radius (ε). Based on this classification, the clusters are formed. The parameters ε and minPts were chosen as 0.5 m and 3, respectively. The bounding boxes are linked to the DBSCAN classes, a minimal bounding box is determined for each cluster class, and the centroid of the new bounding box is calculated. To obtain the number of parked vehicles, the points (centroids of minimal bounding boxes of clusters) within the parking polygons are counted (Figure 7). In combination with the known number of parking spaces per polygon, it is straightforward to create statistics on the occupancy of parking spaces.

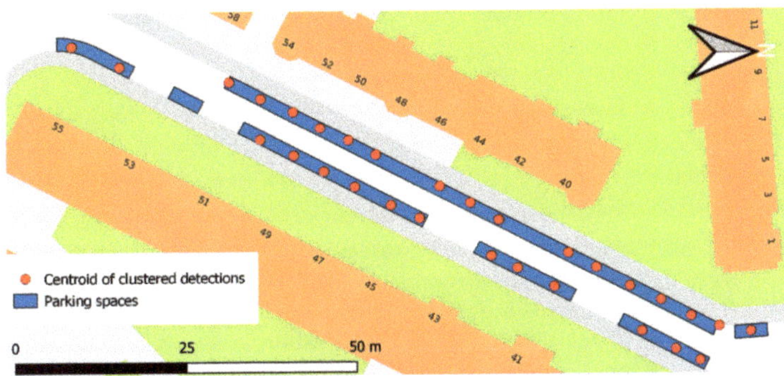

Figure 7. Result of the GIS analysis. Centroids of the clustered and filtered vehicle detections (red dots) plotted with the parking spaces (blue areas) (data source for background map: Geodaten Kanton Basel-Stadt).

Since we count the centroids of clustered vehicle detections within the parking spaces, moving vehicles on the streets do not influence the result. Hence, we do not need to perform a removal of moving vehicles, such as in Bock et al. [4]. Furthermore, we assume that while the parked cars are recorded, they are static. Among other things, this assumption can also be made because the passing MM vehicle prevents other vehicles from leaving or entering parking spaces during capturing.

4. Experiments and Results

Our proposed low-cost MMS and object detection approach consists of three main components determining the capabilities and performance of the overall system: the low-cost navigation unit, the low-cost 3D cameras, and the AI-based object detection algorithms. This section contains a short introduction to the study areas and the test data (Figure 8), followed by a description of the three main experiments aimed at evaluating the main components and the overall system performance:

- Georeferencing investigations in demanding urban environments
- 3D camera performance evaluation in indoor and outdoor settings
- AI-based 3D vehicle detection experiments in a representative urban environment with different parking types.

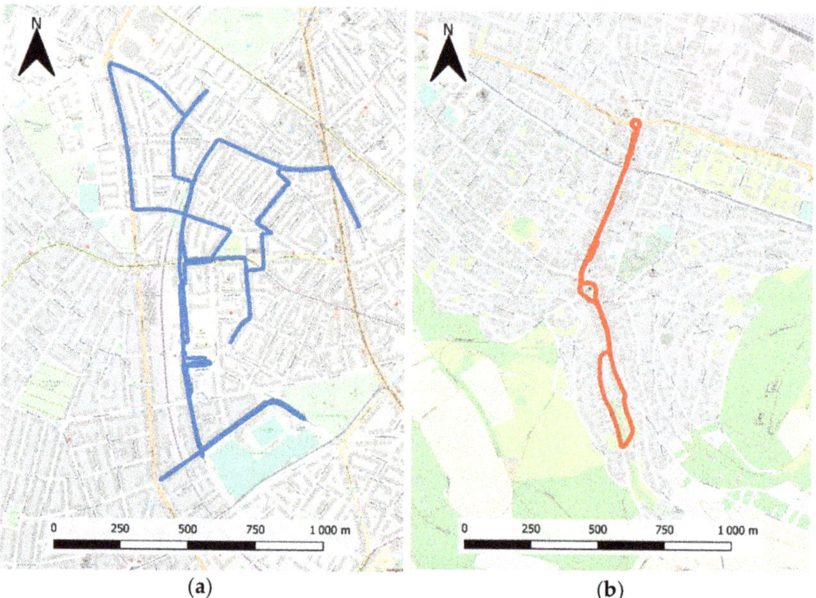

Figure 8. Maps showing the mobile measurement campaigns conducted in the area of Basel, Switzerland: (**a**) Basel city, 11 December 2020, (**b**) Muttenz, 3 April 2021. Background map: © OpenStreetMap contributors.

4.1. Study Areas and Data

Our proposed system was evaluated using datasets from two mobile mapping campaigns in demanding urban and suburban environments (see Figure 8 and Table 2). The test areas are representative of typical European cities and suburbs in terms of building heights, streets widths, vegetation/trees, and a variety of on-street parking types.

Table 2. Overview of two mobile mapping campaigns with their main purpose, characteristics, and the resulting datasets.

Test Campaign Characteristics	Basel City	Muttenz
Main purpose	Evaluation of 3D vehicle detection for different on-street parking types	Georeferencing investigations with parallel operation of the low-cost MMS and a high-end reference MMS
Payload/Sensors	Front: Low-cost MM payload with single RealSense D455	Front: Low-cost MM payload with dual RealSense D455 Back: High-end BIMAGE Backpack as a reference system (for configuration, see Figure 2a)
Characteristics of test area	Residential district, west of the city center of Basel; roads often lined by multi-story buildings and trees; route mostly flat, selected to encompass a large variety of parking slot types	Suburban town located southeast of Basel; route passing shopping district and historical center; route partially lined by large trees; partially steep roads and large elevation differences
Acquisition date	11 December 2020	3 April 2021
Trajectory length	9.7 km	3.4 km
Average acquisition speed	9 km/h	10 km/h
GNSS epochs	~3500	~3500
Image capturing frame rates	5 fps (D455)	5 fps (D455)/1 fps (BIMAGE)
Number of RGB-D images	22,382	5443

Investigated On-Street Parking Types

Four different types of parking slots that can be found along the campaign route "Basel city" were included, as depicted in Figure 8a: parallel, perpendicular, angle, and 2 × 2 block parking. This allowed us to develop and test a workflow capable of handling each type. With each parking type also comes a unique set of challenges to accurately detect parked vehicles. In the case of several densely perpendicularly parked vehicles (Figure 9b), only a small section of the vehicle facing the road is seen by the cameras. Similarly, with angled parking spaces such as that shown in Figure 9c, a large car (e.g., minivan or SUV) can obstruct the view of a considerably smaller vehicle parked in the following space, leading to fewer or no detections. Lastly, one of the residential roads contains a rather unique parking slot type: an arrangement of 2 × 2 parking spaces on one side of the road (Figure 9d). The main challenge with this type of on-street parking is to be able to correctly identify a vehicle parked in the second, more distant row when all four spaces are occupied.

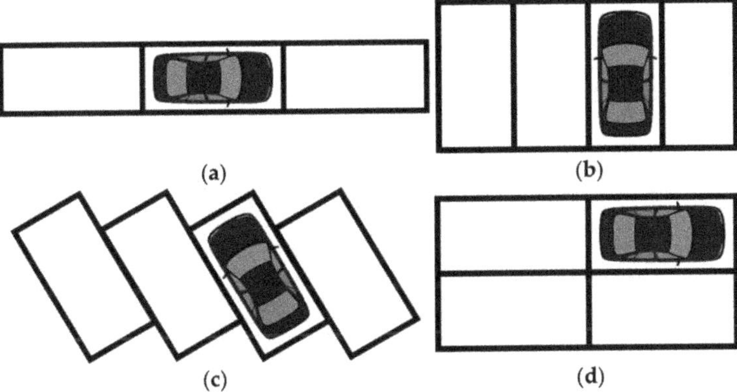

Figure 9. Types of parking slots encountered in the city of Basel. (**a**) (Slotted) parallel parking; (**b**) perpendicular parking; (**c**) angle parking; (**d**) 2 × 2 parking slot clusters.

4.2. Georeferencing Investigations

As outlined in Section 3.2.2., a high-end backpack MMS (Figure 2a, II) was installed on the tricycle, serving as a kinematic reference for the low-cost MM payload. In this case, the investigated low-cost system was mounted on the front payload rack of the vehicle and the high-end reference system at the back, resulting in a fixed lever-arm between the two positioning systems. Both systems have independent GNSS and IMU-based navigation sensors. While the sensors of the low-cost MM sensor configuration fall into the consumer grade category, the backpack MMS navigation sensors have tactical grade performance. Since both navigation systems are precisely synchronized using the GNSS time, poses of both systems are accurately time-stamped. Thus, poses from both processed trajectories can be compared, e.g., to estimate the lever arm between both navigation coordinate frames.

$$
{}^{b_r}T_{b_{lc}} = \left({}^{w}T_{b_r}\right)^{-1} * {}^{w}T_{b_{lc}}
\tag{1}
$$

Equation (1) describes the lever arm estimation using transformation matrices for homogeneous coordinates. Here, we consider the poses of the navigation systems as transformations from the respective body frame b to the global navigation coordinate frame w. By concatenating the inverse navigation sensor pose of the reference system (backpack MMS) ${}^{w}T_{b_r}$ with the navigation sensor pose of the low-cost MMS ${}^{w}T_{b_{lc}}$, we obtain the transformation ${}^{b_r}T_{b_{lc}}$ from the body frame of the low-cost navigation system b_{lc} to the body frame of the reference (backpack MMS) navigation system b_r, from which we extract the lever arm.

To investigate the magnitudes of the position deviations along-track and cross-track, we subtracted the mean lever arm from all estimated lever arm estimations for each acquisition period. Since the georeferencing performance of the high-end BIMAGE backpack—serving as a reference—has been well researched [42,43], we can safely assume that position deviations can be primarily attributed to the investigated low-cost system.

For further 3D vehicle detection and subsequent GIS analysis, an accurate 2D position is crucial. By contrast, the accuracy of the absolute height is negligible due to the back projection onto the 2D xy-plane for the GIS analysis. Therefore, we considered the 2D position and the height separately in the following investigations.

We investigated two different acquisition periods from our dataset from the test site in Muttenz, and we used one pose per second for our examinations. The first "static" period was shortly before moving into the town center, when the system had already initialized but was motionless for 270 s and had good GNSS coverage. By contrast, the second "dynamic" period was from the data acquisition in the village center during 1240 s with various qualities of GNSS reception. For both datasets, we evaluated (a) the intrinsic accuracy provided by the trajectory processing software and (b) the absolute position accuracy, using the high-end system with its estimated pose and the mean lever arm as a reference.

The investigations yielded the following results: In the "static" dataset, the mean intrinsic standard deviation of the estimated 2D positions was 0.6 cm for the reference system and 5.1 cm for the low-cost MM sensor components, respectively. In the "dynamic" dataset, the mean intrinsic standard deviation of the 2D positions was 1.1 cm for the reference system and 10.5 cm for the low-cost MM sensor components, respectively. Regarding the intrinsic standard deviation of the altitudes, they amount to 0.7 cm for the reference system and 7.0 cm for the low-cost MM sensor configuration for the "static" dataset. For the "dynamic" dataset, they amount to 1.1 cm for the reference system and 18.0 cm for the low-cost MM sensor configuration.

The investigations of the absolute coordinate deviations between both trajectories resulted in a mean 2D deviation of 8.3 cm in the "static" dataset and 36.4 cm in the "dynamic dataset". The mean height deviation amounted to 9.8 cm in the "static" dataset and 90.9 cm in the "dynamic" dataset. The kinematic 2D deviations in the order of 1 m are somewhat larger than the direct georeferencing accuracies of high-end GNSS/IMU systems in challenging urban areas [56].

However, while the coordinate deviations in the "static" dataset remain in a range of approx. 10 cm during the entire period (Figure 10a), they vary strongly in the "dynamic" dataset with 2D position deviation peaks in excess of 100 cm (Figure 10b).

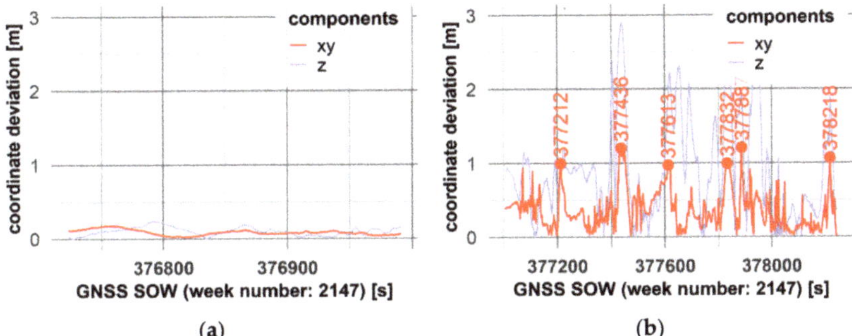

Figure 10. Coordinate deviations between the low-cost data capturing system and the BIMAGE backpack reference systems as a function of time in GNSS Seconds of Week (SOW). Two time periods with different conditions were investigated: (**a**) static data recording with good GNSS reception; (**b**) dynamic data recording with different GNSS coverage.

Furthermore, we investigated the environmental conditions of some peaks in Figure 10b to identify possible causes. At second 377,212, only five GNSS satellites were observed, while a building in the southerly direction covers the GNSS reception at second 377,436. At second 377,613, the GNSS reception was affected by dense trees in the easterly direction and at second 377,832 by a church in the south. At second 377,877, a complete loss of satellites occurred, and finally at second 378,218, a tall building in the south interfered with satellite reception.

4.3. 3D Camera Performance Evaluation

In earlier investigations by Frey [57], different RealSense 3D camera types, including the solid-state LiDAR model L515 and the active stereo-based models D435 and D455, were compared in indoor and outdoor environments. All 3D cameras performed reasonably well in indoor environments. However, in outdoor environments and under typical daylight conditions, the maximum range of the L515 LiDAR sensor was limited to 1–2 m and that of the D435 to 2–3 m. The latest model D455, however, showed a clearly superior performance in outdoor environments with maximum ranges of more than 5–7 m [57]. Based on these earlier trials, the D455 was chosen for this project and subsequently evaluated in more detail.

4.3.1. Distance-Dependent Depth Estimation

In a first series of experiments, we evaluated bias and precision of the depth measurements with the Intel RealSense D455 in a controlled indoor environment. The evaluation was conducted in line with the methodology and metrics proposed by Helmetschlager-Funek et al. [25]. We examined three units (cam1, cam2, and cam3). The experimental setup consisted of a fixed 3D camera and a reference plane oriented orthogonally to the main viewing direction (Figure 11a,b). This plane was re-positioned in one-meter intervals at measuring distances from 1 to 12 m, leading to a total of 12 evaluated distances. At each interval, 100 frames were captured to obtain a sufficiently large number of measuring samples. Ground-truth measurements were obtained with a measuring tape with an accuracy of approx. 1 cm. The analyses were performed on a 20 × 20 pixel window, and we calculated the average difference to the reference distance (bias) and the precision (standard deviation) over the 100 frames per position.

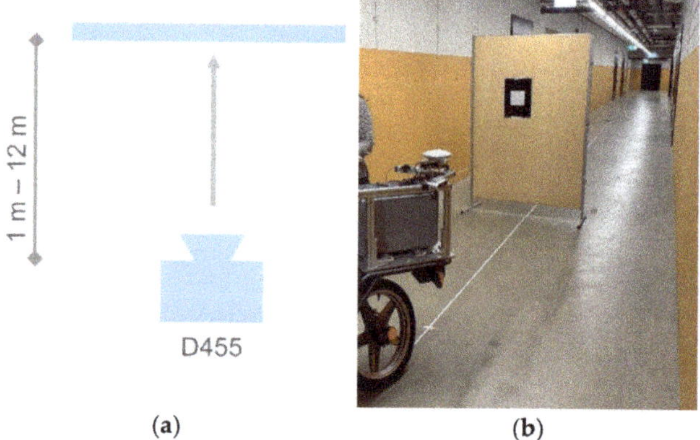

Figure 11. Indoor experimental setup for the performance evaluation of depth measurements: (**a**) illustration of the experimental concept; (**b**) practical setup for the evaluation of the left-facing 3D camera.

Table 3 shows the resulting bias and precision values of the three camera units (cam 1–3) at selected distances of 4 m and 8 m. According to the manufacturer specifications, the D455 should have a depth accuracy (bias) of $\leq$2% and a standard deviation (precision) of $\leq$2% for ranges up to 4 m. In our experiments, only cam 2 showed a bias well within the specifications, while the bias of cam 3 was more than double the expected value.

Table 3. Bias and precision values of depth measurements at 4 m and 8 m with three different RealsSense D455 units with respective colors as shown in Figure 12. Percentage values show the respective metric bias and precision values in relation to the respective range.

	Cam 1 (Blue)		Cam 2 (Red)		Cam 3 (Black)	
	Bias	**Precision**	**Bias**	**Precision**	**Bias**	**Precision**
4 m	−11 cm (2.8%)	5.7 cm (1.4%)	−1 cm (0.3%)	5.6 cm (1.4%)	17 cm (4.3%)	8.0 cm (2.0%)
8 m	−27 cm (3.4%)	26.8 cm (3.4%)	0 cm (0.0%)	27.2 cm (3.4%)	106 cm (13.3%)	9.4 cm (1.2%)

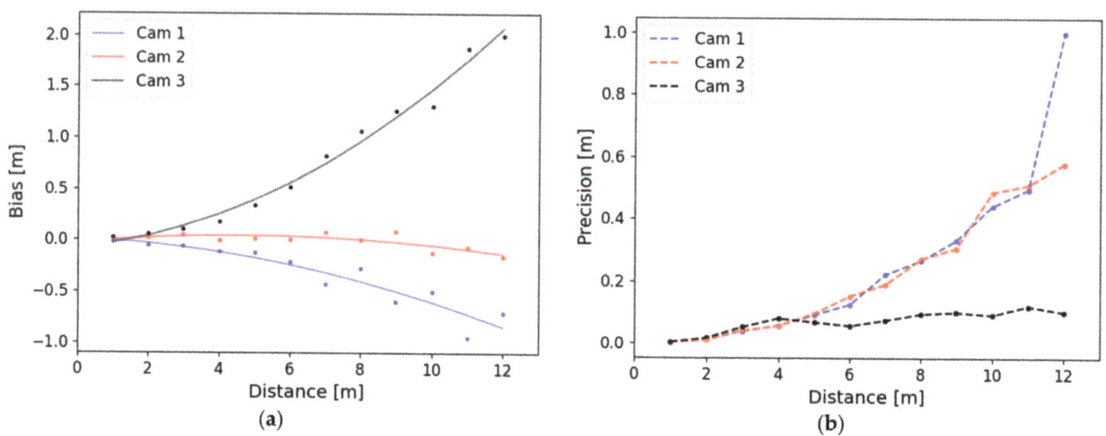

Figure 12. Performance evaluation of depth measurements by three different D455 units (cam 1–3): (**a**) bias in relation to measuring distance; (**b**) precision in relation to measuring distance.

Figure 12a,b show the bias and precision values for all the evaluated measuring distances from 1 to 12 m. Bias and precision show a roughly linear behavior for distances up to 4 (to max. 6 m). For longer ranges, all cameras exhibited a non-linear distance-dependent bias and precision—except cam 3 with an exceptional, nearly constant value for precision. The exponential behavior of the precision can be expected from a stereo system with a small baseline of only 95 mm and a very small b/h ratio for longer measuring distances. The large bias of cam 3 over 4 m and the exponential behavior of its distance-dependent bias values indicate a calibration problem.

4.3.2. Influence of Lighting Conditions on Outdoor Measurements

Outdoor environments pose several challenges for (3D) cameras, including large variations in lighting conditions and in the radiometric properties of objects to be mapped. The main question of interest was whether the D455 cameras could be operated in full daylight, at dusk, or even at night with only artificial street lighting available. For this purpose, we investigated the influence of lighting conditions on the camera performance.

In this experiment, cam 3 was placed in front of a staircase in the park of the FHNW Campus in Muttenz (Figure 13). Selected steps of the stair featured targets so that three different distances (3.81 m, 5.13 m, 6.09 m) could be evaluated. Additionally, a luxmeter was used for measuring light intensity and a tachymeter for reference distance measurements. The experiment was performed in full daylight on a sunny day with a maximum light intensity of 1770 lux and during sunset until dusk with a minimal intensity of only 2 lux. For comparison, typical indoor office lighting has an intensity of approx. 500 lux.

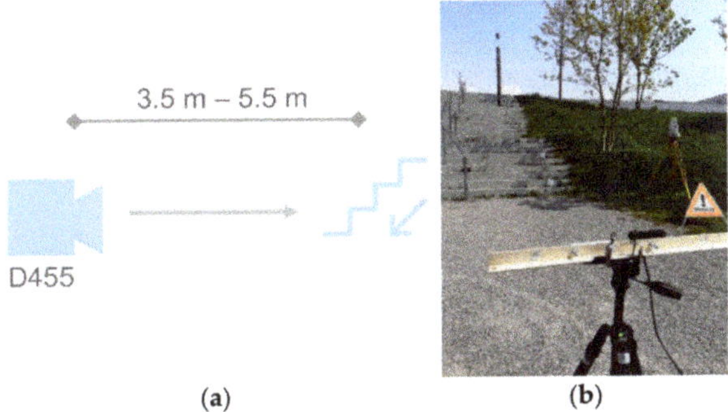

(a) (b)

Figure 13. Outdoor experimental setup for the performance evaluation of depth measurements under different lighting conditions: (**a**) schematic view of the experiment; (**b**) illustration of the outdoor test field.

Data capturing and analysis were carried out in analogy to the indoor experiments outlined above, but this time with the capture of 100 frames per lighting condition and with the evaluation of 20 × 20 pixel windows. Bias values for the three targets were derived by comparing the observed average distance with the respective reference distance.

The results for bias and precision under different lighting conditions are shown in Figure 14a,b. It can be seen in Figure 14a that the bias values for cam 3 exhibit the same distance-based scale effect as already shown in the indoor tests above. Bias and precision values proved to be relatively stable under medium light intensities between 330 and 1000 lux. Slight increases in bias can be observed in very dark and very bright conditions—with a sharp increase for the target at 6.09 m under 1750 lux, for which we currently have no plausible explanation. The highest bias of 0.95 m is with 2 lux at a distance of 6.09 m. This corresponds to the end of the sunset in almost complete darkness. The precision values for the two shorter distances were very stable and consistently below 10 cm, varying only by ±2.4 cm across all investigated illuminations. So, rather surprisingly, lighting conditions seem to have no significant influence on measuring precision. The larger values and variations in precision at a distance of 6.09 m are likely caused by the fact that this target was brown while the others were white.

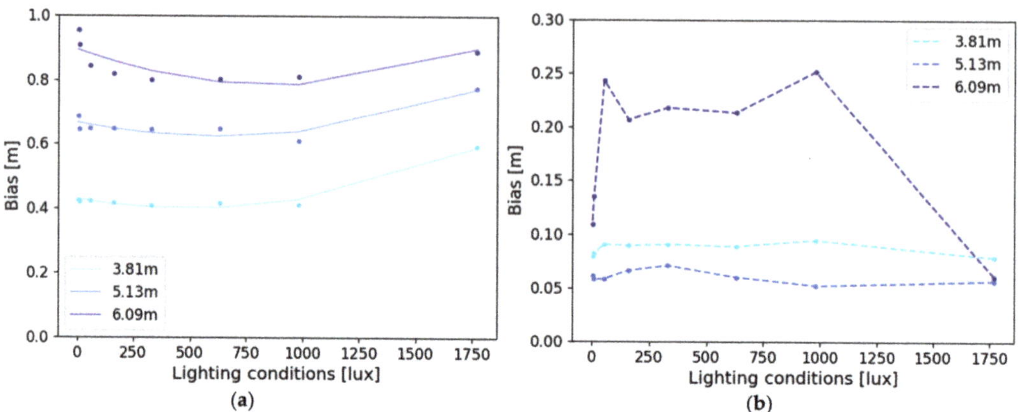

Figure 14. Performance evaluation of depth measurements by D455 (Cam 3): (**a**) bias and (**b**) precision in relation to the three measuring distances (light, middle, and dark blue values) and lighting intensity.

4.4. AI-Based 3D Vehicle Detection

To evaluate the AI-based 3D vehicle detection algorithm used, images of seven streets in the test area Basel were processed according to the workflow described in Section 3.4. The streets were selected so that all parking types were represented. However, since the types of parking spaces occur with different frequencies, their number in the evaluation varies greatly. In total, the selected test streets include 350 parking spaces, of which 283 were occupied. The vehicle detections were manually counted and verified.

Initial investigations using the point cloud-based PointRCNN 3D object detector as part of OpenPCDet yielded a recall of 0.87, whereby no tall vehicles such as family vans, camper vans, or delivery vans were detected. Further investigations revealed that the pre-trained networks provided in OpenPCDet can only detect the classes car, pedestrian, and cyclist and do not yet include classes such as van, truck, or bus. In order to correctly assess the detection capabilities for a vehicle class with existing training data, subsequently only the class "car" representing standard cars (including hatchbacks, sedans, station wagons, etc.) was considered. Thus, all 37 tall vehicles were removed from the dataset. For the evaluation, a total of 313 parking spaces were considered, 246 of which were occupied by cars. Our 3D detection approach achieved an average precision of 1.0 (100%) and an average recall of 0.97 (97%) for all parking types (Table 4). Parallel parking showed the best detection rate with a recall of 1.0, followed by angle parking with 0.96 and perpendicular parking with 0.94. Cars parked in 2 × 2 parking slots were significantly harder to detect, resulting in a recall of only 0.50. However, the sample for this type is too small for a statistically sound evaluation.

Table 4. AI-based detection results for cars, broken down by type of parking space.

	Parallel	**Perpendicular**	**Angle**	**2 × 2**	**Overall**
True positive	163	49	24	3	239
True negative	36	26	4	1	67
False positive	0	0	0	0	0
False negative	0	3	1	3	7
Precision	1.00	1.00	1.00	1.00	1.00
Recall	1.00	0.94	0.96	0.50	0.97

5. Discussion

5.1. Georeferencing Investigations

Creating valid on-street parking statistics requires absolute object localization accuracies roughly at the meter level if the occupation of individual parking slots is to be correctly determined. As shown in several studies [42,56], georeferencing in urban areas is very challenging, even with high-end MMS. In our georeferencing investigations, we equipped our electrical tricycle with our low-cost mobile mapping payload and with the well-researched high-performance BIMAGE backpack as the second payload, which subsequently served as a reference for static and kinematic experiments. By estimating a lever arm between both systems for each measurement epoch and comparing these estimates with the lever arm fixed to the MMS body frame, the georeferencing performance could successfully be evaluated. The direct georeferencing comparisons yielded a good average 2D deviation between the low-cost and the reference system for the static case of approx. 10 cm. The kinematic tests resulted in an average 2D deviation of 0.9 m, however with peaks of more than 1 m. These peaks can be attributed to locations with extended GNSS signal obstructions caused by buildings or large trees. The georeferencing tests showed that the current low-cost system fulfills the sub-meter accuracy requirements in areas with few to no GNSS obstructions. However, in demanding urban environments it does not yet fulfill these strict requirements, mainly due to its low-quality IMU, and complementary positioning strategies will likely be required. Future research could include the evaluation of higher-grade and possibly redundant low-cost IMUs, and the development of novel map matching approaches offered by 3D cameras. However, the current system shortcomings only limit the absolute localization and automated mapping capabilities of the system; they do not affect the 3D vehicle detection performance and the object clustering, which mainly rely on relative accuracies.

5.2. 3D Camera Performance

In our experiments, we investigated the depth accuracy and precision of RealSense D455 low-cost 3D cameras. In a previous evaluation, this camera type had proven to be the first 3D camera suitable for outdoor applications supporting measuring ranges beyond 4 m. In a first experiment, we investigated three units for depth accuracy (bias) and depth precision (standard deviation) at different measuring distances, ranging from 1 to 12 m. Two of the three units exhibited a depth accuracy and a depth precision with a significant non-linear dependency on measuring range, which confirmed the findings of Halmetschlager et al. [25]. The manufacturer specification [40] of $\leq 2\%$ accuracy for ranges up to 4 m was only met by one unit. The specification of $\leq 2\%$ precision was met by all units. The investigations showed that 3D cameras should be tested—and if necessary re-calibrated—if they are to be used for long-range measurements. By limiting the depth measurements to a max. range of 8 m, we limited the localization error contribution of the depth bias to max. 0.3–0.4 m. The second experiment demonstrated that ambient lighting conditions have no significant effect on the depth bias and precision of the RealSense D455—with only a minor degradation in very dark and very bright conditions. This robustness towards different lighting conditions and the capability to reliably operate in very dark and very bright environments is an important factor for our application.

5.3. 3D Object Detection Evaluation

In the last experiment, we addressed the challenge of reliably detecting and localizing vehicles in the point clouds derived from the low-cost 3D cameras, which are significantly noisier and have more data gaps than LiDAR-based point clouds. In an evaluation of five candidate 3D object detection algorithms, PointRCNN clearly outperformed all the others. On our dataset in the city of Basel with four different parking types (parallel, perpendicular, angle, and 2 × 2 blocks) and a total of 313 parking spaces, we obtained an average precision of 100% and an initial average recall of 97%. These detection results apply for the class "car", representing typical standard cars such as sedans, hatchbacks, etc. It should be noted that our current detection framework has not yet been trained for other vehicle classes such as vans, buses, or trucks. Consequently, tall vehicles such as family or delivery vans are not yet detected. For the supported class of cars, our approach outperformed the other approaches (Mathur et al. [3], Bock et al. [4], Grassi et al. [11], and Fetscher [5]) in precision and was equal or better for recall. When broken down by the type of parking space, all cars in parallel parking spaces were detected with slightly inferior performance for angle and perpendicular parking (see Table 4). In conclusion, PointRCNN, which was originally developed for and trained with point clouds from high-end LiDAR data, can be successfully applied to depth data from low-cost RGB-D cameras. Ongoing and future research includes the training of PointRCNN for additional vehicle classes, in particular "vans", which according to our preliminary investigations account for about 10% of the vehicles in the test area. The intention is to use the labeled object classes from the KITTI 3D Object detection benchmark [33] and to complement the training data with RGB-D data from mobile mapping missions with our own system.

5.4. Overall Capabilities and Performance

Table 5 shows a comparison of our method with the state-of-the-art methods introduced in Section 2. It shows that our low-cost system has the potential for the necessary revisit frequencies, e.g., for temporal parking analyses over the course of the day. The comparison also shows that only our approach and the one by Fetscher [5] support the three main parking types of parallel, angle, and perpendicular parking. However, the latter method by Fetscher is not suitable for practical statistics due to its reliance on street-level imagery. One of the advantages of our approach is the availability of the complementary nature of the co-registered RGB imagery and depth data from the 3D cameras. This not only allows for the visual inspection of the detection results, but it will also facilitate future retraining of existing vehicle classes or the training of new, currently unsupported vehicle classes in 3D object detection, e.g., by exploiting labels from 2D object detection and segmentation [13].

Table 5. Comparison of our method with the main methods for deriving on-street parking statistics from mobile sensors.

	Mathur et al. [3]	Bock et al. [4]	Grassi et al. [11]	Fetscher [5]	Ours
Mapping platform	Probe vehicles (e.g., taxis)	High-end MLS vehicle	Vehicle (dashboard-mounted)	High-end multi-view stereo MMS	Electric Tricycle
Mapping sensors/ mapping data	Ultrasonic range finder/range profiles	Dual LiDAR/3D point clouds	Smartphone camera/2D imagery	High-end stereo cameras/ RGB-D imagery	Low-cost 3D camera/ RGB-D imagery
Revisit frequency	potentially high	on-demand	potentially high	low	potentially high
Supported parking types	parallel only	parallel, perpendicular	parallel only	parallel, angle, perpendicular, 2 × 2 clusters	parallel, angle, perpendicular, 2 × 2 clusters
Detection type	gaps (in range profiles)	object segmentation and classification (RF)	image-based car detection	corner detection or clustering	AI-based 3D object detection (PointRCNN)
Sample size (# of slots or vehicles)	57	717	8176	184	313
Detection accuracy	~90% (of free spaces)	Recall 93.7% Precision 97.4%	~90%	97.0–98.3%	Recall 97% Precision 100%

6. Conclusions and Future Work

In this article, we introduced a novel system and approach for creating on-street parking statistics by combining a low-cost mobile mapping payload featuring RGB-D cameras with AI-based 3D vehicle detection algorithms. The new payload integrates two active stereo RGB-D consumer cameras of the latest generation, an entry-level GNSS/INS positioning system, and an embedded single-board computer using the Robot Operating

System (ROS). Following direct georeferencing and anonymization steps, the RGB-D imagery is converted to 3D point clouds. These are subsequently used by PointRCNN for detecting vehicle location, size, and orientation—subsequently represented by 3D bounding boxes. These vehicle detection candidates and their scores are subsequently used for a GIS-based creation of parking statistics. The new automated mobile mapping and 3D vehicle detection approach yielded a precision of 100% and an average recall of 97% for cars and can support all common parking types, including perpendicular and angle parking. To our knowledge, this is one of the first studies successfully using low-cost 3D cameras for kinematic mapping purposes in an outdoor environment.

In our study, we investigated several critical components of a mobile mapping solution for automatically detecting parked cars and creating parking statistics, namely, georeferencing accuracy, 3D camera performance, and AI-based vehicle detection.

The methods and results described in this paper leave room for further improvement. One of the first goals is the training of the PointRCNN detector to allow the detection of tall vehicles and vans. For this purpose, a reasonably large training and validation dataset will be collected in further mobile mapping missions. Another goal is to employ edge computing for onboard 3D object detection. This would eliminate the current need for massive onboard data storage performance and capacity, would avoid privacy issues, and would dramatically reduce the post-processing time. Finally, we will investigate novel map matching strategies, which will be offered by the RGB-D data and which will avoid some of the current difficulties with direct georeferencing.

Author Contributions: S.N. designed and supervised the overall research and wrote the majority of the paper. J.M. designed, implemented, and analyzed the AI-based 3D vehicle detection and designed and implemented the anonymization workflow. S.B. designed and developed the low-cost mobile mapping system, integrated the 3D cameras, and evaluated the georeferencing experiments. M.A. designed, performed, and evaluated the 3D camera performance evaluation experiments and operated the mobile mapping system during the Muttenz campaign. S.R. was responsible for the planning, execution, processing, and documentation of the field test campaigns. All authors have read and agreed to the published version of the manuscript.

Funding: This research was partly funded by the Amt für Mobilität, Bau-und Verkehrsdepartement des Kantons Basel-Stadt as part of the pilot study "Bildbasiertes Parkplatzmonitoring".

Data Availability Statement: Not applicable.

Acknowledgments: We would like to thank our project partners Luca Olivieri, Mobility Strategy, Amt für Mobilität Basel-Stadt and Alex Erath, Institute of Civil Engineering, FHNW University of Applied Sciences and Arts Northwestern Switzerland for making this project possible and for their numerous and important inputs from the perspective of mobility experts. We would further like to acknowledge the hints and support of Pia Bereuter in establishing the GIS workflow. We also thank Simon Fetscher and Jasmin Frey for their earlier studies and for contributing valuable experiments and data.

Conflicts of Interest: The authors declare no conflict of interest.

References

1. Shoup, D.C. Cruising for Parking. *Transp. Policy* **2006**, *13*, 479–486. [CrossRef]
2. Rapp Trans AG Basel-Stadt. *Erhebung Parkplatzauslastung Stadt Basel 2019*; Version 1; RAPP Group: Basel, Switzerland, 2019.
3. Mathur, S.; Jin, T.; Kasturirangan, N.; Chandrasekaran, J.; Xue, W.; Gruteser, M.; Trappe, W. ParkNet. In Proceedings of the 8th International Conference on Mobile Systems, Applications and Services—MobiSys'10, San Francisco, CA, USA, 15–18 June 2010; p. 123. [CrossRef]
4. Bock, F.; Eggert, D.; Sester, M. On-Street Parking Statistics Using LiDAR Mobile Mapping. In Proceedings of the 2015 IEEE 18th International Conference on Intelligent Transportation Systems, Gran Canaria, Spain, 15–18 September 2015; pp. 2812–2818. [CrossRef]
5. Fetscher, S. Automatische Analyse von Streetlevel-Bilddaten Für das Digitale Parkplatzmanagement. Bachelor's Thesis, FHNW University of Applied Sciences and Arts Northwestern Switzerland, Muttenz, Switzerland, 2020.

6. Polycarpou, E.; Lambrinos, L.; Protopapadakis, E. Smart Parking Solutions for Urban Areas. In Proceedings of the 2013 IEEE 14th International Symposium on "A World of Wireless, Mobile and Multimedia Networks" (WoWMoM), Madrid, Spain, 4–7 June 2013; pp. 1–6. [CrossRef]

7. Paidi, V.; Fleyeh, H.; Håkansson, J.; Nyberg, R.G. Smart Parking Sensors, Technologies and Applications for Open Parking Lots: A Review. *IET Intell. Transp. Syst.* **2018**, *12*, 735–741. [CrossRef]

8. Barriga, J.J.; Sulca, J.; León, J.L.; Ulloa, A.; Portero, D.; Andrade, R.; Yoo, S.G. Smart Parking: A Literature Review from the Technological Perspective. *Appl. Sci.* **2019**, *9*, 4569. [CrossRef]

9. Houben, S.; Komar, M.; Hohm, A.; Luke, S.; Neuhausen, M.; Schlipsing, M. On-Vehicle Video-Based Parking Lot Recognition with Fisheye Optics. In Proceedings of the 16th International IEEE Conference on Intelligent Transportation Systems (ITSC 2013), The Hague, The Netherlands, 6–9 October 2013; pp. 7–12. [CrossRef]

10. Suhr, J.; Jung, H. A Universal Vacant Parking Slot Recognition System Using Sensors Mounted on Off-the-Shelf Vehicles. *Sensors* **2018**, *18*, 1213. [CrossRef]

11. Grassi, G.; Jamieson, K.; Bahl, P.; Pau, G. Parkmaster: An in–Vehicle, Edge–Based Video Analytics Service for Detecting Open Parking Spaces in Urban Environments. In Proceedings of the Second ACM/IEEE Symposium on Edge Computing, San Jose, CA, USA, 12–14 October 2017; pp. 1–14. [CrossRef]

12. Nebiker, S.; Cavegn, S.; Loesch, B. Cloud-Based Geospatial 3D Image Spaces—A Powerful Urban Model for the Smart City. *ISPRS Int. J. Geo-Inf.* **2015**, *4*, 2267–2291. [CrossRef]

13. Wu, Y.; Kirillov, A.; Massa, F.; Lo, W.-Y.; Girshick, R. Detectron2. 2019. Available online: https://github.com/facebookresearch/detectron2 (accessed on 3 August 2021).

14. Ulrich, L.; Vezzetti, E.; Moos, S.; Marcolin, F. Analysis of RGB-D Camera Technologies for Supporting Different Facial Usage Scenarios. *Multimed. Tools Appl.* **2020**, *79*, 29375–29398. [CrossRef]

15. Kuznetsova, A.; Leal-Taixe, L.; Rosenhahn, B. Real-Time Sign Language Recognition Using a Consumer Depth Camera. In Proceedings of the 2013 IEEE International Conference on Computer Vision Workshops, Sydney, Australia, 2–8 December 2013; pp. 83–90. [CrossRef]

16. Jain, H.P.; Subramanian, A.; Das, S.; Mittal, A. Real-Time Upper-Body Human Pose Estimation Using a Depth Camera. In Proceedings of the International Conference on Computer Vision/Computer Graphics Collaboration Techniques and Applications, Rocquencourt, France, 10–11 October 2011; pp. 227–238.

17. Zollhöfer, M.; Stotko, P.; Görlitz, A.; Theobalt, C.; Nießner, M.; Klein, R.; Kolb, A. State of the Art on 3D Reconstruction with RGB-D Cameras. *Comput. Graph. Forum* **2018**, *37*, 625–652. [CrossRef]

18. Holdener, D.; Nebiker, S.; Blaser, S. Design and Implementation of a Novel Portable 360° Stereo Camera System with Low-Cost Action Cameras. *Int. Arch. Photogramm. Remote Sens. Spat. Inf. Sci.* **2017**, *42*, 105–110. [CrossRef]

19. Hasler, O.; Blaser, S.; Nebiker, S. Performance evaluation of a mobile mapping application using smartphones and augmented reality frameworks. *ISPRS Ann. Photogramm. Remote Sens. Spat. Inf. Sci.* **2020**, *5*, 741–747. [CrossRef]

20. Torresani, A.; Menna, F.; Battisti, R.; Remondino, F. A V-SLAM Guided and Portable System for Photogrammetric Applications. *Remote Sens.* **2021**, *13*, 2351. [CrossRef]

21. Henry, P.; Krainin, M.; Herbst, E.; Ren, X.; Fox, D. RGB-D Mapping: Using Kinect-Style Depth Cameras for Dense 3D Modeling of Indoor Environments. *Int. J. Rob. Res.* **2012**, *31*, 647–663. [CrossRef]

22. Henry, P.; Krainin, M.; Herbst, E.; Ren, X.; Fox, D. RGB-D Mapping: Using Depth Cameras for Dense 3D Modeling of Indoor Environments. In *Experimental Robotics*; Springer: Berlin/Heidelberg, Germany, 2014; pp. 477–491.

23. Brahmanage, G.; Leung, H. Outdoor RGB-D Mapping Using Intel-RealSense. In Proceedings of the 2019 IEEE SENSORS, Montreal, QC, Canada, 27–30 October 2019; pp. 1–4. [CrossRef]

24. Iwaszczuk, D.; Koppanyi, Z.; Pfrang, J.; Toth, C. Evaluation of a mobile multi-sensor system for seamless outdoor and indoor mapping. *Int. Arch. Photogramm. Remote Sens. Spat. Inf. Sci.* **2019**, *XLII-1/W2*, 31–35. [CrossRef]

25. Halmetschlager-Funek, G.; Suchi, M.; Kampel, M.; Vincze, M. An Empirical Evaluation of Ten Depth Cameras: Bias, Precision, Lateral Noise, Different Lighting Conditions and Materials, and Multiple Sensor Setups in Indoor Environments. *IEEE Robot. Autom. Mag.* **2019**, *26*, 67–77. [CrossRef]

26. Lourenço, F.; Araujo, H. Intel RealSense SR305, D415 and L515: Experimental Evaluation and Comparison of Depth Estimation. In Proceedings of the 16th International Joint Conference on Computer Vision, Imaging and Computer Graphics Theory and Applications (VISIGRAPP 2021), Online Streaming, 8–10 February 2021; Volume 4, pp. 362–369. [CrossRef]

27. Vit, A.; Shani, G. Comparing RGB-D Sensors for Close Range Outdoor Agricultural Phenotyping. *Sensors* **2018**, *18*, 4413. [CrossRef]

28. Zhao, Z.Q.; Zheng, P.; Xu, S.T.; Wu, X. Object Detection with Deep Learning: A Review. *IEEE Trans. Neural Netw. Learn. Syst.* **2019**, *30*, 3212–3232. [CrossRef]

29. Friederich, J.; Zschech, P. Review and Systematization of Solutions for 3d Object Detection. In Proceedings of the 15th International Conference on Business Information Systems, Potsdam, Germany, 8–11 March 2020. [CrossRef]

30. Arnold, E.; Al-Jarrah, O.Y.; Dianati, M.; Fallah, S.; Oxtoby, D.; Mouzakitis, A. A Survey on 3D Object Detection Methods for Autonomous Driving Applications. *IEEE Trans. Intell. Transp. Syst.* **2019**, *20*, 3782–3795. [CrossRef]

31. Shi, S.; Wang, X.; Li, H. PointRCNN: 3D Object Proposal Generation and Detection from Point Cloud. In Proceedings of the IEEE Computer Society Conference on Computer Vision and Pattern Recognition, Long Beach, CA, USA, 16–20 June 2019; pp. 770–779. [CrossRef]
32. Shi, S.; Guo, C.; Jiang, L.; Wang, Z.; Shi, J.; Wang, X.; Li, H. PV-RCNN: Point-Voxel Feature Set Abstraction for 3D Object Detection. In Proceedings of the Conference on Computer Vision and Pattern Recognition, Seattle, WA, USA, 14–19 June 2020; pp. 10526–10535. [CrossRef]
33. KITTI 3D Object Detection Online Benchmark. Available online: http://www.cvlibs.net/datasets/kitti/eval_object.php?obj_benchmark=3d (accessed on 29 July 2021).
34. Zheng, W.; Tang, W.; Jiang, L.; Fu, C.-W. SE-SSD: Self-Ensembling Single-Stage Object Detector from Point Cloud. *arXiv* **2021**, arXiv:2104.09804.
35. Li, Z.; Yao, Y.; Quan, Z.; Yang, W.; Xie, J. SIENet: Spatial Information Enhancement Network for 3D Object Detection from Point Cloud. *arXiv* **2021**, arXiv:2103.15396.
36. Deng, J.; Shi, S.; Li, P.; Zhou, W.; Zhang, Y.; Li, H. Voxel R-CNN: Towards High Performance Voxel-Based 3D Object Detection. *arXiv* **2020**, arXiv:2012.15712.
37. Qi, C.R.; Liu, W.; Wu, C.; Su, H.; Guibas, L.J. Frustum PointNets for 3D Object Detection from RGB-D Data. In Proceedings of the Conference on Computer Vision and Pattern Recognition, Salt Lake City, UT, USA, 18–23 June 2018; pp. 918–927. [CrossRef]
38. Wang, Z.; Jia, K. Frustum ConvNet: Sliding Frustums to Aggregate Local Point-Wise Features for Amodal. In Proceedings of the IEEE International Conference on Intelligent Robots and Systems, Macau, China, 3–8 November 2019; pp. 1742–1749. [CrossRef]
39. Swift Navigation, I. PiksiMulti GNSS Module Hardware Specification. 2019, p. 10. Available online: https://www.swiftnav.com/latest/piksi-multi-hw-specification (accessed on 3 August 2021).
40. Intel Corporation. Intel Product Family D400 Series: Datasheet. 2020, p. 134. Available online: https://dev.intelrealsense.com/docs/intel-realsense-d400-series-product-family-datasheet (accessed on 3 August 2021).
41. nVidia Developers. Jetson TX2 Module. Available online: https://developer.nvidia.com/embedded/jetson-tx2 (accessed on 29 July 2021).
42. Blaser, S.; Meyer, J.; Nebiker, S.; Fricker, L.; Weber, D. Centimetre-accuracy in forests and urban canyons—Combining a high-performance image-based mobile mapping backpack with new georeferencing methods. *ISPRS Ann. Photogramm. Remote Sens. Spat. Inf. Sci.* **2020**, *1*, 333–341. [CrossRef]
43. Blaser, S.; Cavegn, S.; Nebiker, S. Development of a Portable High Performance Mobile Mapping System Using the Robot Operating System. *ISPRS Ann. Photogramm. Remote Sens. Spat. Inf. Sci.* **2018**, *4*, 13–20. [CrossRef]
44. Quigley, M.; Conley, K.; Gerkey, B.; Faust, J.; Foote, T.; Leibs, J.; Berger, E.; Wheeler, R.; Mg, A. ROS: An Open-Source Robot Operating System. *ICRA Workshop Open Source Softw.* **2009**, *3*, 5.
45. Dorodnicov, S.; Hirshberg, D. Realsense2_Camera. Available online: http://wiki.ros.org/realsense2_camera (accessed on 29 July 2021).
46. NovAtel. Inertial Explorer 8.9 User Manual. 2020, p. 236. Available online: https://docs.novatel.com/Waypoint/Content/PDFs/Waypoint_Software_User_Manual_OM-20000166.pdf (accessed on 3 August 2021).
47. understand.ai. Anonymizer. Karlsruhe, Germany, 2019. Available online: https://github.com/understand-ai/anonymizer (accessed on 3 August 2021).
48. OpenPCDet Development Team. OpenPCDet: An Open-source Toolbox for 3D Object Detection from Point Clouds. Available online: https://github.com/open-mmlab/OpenPCDet (accessed on 29 July 2021).
49. Lang, A.H.; Vora, S.; Caesar, H.; Zhou, L.; Yang, J.; Beijbom, O. Pointpillars: Fast Encoders for Object Detection from Point Clouds. In Proceedings of the Conference on Computer Vision and Pattern Recognition (CVPR), Long Beach, CA, USA, 16–20 June 2019; pp. 12689–12697. [CrossRef]
50. Yan, Y.; Mao, Y.; Li, B. Second: Sparsely Embedded Convolutional Detection. *Sensors* **2018**, *18*, 3337. [CrossRef]
51. Shi, S.; Wang, Z.; Shi, J.; Wang, X.; Li, H. From Points to Parts: 3D Object Detection from Point Cloud with Part-Aware and Part-Aggregation Network. *IEEE Trans. Pattern Anal. Mach. Intell.* **2020**, *1*, 2977026. [CrossRef] [PubMed]
52. Geiger, A.; Lenz, P.; Urtasun, R. Are We Ready for Autonomous Driving? The KITTI Vision Benchmark Suite. In Proceedings of the 2012 IEEE Conference on Computer Vision and Pattern Recognition, Providence, RI, USA, 16–21 June 2012; pp. 3354–3361. [CrossRef]
53. Butler, H.; Daly, M.; Doyle, A.; Gillies, S.; Hagen, S.; Schaub, T. The GeoJSON Format. RFC 7946. *IETF Internet Eng. Task Force* **2016**. Available online: https://www.rfc-editor.org/info/rfc7946 (accessed on 30 June 2021). [CrossRef]
54. QGIS Development Team. QGIS Geographic Information System. *Open Source Geospatial Foundation.* 2009. Available online: https://qgis.org/en/site/ (accessed on 3 August 2021).
55. Ester, M.; Kriegel, H.-P.; Sander, J.; Xu, X. A Density-Based Algorithm for Discovering Clusters in Large Spatial Databases with Noise. In Proceedings of the Second International Conference on Knowledge Discovery and Data Mining (KDD-96), Portland, OR, USA, 2–4 August 1996; Volume 96, pp. 226–231.
56. Cavegn, S.; Nebiker, S.; Haala, N. A Systematic Comparison of Direct and Image-Based Georeferencing in Challenging Urban Areas. *Int. Arch. Photogramm. Remote Sens. Spat. Inf. Sci.* **2016**, *41*, 529–536. [CrossRef]
57. Frey, J. Bildbasierte Lösung für das mobile Parkplatzmonitoring. Master's Thesis, FHNW University of Applied Sciences and Arts Northwestern Switzerland, Muttenz, Switzerland, 2021.

Article

Robust Loop Closure Detection Integrating Visual–Spatial–Semantic Information via Topological Graphs and CNN Features

Yuwei Wang [1], Yuanying Qiu [1,*], Peitao Cheng [2] and Xuechao Duan [1]

1   Key Laboratory of Electronic Equipment Structure Design of Ministry of Education, Xidian University, Xi'an 710071, China; yweiwang@stu.xidian.edu.cn (Y.W.); xchduan@xidian.edu.cn (X.D.)
2   School of Mechano-Electronic Engineering, Xidian University, Xi'an 710071, China; ptcheng@xidian.edu.cn
*   Correspondence: yyqiu@mail.xidian.edu.cn

Received: 11 October 2020; Accepted: 23 November 2020; Published: 27 November 2020

Abstract: Loop closure detection is a key module for visual simultaneous localization and mapping (SLAM). Most previous methods for this module have not made full use of the information provided by images, i.e., they have only used the visual appearance or have only considered the spatial relationships of landmarks; the visual, spatial and semantic information have not been fully integrated. In this paper, a robust loop closure detection approach integrating visual–spatial–semantic information is proposed by employing topological graphs and convolutional neural network (CNN) features. Firstly, to reduce mismatches under different viewpoints, semantic topological graphs are introduced to encode the spatial relationships of landmarks, and random walk descriptors are employed to characterize the topological graphs for graph matching. Secondly, dynamic landmarks are eliminated by using semantic information, and distinctive landmarks are selected for loop closure detection, thus alleviating the impact of dynamic scenes. Finally, to ease the effect of appearance changes, the appearance-invariant descriptor of the landmark region is extracted by a pre-trained CNN without the specially designed manual features. The proposed approach weakens the influence of viewpoint changes and dynamic scenes, and extensive experiments conducted on open datasets and a mobile robot demonstrated that the proposed method has more satisfactory performance compared to state-of-the-art methods.

Keywords: loop closure detection; visual SLAM; semantic topology graph; graph matching; CNN features; deep learning

1. Introduction

Simultaneous localization and mapping (SLAM) [1] is of great importance in autonomous robots and has become a hotspot in robotics research [2,3]. SLAM mainly solves the problem of robot localization and map establishment in an unknown environment, relying on external sensors to work. Since the camera can capture a wealth of information, it is currently widely used in visual SLAM systems. Loop closure detection is an important module of visual SLAM, because its role is to determine whether the robot returns to its previous environment [4] and then to correct the localization errors accumulated over time to construct an accurate and global-consistent map. In addition, loop closure detection can create new edge constraints between revisited pose nodes [5–7] for visual SLAM based on pose graphs. These additional constraints are optimized by bundle adjustment [8] in the backend of a visual SLAM to get more accurate estimation results [9].

Traditional appearance-based methods have nothing to do with the frontend and backend of visual SLAM, as they only detect loops based on the similarity of image pairs. They are mostly based on the bag of words (BoW) model [10], which clusters visual features such as SIFT and SURT to generate words and then construct a dictionary. In that way, images can be characterized by word vectors according to the dictionary, and the loops can be detected according to the vector difference between the images. They can effectively work in different scenarios and have become the mainstream method in visual SLAM [11]. Among them, the loop closure detection methods based on local features utilize SIFT [12], SURF [13], and ORB [14] to describe an image. For example, Angeli et al. [15] used SIFT features for loop closure detection, FAB-MAP [10] employed SURF features, RTAB-Map SLAM [16] utilized SIFT and SURF features, and ORBSLAM [17] exploited ORB features. These works have yielded gratifying results. In addition, there have been many methods based on global features. Sünderhauf et al. [18] applied GIST [19] to place recognition, encoding the response of the image in different directions and scales as a global description through Gabor filters. Additionally, Naseer et al. [20] used HOG descriptors to characterize the holistic environment for image recognition. However, in the above-mentioned methods, the features are artificially designed and can only cope with limited scene changes. Moreover, they only contain low-level information and cannot express complex structural information, so it is difficult to deal with drastic appearance changes.

The sequence-based approach has achieved great success in dealing with appearance changes. SeqSLAM [21] considers a short image sequence instead of a single image to solve perceptual aliasing. It uses correlation matching to find the local best match for each query image in all short image sequences. Abdollahyan et al. [22] proposed a sequence-based method for visual localization that employed a directed acyclic graph to model an image sequence to form a string, and then they exploited the partial order kernel to compare strings. Naseer et al. [20] modeled image matching as a minimum cost flow problem in a data association graph and used the HOG descriptor of the image to match the image pair. SMART [23] applied a query image sequence to match a dataset image sequence by calculating the similarity in the downsampled and patch-normalized image sequences. Hansen et al. [24] used the Hidden Markov Model to retrieve the image sequence of a dataset matching the query image sequence by calculating an image similarity probability value matrix. However, these methods do not consider the spatial geometric relationship of the objects in the image, and they are difficult to use in the face of changes in the viewpoint.

With the rise of deep learning in computer vision fields such as image recognition and classification, researchers have begun to apply the deep convolutional neural networks (CNNs) for loop closure detection. A multi-layer neural network automatically learns inherent feature expression directly from raw data and expresses an image as a global feature [25]. This has become an effective way to solve the loop closure detection problem of visual SLAM. Hou et al. [26] used the output of the intermediate layer of a pre-trained CNN to construct feature descriptions for loop closure detection, and it was proved that the output effects of the third convolutional layer and the fifth pooling layer were better than those of other layers. Sünderhauf et al. [27] comprehensively evaluated the application of three advanced CNNs in loop closure detection and found that the output of the low-level network was robust to appearance changes. Moreover, the output of the high-level network was found to contain more semantic information that was robust to changes in viewpoint. Arroyo et al. [28] combined the output of each layer of a CNN and expressed it as a separate feature vector. It was found that this vector had strong appearance and viewpoint robustness. Gao et al. [29] adopted the stacked denoising auto-encoder method to automatically learn the compressed representation of an image in an unsupervised learning manner. Sünderhauf et al. [30] proposed a method based on CNN landmarks that effectively integrated global and local features. However, deep learning automatically learns the global features of an image while ignoring local features, so it cannot cope with drastic changes in viewpoint.

In order to achieve strong robustness to viewpoint changes, many spatial-based methods have been proposed in recent years. Cascianelli et al. [31] proposed a method based on a co-visibility graph—that is, if the underlying landmark was co-observed in the image, the two nodes were connected and the image was modeled as a graph structure of nodes and edges for place recognition. Finman et al. [32]

performed a convolution operation on an RGB-D-dense map to detect an object and then connected the objects to construct a sparse object graph for place recognition. Oh et al. [33] represented an object-based place recognition method that characterized the objects by the center of the position and connected them by edges. Then, the objects and the edges were used to measure the similarity for loop closure detection. Pepperell et al. [34] used roads as directed edges connecting intersections, which promoted the sequence matching of locations. Stumm et al. [35] applied an adjacency matrix to encode the spatial relationship of landmarks. Gawel et al. [36] utilized a graph structure to encode the spatial relationship of landmark regions, and their model had strong robustness against viewpoint changes. Furthermore, some techniques have been used to encode graph structure information into a vector space for similarity calculation. Graph kernels were used to calculate the similarity between a query and candidate image for pace recognition [37]. Han et al. [38] proposed an unsupervised learning method to learn a projection from landmarks in a scene to low-dimensional space that preserved the local consistency, i.e., the distance information between the landmarks of the original data was retained in the projection space. A random walk descriptor was applied to describe graph structure [36]. Chen et al. [39] employed a feature-encoding method based on convolutional layer activations to handle viewpoint changes. Schönberger et al. [40] obtained three-dimensional descriptors for visual localization by encoding spatial and semantic information. In addition to vector-based descriptors, Gao et al. [41] proposed a multi-order graph matching method for loop closure detection. Though these methods have achieved good results, they have not effectively integrated visual, spatial, and semantic information, so they are difficult to use in drastic viewpoint changes and dynamic scenes.

In this paper, a robust loop closure detection approach integrating visual–spatial–semantic information is proposed by using topological graphs and CNN features; this approach makes effective use of appearance-invariant CNN features and viewpoint-invariant landmark regions to improve robustness in the face of viewpoint changes and dynamic scenes. The approach consists of two parts: the construction of the semantic topology graphs and loop closure detection. Firstly, the algorithm of semantic topological graph performs semantic segmentation on the image to extract landmark regions. At the same time, the distinctive landmarks are selected for loop closure detection after eliminating dynamic landmarks. Then, acquired landmarks are input into a pre-trained AlexNet network, and the third convolution layer output is used as the global feature of landmarks. Finally, the image is constructed as a semantic topology graph of nodes and edges to represent the spatial relationship of landmarks, and a random walk descriptor is used to represent the graph structure. The algorithm of loop closure detection first quickly retrieves candidate images based on the semantic information of landmarks by using shared nodes of the same category. Furthermore, the appearance similarity of the landmark pair is calculated according to the CNN and contour features, and the random walk descriptor is used to calculate the geometric similarity between images. Then, loop closure detection is organized according to the overall similarity of the appearance and spatial information. Experiments conducted on public datasets demonstrated the superiority of the proposed method over other state-of-the-art methods. To verify the robustness of the approach in viewpoint changes and dynamic scenes, further experiments were performed on a mobile robot in outdoor scenes, and satisfactory results were obtained.

In short, the main contributions of this work are as follows:

- A robust loop closure detection approach that combines visual, spatial, and semantic information to improve the robustness for changes in viewpoint and dynamic scenes is proposed.
- A pre-trained semantic segmentation model is used to segment landmarks and a pre-trained AlexNet network is employed to extract CNN features that can be used without specific scene training. In addition, the semantic segmentation model and feature extraction network can be replaced by other models.

The remainder of this paper is organized as follows: Section 2 describes the proposed loop closure detection method. Section 3 gives experimental details and comparison results. Finally, conclusions are presented in Section 4.

2. Materials and Methods

As shown in Figure 1, the algorithm of loop closure detection proposed in this paper includes the following key modules:

(1) The extraction of the semantic landmarks.
(2) The elimination of the dynamic landmarks and selection of the distinctive landmarks.
(3) The calculation of the CNN features in landmark regions and dimensionality reduction processing on CNN features.
(4) The construction of the semantic topological graphs and expression of random walk descriptors.
(5) The calculation of geometric similarity with random walk descriptors.
(6) The calculation overall similarity for loop closure detection.

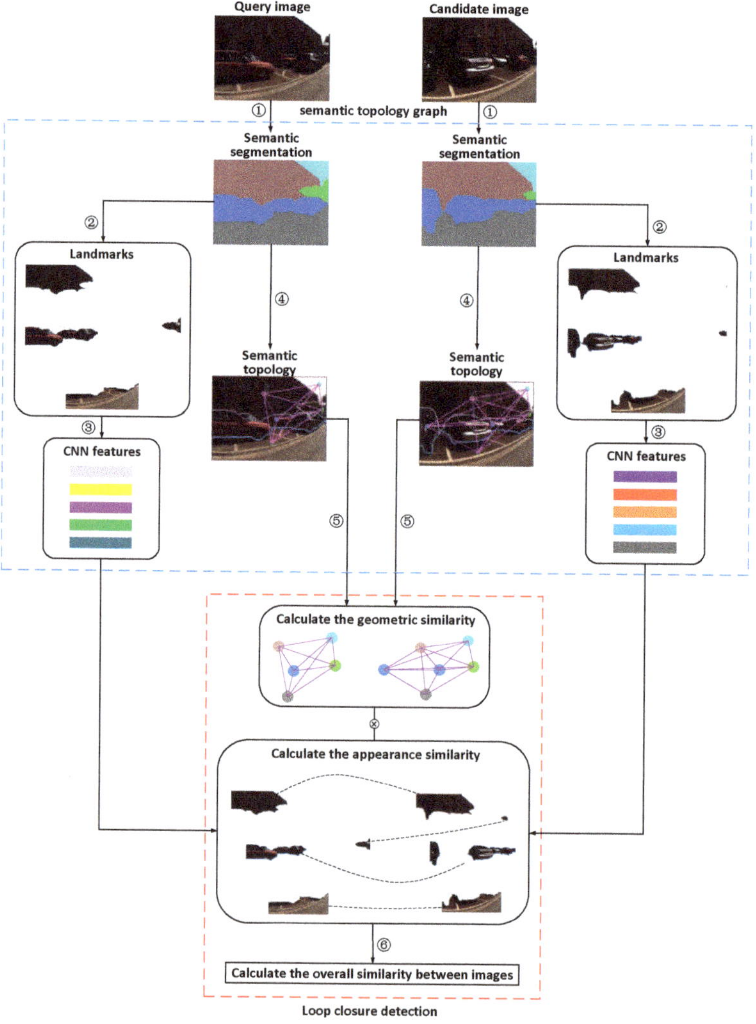

Figure 1. Overview of the semantic loop closure detection algorithm based on topology graphs and convolutional neural network (CNN) features.

Among the key steps, steps (1)–(4) are algorithms for constructing semantic topological graphs, which are given in Section 2.1, and steps (5) and (6) are loop closure detection algorithms, which are provided in Section 2.2.

The method presented in this paper is different from previous methods as follows:

- The extraction of the landmarks uses a pre-trained semantic segmentation network.
- It utilizes semantic information to eliminate dynamic landmarks and select of the distinctive landmarks.
- It adopts random walk descriptors to represent the topological graphs for graph matching.
- It adds geometric constraints on the basis of appearance similarity.

2.1. Semantic Topology Graph

The construction of the semantic topology graphs is the basis of loop closure detection, so this section first introduces the construction process of the semantic topology graphs (see Figure 2).

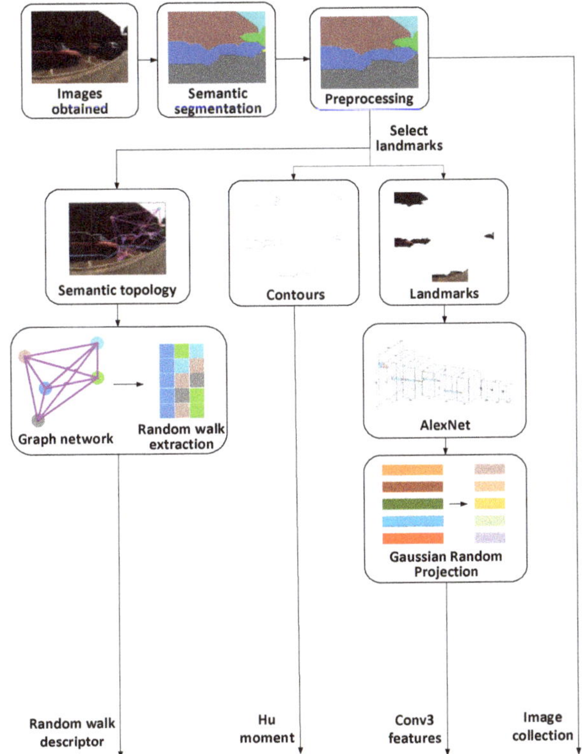

Figure 2. Flow chart for constructing semantic topology graph.

For each obtained image, semantic segmentation is performed to extract landmarks. After preprocessing, the image is divided into landmark regions and contours. The obtained landmarks are selected and sent to AlexNet to extract third convolutional layer (Conv3) features. Then, Gaussian random projection is used to reduce the dimensionality of the feature vectors, and low-dimensional feature vectors are output. In addition, the Hu moment is calculated according to the obtained contours. At the same time, the preprocessed semantic segmentation images are used to extract the center of the landmark region as a node, and the landmarks seen from the same viewpoint are connected by undirected edges to establish a semantic topology graph according to the co-visibility information.

Finally, a random walk descriptor is exploited to describe the topological graph structure. Here, the images obtained before the loop closure detection are called dataset images.

2.1.1. Landmark Extraction

Previous methods [30,31] have employed object proposal [42] to extract landmark regions even though it contains a lot of irrelevant feature information. In particular, the proposed method adopts semantic segmentation to extract landmarks, and this can accurately obtain the range of landmark regions.

DeepLabV3 + [43] is one of the most influential semantic segmentation models. It is better than FCN [44], U-Net [45], and SegNet [46] for some datasets, and it is widely used in the field of engineering technology. The ADE20K dataset [47,48] covers a wide range of scenes and object categories. Furthermore, it provides dense annotations, so it is used to train the DeepLabV3 +. The pre-trained DeepLabV3 + model can be applied for extracting landmarks.

Here, DeepLabV3 + was used to fuse the shallow features outputted by the encoder with the deep features generated from the ASPP module so that it could produce high-precision semantic segmentation results.

2.1.2. Landmark Selection

Due to the effects of illumination and dynamic disturbance in the images obtained by the robot, as well as the inherent defects of the semantic segmentation model, there was a lot of noise, as well as dynamic and secondary landmarks, in the semantic segmentation image that was obtained via the model discussed in Section 2.1.1. To overcome these problems, the landmarks were preprocessed to obtain significant landmark regions. Then, the dynamic regions from the landmarks obtained by preprocessing were removed, and the distinctive patches were selected.

As shown in Figure 3, the semantic segmentation image (see Figure 3b) was filtered to remove the regions; its area was less than the specified threshold (the threshold was 100 in this paper). Figure 3c was obtained by merging region filtered out with the surrounding area. Through the above-discussed procedures, the secondary landmarks and holes were filtered out to obtain the obvious landmark region with clear boundaries.

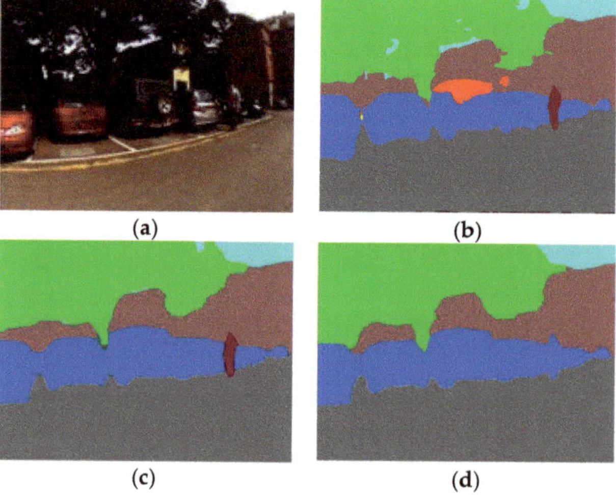

Figure 3. Selection of the landmarks: (**a**) raw image, (**b**) semantic segmentation image, (**c**) the result of filtering, and (**d**) the result of eliminating pedestrian dynamic landmarks.

In order to overcome the impact of dynamic scenes, the semantic information of landmarks was then used to eliminate the pedestrian dynamic landmarks. At the same time, the pedestrians and long-term parking car region were merged, and the merged area could be used as car landmarks in subsequent work. Figure 3d was obtained by the above-mentioned operations. After excluding dynamic landmarks, in the follow-up loop closure detection, the dynamic landmarks were no longer matched. Furthermore, the number of pixels could be calculated for the landmark region. The distinctive landmarks were selected according to the number of landmark pixels and semantic information combined with experimental scenes for loop closure detection. In addition, dynamic landmarks were determined by scene content and landmark semantic information. In other words, according to the movement status of the landmarks in each experimental scene, we removed the moving landmarks in the dataset images and prevented them from participating in subsequent experiments. Formally, we denoted t as the number of distinctive landmarks selected in the image. (t was 5 or 10 in this work).

2.1.3. CNN Features

CNN features have appearance invariance, so they far surpass manual features in the field of image retrieval and classification. AlexNet [49] won the 2012 ImageNet competition champion, and the network was pre-trained for object recognition tasks on the ILSVRC dataset [50].

The AlexNet network architecture had 8 layers, of which the first 5 layers were convolutional layers and the last 3 layers were fully connected layers. There was a pooling layer after the 1st, 2nd, and 5th convolutional layers, but there was none after the 3rd and 4th convolutional layers. Each convolutional layer had activation function ReLU and normalization. The input of the network was a 227×227 3-channel image, and the output feature of the third convolutional layer was $13 \times 13 \times 384 = 64.896$. According to the research of [27], the output features of the third convolutional layer of AlexNet perform best under appearance changes. We found that the output features of the fully connected layer had strong semantic information that was robust to viewpoint changes but poor for appearance changes. At the same time, it was proved that the AlexNex obtained by pre-training in the object recognition task was better than the CNN model based on place recognition training when considering the characteristics of the entire image under the viewpoint changes. Other advanced networks such as VGG, ResNet, and DenseNet have complex architectures, as well as a lack of research and utilization in the field of loop closure detection. Therefore, this article used the relatively lightweight and mature AlexNet in the loop closure detection field to extract CNN features. Based on the above research, the proposed method employed the output of the Conv3 of the AlexNet as the global feature in the landmark region.

The landmark proposal extracted by the object proposal method contained a large amount of irrelevant feature information. This led to a certain amount of noise influence in the CNN feature description. However, the landmark area extracted by semantic segmentation in this paper only contained the landmark feature and no other unrelated features. Figure 4a–c is introduced in Section 2.1.2, this section explains the landmark area and contour extraction. The landmarks from the filtering result (see Figure 4c) were selected to get Figure 4d, and the contour binary (Figure 4e) of the corresponding landmark was obtained by a Canny operator. Then, the landmarks (see Figure 4d) were resized to 227×227 pixels and input to the pre-trained AlexNet to extract features. As a result, the features of each landmark could be represented by a 64.896-dimensional vector. In order to keep the original size information of the landmark, this paper added the Hu moment of the contour (see Figure 4e) to the CNN feature to describe the landmark.

The obtained high-dimensional vector contained redundant landmark feature information, and a large amount of computational cost was required to calculate landmark similarity. Due to the real-time requirements of visual SLAM, the Gaussian random projection method [51] utilized in [30] was employed to reduce the dimensionality of the feature vector to 2048 dimensions.

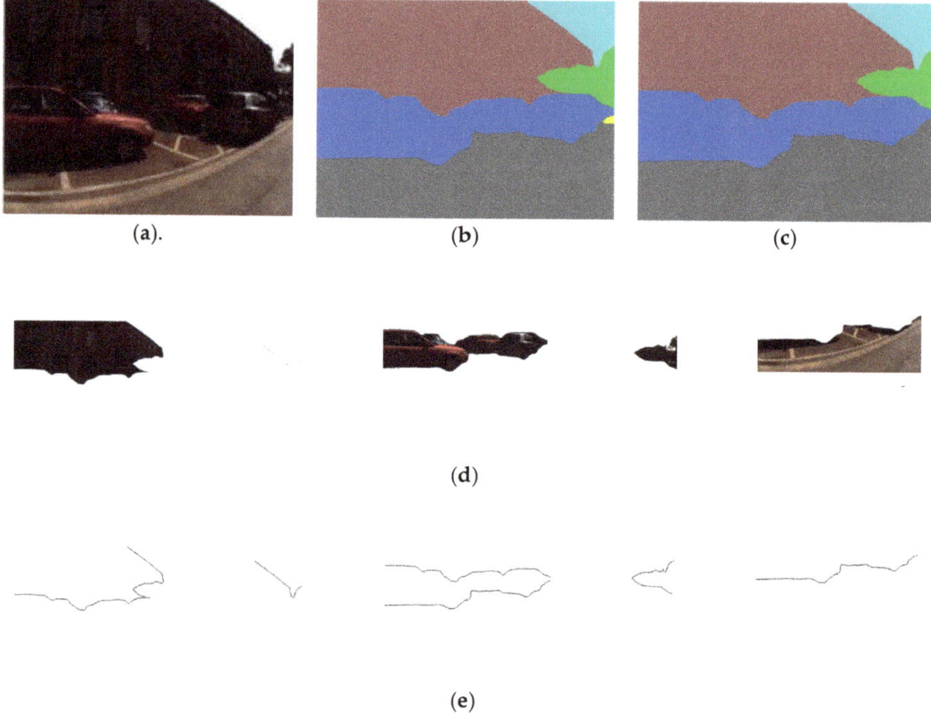

Figure 4. Extraction of landmark regions and contours: (**a**) raw image, (**b**) semantic segmentation image, (**c**) the result of filtering, (**d**) the selected landmark regions, and (**e**) the contours of the selected landmarks.

2.1.4. Graph Representation

In order to preserve the spatial relationship of the scene, a semantic topology graph was constructed from a single image obtained by the robot. When the robot was initialized, the camera captured the first image and started to create the semantic topology graph. In this paper, each landmark was abstracted as a node containing category and pixel number information. Additionally, the node was located at the center of the landmark region.

The truncated random walk proposed by Perozzi et al. [52] was used to describe the semantic topological graph and represented each node as a fixed-length embedding vector. In order to enrich the feature expression, the node index and the number of pixels were adopted to describe the nodes. The node index was obtained according to the 150 semantic categories of the ADE20K dataset, and the number of pixels was acquired from the landmark region.

Then, the random walk descriptor of each node was calculated, and each node was used as the target node. In the semantic topological graph, the next adjacent node was randomly selected until the walking depth n was reached and a random walking path was obtained. In this paper, m random walks were performed on each node to obtain m random walk paths. Finally, each target node could be expressed as a matrix $M = \{m_{ij}\} \in \mathbb{R}^{m \times n}$. Since each node contained information about the index and the number of pixels, a matrix $M = \{m_{ij}\} \in \mathbb{R}^{m \times 2n}$ was finally obtained. In addition, the random walks followed certain rules, i.e., they would not repeat the same path and would not return to the previous nodes during each walk. The selection of m and n was related to the number of nodes in the graph structure. Based on the research of Gawel et al. [36] and the number of landmarks selected in Section 2.1.2, m was selected as 10, 20, or 50, and n was 3 or 5 for experiments.

Figure 5 shows an example of a constructing descriptor. After the image in Figure 5a was obtained by the robot, the landmark nodes were extracted to obtain Figure 5b. Then, the nodes in Figure 5b were connected by undirected edges according to the co-visibility information to get Figure 5c. The semantic topology graph (see Figure 5d) was used to describe the geometric connection of the nodes in Figure 5a. Furthermore, a random walk graph descriptor (see Figure 5e) was constructed according to the semantic topology graph (see Figure 5d). The blue node (car) was used here to construct a descriptor for the target node. For the purpose of illustration, a graph description matrix $M \in \mathbb{R}^{5 \times 6}$ was made to represent the geometric characteristics of the image by five random walks with a depth of 3 each time. The last row of Figure 5e corresponds to the random walk path shown by the black arrow in Figure 5d. Figure 5f used a matrix to quantify the description of Figure 5e, where the red box represents the index of nodes and the number of pixels. The index of the target node car was 21, and 68,192 was the number of pixels contained in the car landmark. In the same way, 2 and 112,918 were the index and pixel number of the building node, respectively; 3 and 13,488 were the index and pixel number of the sky node, respectively; 5 and 10,697 were the index and pixel number of the tree node, respectively; and 7 and 101,572 were the index and pixel number of the road node, respectively. For visualization, some node information was omitted.

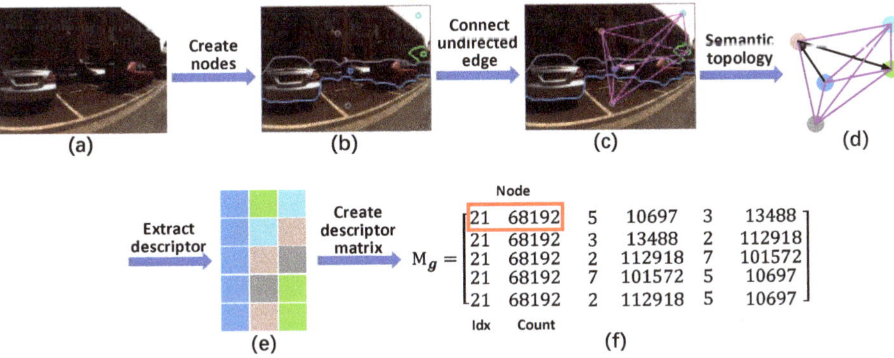

$$M_g = \begin{bmatrix} 21 & 68192 & 5 & 10697 & 3 & 13488 \\ 21 & 68192 & 3 & 13488 & 2 & 112918 \\ 21 & 68192 & 2 & 112918 & 7 & 101572 \\ 21 & 68192 & 7 & 101572 & 5 & 10697 \\ 21 & 68192 & 2 & 112918 & 5 & 10697 \end{bmatrix}$$

Figure 5. Construction of the topology graph and extraction of the descriptor: (**a**) raw image, (**b**) creation of nodes, (**c**) connection of undirected edges, (**d**) construction of topology graph, (**e**) random walk descriptor, and (**f**) descriptor matrix.

2.2. Loop Closure Detection

This section introduces the algorithm of loop closure detection, and its flowchart is shown in Figure 6. Firstly, dataset images (candidate images) that matched the current image (query image) were retrieved. Then, the appearance similarity was calculated according to the CNN and contour features between images. In addition, geometric similarity was obtained by using the random walk descriptor. Finally, loop closure detection was performed according to the overall similarity.

2.2.1. Obtain Candidate Images

When the robot enters a previous environment again, it needs to retrieve the candidate images from the dataset images. In this article, by controlling the number of landmarks that shared the same label between the query image and each image in the dataset, candidate images of the current query image were obtained. The smaller the number of shared nodes, the more candidate images and the longer the retrieval time, and vice versa. A reasonable setting of the number of shared nodes can improve the speed and accuracy of loop closure detection, so the number of shared nodes was set as 1 to obtain candidate images. In other words, when both the query image and an image in the dataset images have a landmark with the same label, the dataset image is considered to be one candidate image.

According to the same principle, each query image can obtain a candidate image set, and each image in the candidate image set has at least one landmark (node) with the same label as the query image.

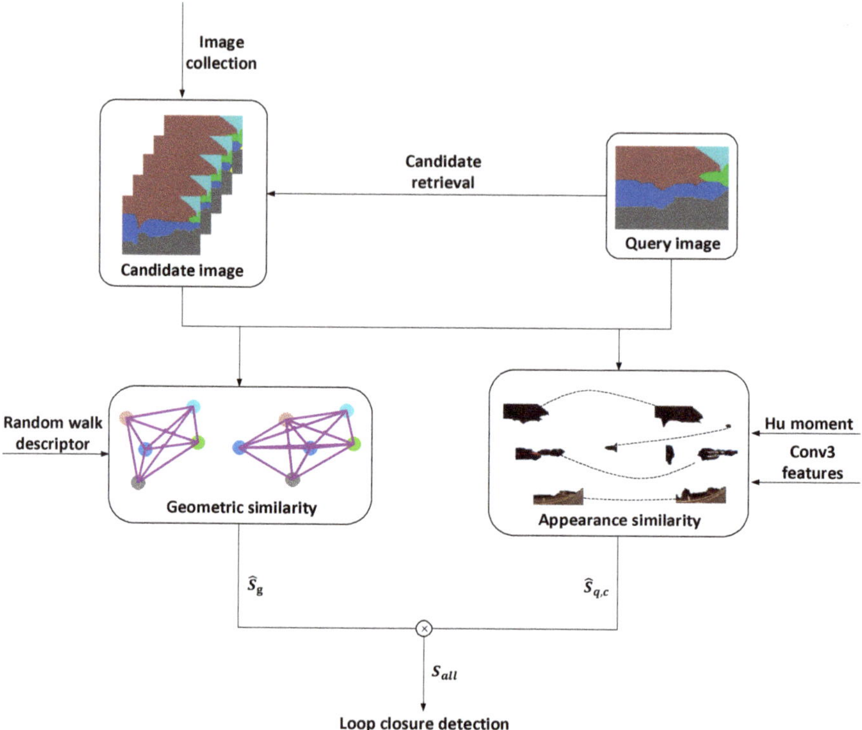

Figure 6. Flow chart of loop closure detection.

2.2.2. Appearance Similarity

To calculate the appearance similarity between the candidate image and the current query image, it is necessary to match the landmark of the query image with all landmarks of the candidate image. By using the semantic information of landmarks, we employed the nearest neighbor search based on the cosine distance (see Equation (1)) of CNN features to match the landmark pairs of the same label in the two images so that only the landmarks of the same category were matched to speed up the matching process. In the matching process, we used a bidirectional matching method, i.e., landmark pairs were accepted only if they were mutual matches.

$$d_{ij}^{cosine} = \frac{1}{2}\left(1 - \frac{v_i^q \cdot v_j^c}{\| v_i^q \|_2 \| v_j^c \|_2}\right) \tag{1}$$

where v_i^q denotes the feature vector of the i-th landmark of the query image and v_j^c describes the feature vector of the j-th landmark of the candidate image.

While calculating the similarity of the CNN features, the geometric shape of the landmark was introduced as a penalty factor to eliminate the false positive phenomenon, i.e., the CNN features were similar but the contours were different. The references [30,31,53] used the difference between the long side and the wide side of the region proposal of the landmark pair to measure the shape difference. However, because they only used the long side and wide side difference of the bounding box, the influences of the scale and rotation was omitted. When the viewpoint changed drastically, the rotation of the landmark caused a large change in the aspect ratio of the bounding box. In the end, the shape penalty factor was too large, resulting in a low appearance similarity.

In order to solve the above problems, Hu moments [54] were used to describe the irregular contour features of landmarks, which possessed invariance about rotation, translation, and scale. Due to the wide range of Hu moments, the logarithm method was used for data compression in order to facilitate comparison. At the same time, considering that the Hu moment may have a negative value, absolute value was taken before the logarithm, as shown in Equation (2):

$$c_i = sign(hu_i)log|hu_i| \quad i = 1, 2, \cdots, 7 \tag{2}$$

where $sign(x)$ is the sign function.

Hu [54] constructed seven invariant moments to describe geometric shape. Therefore, each landmark contour could be expressed as a feature vector by seven Hu moment values through Equation (3):

$$C = (c_1, c_2, c_3, c_4, c_5, c_6, c_7) \tag{3}$$

Through the contour feature vector, the shape difference of landmark contour between query image and candidate image could be calculated by Equation (4):

$$\gamma_{ij} = exp\left(\max_{m=1...7} \frac{\left| c_m^{qi} - c_m^{cj} \right|}{\left| c_m^{qi} \right|} \right) \tag{4}$$

where c_m^{qi} and c_m^{cj} denote the m-th Hu moment of the i-th landmark contour in query image and the m-th Hu moment of the j-th landmark contour in candidate image, respectively.

According to cosine distance of the CNN feature and shape similarity obtained by the above calculation, the appearance similarity of the landmark pair between the query and the candidate images could be obtained by Equation (5):

$$d_{ij} = 1 - d_{ij}^{Cosine} \cdot \gamma_{ij} \tag{5}$$

In Equation (5), when the contour shape of the landmark is close, γ_{ij} is close to 1. If γ_{ij} is larger, it indicates that the contour difference of the landmark is large. In addition, when d_{ij} is close to 1, it means that the landmarks both have similar CNN features and geometric shapes. Furthermore, when d_{ij} is a negative number, it indicates that the geometric shapes of the landmarks differ greatly. If d_{ij} is small, it reveals that there may be differences in the CNN features or geometric shapes.

2.2.3. Geometric Similarity

In visual SLAM, the accuracy of loop closure detection is particularly important. Therefore, it is necessary to consider both appearance similarity and geometric similarity during loop closure detection. Thus, the random walk graph descriptor proposed in Section 2.1.4 was used to calculate geometric similarity for graph matching. Denote that the vectorized form of the descriptor matrix $M = \{m_{ij}\} \in \mathbb{R}^{m \times 2n}$ using $G \in \mathbb{R}^{2mn}$ is a concatenation of the columns of M into a vector.

In Section 2.1.4, we obtained the random walk descriptor of the dataset images and only needed to construct the semantic topology graph to extract the descriptor for the query image. Since the number of pixels was much larger than the node index in value, the absolute size of the feature vector of the description changed greatly. Therefore, it was more appropriate to use cosine similarity to express the relative difference of graph descriptors by Equation (6):

$$S_g(G_q, G_c) = \frac{G_q \cdot G_c}{\| G, q \|_2 \cdot \| G_c \|_2} \tag{6}$$

where G_q and G_c denote the feature vector of the random walk descriptor in query image and candidate image, respectively. The denominator is the product of the corresponding vector modulus length. After getting a similarity score, it needed to be normalized with Equation (7):

$$S_g = \frac{1}{2} + \frac{1}{2} S_g\left(G_q, G_c\right) \tag{7}$$

Through Equation (7), the similarity score in the range of $[0, 1]$ could be obtained.

2.2.4. Overall Similarity

This section discusses the calculation of the overall similarity between the query image and the candidate image. We not only considered the appearance characteristics of the image but also added geometric constraints. Sections 2.2.2 and 2.2.3 obtained the appearance and geometric similarities of a single landmark pair. Through Equations (8) and (9), we scored each best matched landmark pair $\left(I_q^i, I_c^j\right)$ between the query image I_q and the candidate image I_c. The similarity score between each landmark i in the query image and the most similar landmark j selected by the nearest neighbor search method in the candidate image was first computed, and then the scores were assigned to the candidate image as the mean value of individual scores of its landmarks. Finally, through Equation (10), the overall similarity score of each candidate image was obtained.

$$\hat{S}_{q,c} = \frac{1}{t} \sum_{i,j} d_{ij} \tag{8}$$

$$\hat{S}_g = \frac{1}{t} \sum_{i,j} S_g \tag{9}$$

where t denotes the number of the landmarks in the candidate image I_c (including unmatched landmarks), i represents the i-th landmark of the current query image, and j is the most similar landmark selected by the nearest neighbor search method in the candidate image. Moreover, the sum is done only on the best matched landmark pairs selected by the nearest neighbor search method.

$$S_{all} = \hat{S}_g \cdot \hat{S}_{q,c} \tag{10}$$

In Equation (10), geometric similarity $\hat{S}_g$ is used as a penalty factor for appearance similarity score to filter candidate images with similar local features but large differences in geometric information. In the experiments, it was normalized to $[0, 1]$. We normalized a set of overall similarity scores between the current query image and all candidate images (in Section 2.2.1). When the overall similarity score set of the current query image be $X = \{x_1, x_2, \ldots, x_m\}$, m denotes the number of candidate images retrieved from the current query image. In addition, x_i ($1 <= i <= m$) denotes the overall similarity score between the current query image and the i-th candidate image. Through Equation (11), each value in X can be normalized to $[0, 1]$:

$$y_i = \frac{x_i - min(X)}{max(X) - min(X)} \tag{11}$$

where $Y = \{y_1, y_2, \ldots, y_m\}$ is the normalized score set and y_i is one of the score values.

After obtaining the normalized similarity score for loop closure detection, it is often necessary to perform time and space consistency verification. In this article, the geometry check was not added. Nevertheless, the proposed method that integrates visual, spatial, and semantic information was still found to improve performance.

3. Results

This section mainly introduces the experimental process and result analysis. In order to evaluate the performance of different components of the proposed method and to compare the performance of the proposed method with other state-of-the-art methods, the following methods were considered:

(1)   A state-of-the-art BoW-based method (named 'DBoW2') that did not need to recreate the vocabulary in different scenarios [11].

(2)   A CNN approach (named 'Conv3') that applied the global Conv3 feature to describe images [27].

(3)   A technique (named 'CNNWL') that combined global and local CNN features but ignored the spatial relationship of landmarks [30].

(4)   An approach (named 'GOCCE') that was based on local features and semi-semantic information [31]. It was closer to the proposed algorithm.

(5)   The proposed complete method (named 'VSSTC') that integrated visual, spatial, and semantic information through topological graphs and CNN features.

(6)   Two simplified versions of the random walk graph descriptor of our method. One version included only label information (named 'VSSTC-Label'), and the other used only pixel number information (named 'VSSTC-Pixel').

(7)   A modified version of our method of landmark segmentation method (named 'VSSTC-OP') that used object proposal instead of semantic segmentation to extract landmark regions, while using the size of the bounding box instead of Hu moments.

(8)   Another reduced version of our method (named 'VSSTC-LS') that did not construct topological graphs and lacks spatial information. In order to compare the performance of the proposed method with that of the above methods, comparative experiments were carried out on the datasets and a mobile robot. Based on the experimental results, the proposed method was fully evaluated.

In the experiments, a precision–recall curve (P–R curve) [55] was used as a quantitative evaluation metric, as it is a standard metric for loop closure detection results. By changing the similarity threshold, the P–R curve could be obtained. In order to further observe the experimental results, the maximum recall rate under the precision of 100% and the area (the value in $[0, 1]$) under the P–R curve (AUC) were used as auxiliary evaluation metrics. The larger the recall rate and the area, the better the performance.

The experiments were carried out on a desktop equipped with GTX 1080Ti GPU. We used a pre-trained DeepLabV3 + based on the TensorFlow [56] for semantic segmentation and AlexNet based on Caffe [57] to extract CNN features.

3.1. Dataset Experiments

3.1.1. Datasets

The performance evaluation was performed through the following four public datasets, which are widely used in the field of loop closure detection and place recognition.

City Centre dataset: This dataset [10] contains 1237 pairs of images in outdoor urban environments. The resolution of each image is 640 × 480. It contains dynamic scenes of pedestrians and vehicles. In addition, there are a lot of scenes with changes in viewpoint caused by lateral displacement and reverse movement. It also includes shadows and spots caused by lighting. Ground truth data are included in the dataset. In the experiment, six types of landmarks were selected: tree, road, sky, building, car, and grass. Furthermore, person and bicycle dynamic landmarks were excluded. The dataset scenario is listed in Figure 7a.

(a)

(b)

(c)

Figure 7. *Cont.*

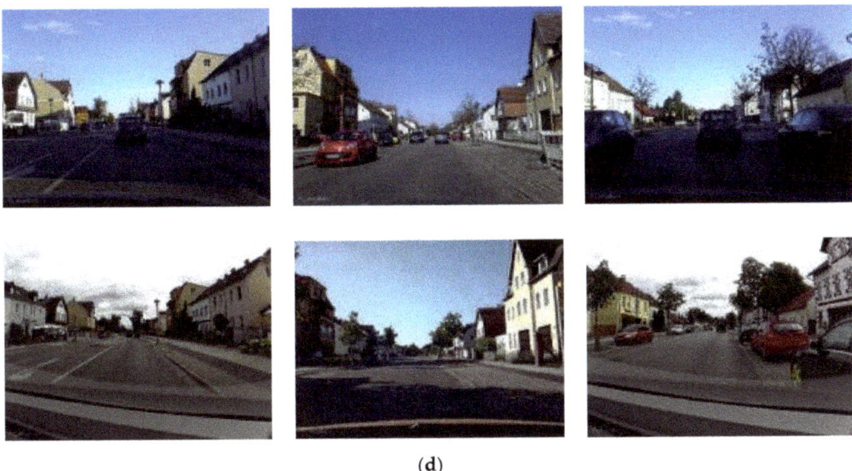

(d)

Figure 7. The example scenes from the used datasets. The query images are placed in the first row, and the matching images are placed in the second row. (**a**) The images in the City Centre dataset. (**b**) The images in the New College dataset. (**c**) The images in the Gardens Point dataset. (**d**) The images in the Mapillary dataset.

New College dataset: This dataset [10] contains 1073 pairs of images of a university campus, which contains a small number of dynamic scenes of pedestrians and vehicles. Additionally, it includes the scenes of viewpoint changes caused by lateral displacement. Furthermore, there are many indoor repetitive structures, such as walls, chairs, and windows. The resolution of each image is 640 × 480. Ground truth data are included in the dataset. In the experiments, eight types of landmarks—sky, wall, chair, building, road, grass, tree, and car—were selected and the person landmark was excluded. The scenes of the dataset images are shown in Figure 7b.

Gardens Point dataset: This dataset has been used in [27] before and is available online (http://tinyurl.com/gardenspointdataset) which contains two traversals on the university campus during the day, one route on the left-hand side of the road and the other on the right-hand side of the road on the same day. In addition, the dataset includes 200 pairs of images, with the left and right sides of the road each containing 200 images. It contains the viewpoint changes caused by walking on the left and right sides of the road, and there are many pedestrian dynamic objects. Ground truth data are included in the dataset. In the experiment, we selected eight types of landmarks: wall, building, sky, road, flooring, door, tree, and ceiling. The person landmark was excluded. The scenes within dataset images are shown in Figure 7c.

Mapillary dataset: The Mapillary datasets were first introduced by [30] and is available online (http://www.mapillary.com) they have street-level imagery and map data from all over the world. We downloaded the Berlin August–Bebel–Straße sequence, as well as ground truth data, to get 318 images. The dataset exhibits significant changes in viewpoints and severe changes in appearance. In addition, it contains a large number of moving vehicle dynamic objects. We selected six types of landmarks—building, road, sky, tree, pole, and signal—and excluded car landmarks. Some images in the dataset are shown in Figure 7d.

3.1.2. Experimental Results and Analysis

The experiments not only illustrated the experimental effects of the proposed approach under different parameter settings but also provided comparison results with other advanced methods. In order to test the topological graph descriptor of the proposed method, the experiments (Figure 8a,b) were conducted by extracting the number of different landmarks ($t = 5$ or 10), using different random walks ($m = 10, 20,$ or 50) and walk depth ($n = 3$ or 5) to change the size of the graph representation. In addition, the proposed complete method was fully compared with other methods (Figure 8). Moreover, we conducted ablation studies (Figure 8c,d) on the components of the proposed system to analyze the performance of each component. Among them, we analyzed the influence of the composition factors of the topological graph descriptors in the proposed method (Figure 8c). The VSSTC-Label method only considered the landmark label, and the VSSTC-Pixel approach only used the number of pixels. These two methods and the proposed VSSTC method were tested on the Gardens Point dataset. In addition, to verify the proposed landmark region extraction method (VSSTC), we replaced the landmark segmentation method with the object proposal method (VSSTC-OP) instead of semantic segmentation to obtain bounding boxes (Figure 8d). At the same time, this experiment verified the effectiveness of Hu moments relative to the size of the bounding box on the Mapillary dataset. Furthermore, to consider the impact of the semantic topology graph on the performance of the proposed method, the removed topology graphs (VSSTC-LS) and the complete method (VSSTC) were analyzed in the mobile robot experiment (see Section 3.2). Figure 8 shows the P–R curves of the experimental results, and Tables 1 and 2 list the maximum recall rate under the precision of 100% and the area under the P–R curve.

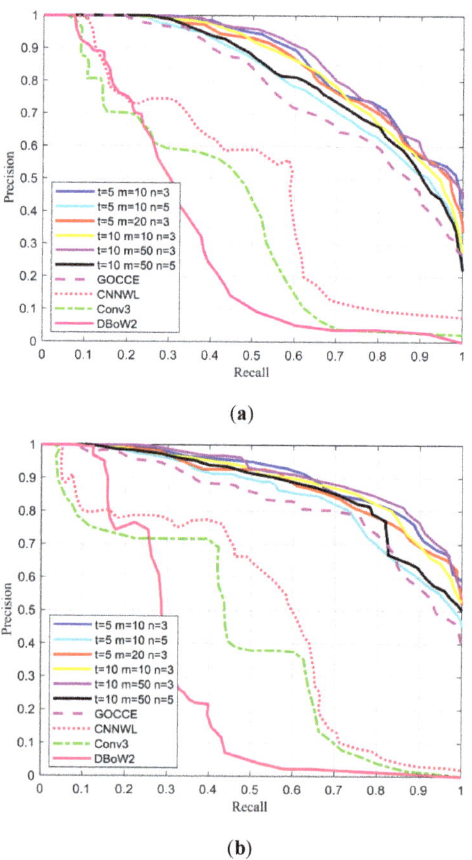

(a)

(b)

Figure 8. *Cont.*

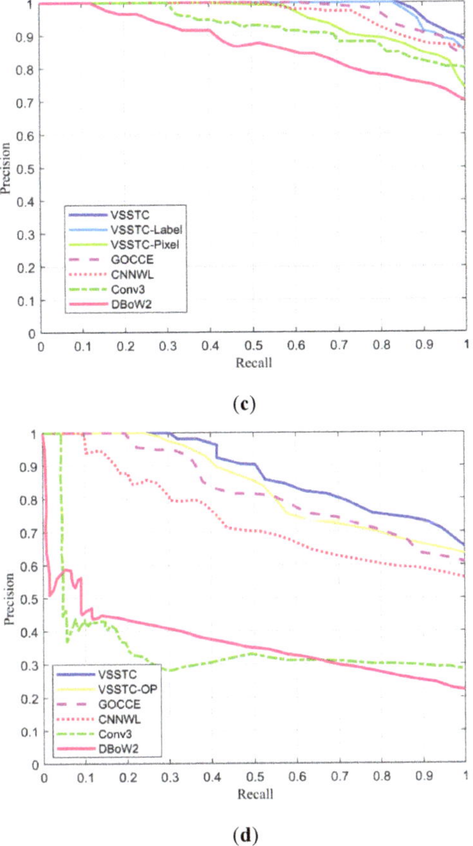

Figure 8. Experimental results for the datasets: (**a**) the precision–recall (P–R) curves of City Centre dataset, (**b**) the P–R curves of New College dataset, (**c**) the P–R curves of Gardens Point dataset, and (**d**) the P–R curves of Mapillary dataset.

Table 1. Experimental results on the City Centre and New College datasets. AUC: area under the curve. Conv3: third convolutional layer.

Methods	City Centre		New College	
	Recall (%)	AUC	Recall (%)	AUC
VSSTC (t = 5 m = 10 n = 3)	29.81	0.8707	20.51	0.9042
VSSTC (t = 5 m = 10 n = 5)	22.44	0.8085	13.14	0.8492
VSSTC (t = 5 m = 20 n = 3)	24.68	0.8512	18.27	0.8848
VSSTC (t = 10 m = 10 n = 3)	21.47	0.8459	15.39	0.8923
VSSTC (t =1 0 m = 50 n = 3)	29.17	0.8636	22.76	0.9109
VSSTC (t = 10 m = 50 n = 5)	25.00	0.8186	14.42	0.8699
GOCCE [31]	19.23	0.7769	8.98	0.8284
CNNWL [30]	6.81	0.4882	5.09	0.5051
Conv3 [27]	5.72	0.3938	3.22	0.4258
DBoW2 [11]	7.36	0.3187	8.04	0.2975

Table 2. Experimental results on the other datasets.

Methods	Gardens Point		Mapillary		Robot	
	Recall (%)	AUC	Recall (%)	AUC	Recall (%)	AUC
VSSTC	84.33	0.9906	30.32	0.8796	30.25	0.7628
VSSTC-Label	82.96	0.9871	- [1]	-	-	-
VSSTC-Pixel	39.52	0.9528	-	-	-	-
VSSTC-OP	-	-	24.36	0.8395	-	-
VSSTC-LS	-	-	-	-	18.14	0.6229
GOCCE [31]	60.99	0.9792	19.59	0.8260	23.49	0.7132
CNNWL [30]	49.57	0.9700	9.37	0.7397	20.09	0.5909
Conv3 [27]	30.15	0.9268	4.43	0.3544	8.35	0.3823
DBoW2 [11]	12.10	0.8730	0	0.3643	0	0.3249

[1] The symbol '-' indicates that the experiment of the method (row) is not performed under the corresponding dataset (column).

From the P–R curves in Figure 8, we can see that in the case of graph descriptor size $m = 10$ and $n = 3$, the maximum recall rate of $t = 5$ was larger than that of $t = 10$. This shows that blindly increasing the number of landmarks can entrain many minor landmarks to participate in loop closure detection, thereby weakening performance. In addition, whether in the case of $t = 5, m = 10$, or $t = 10, m = 50$, the maximum recall rate at $n = 3$ exceeded that at $n = 5$. This indicates that when the walk depth n reached the graph size limit, continuing to increase the walk depth n made the model visit the nodes that had been visited before. This reduced the ability to express the graph descriptor and diminished the loop closure detection performance. Furthermore, in the case of $t = 5$ and $n = 3$, the maximum recall rate when $m = 10$ was larger than that when $m = 20$. However, in the case of $t = 10$ and $n = 3$, the maximum recall rate when $m = 50$ was larger than that when $m = 10$. This demonstrates that the number of random walks m was determined by the size of the semantic topology graph. When the size of the semantic topology graph was small, too large a number of walks reduced the performance of the graph descriptor. When the size of the semantic topology graph was small, a large number of random walk times caused the model to access repeated paths, thereby reducing performance. However, appropriately increasing the number of random walk times according to the size of the semantic topology graph improved the expressive ability of the graph descriptor. In summary, when $t = 5, m = 10$, and $n = 3$, the effect of loop closure detection achieved a good compromise between accuracy and complexity. In order to further clarify the experimental results, it can be observed from Table 1 that the maximum recall rate and AUC value in the City Centre and New College datasets also conformed to the above conclusions.

From the experimental results on the City Centre and New College datasets, it can be seen that the performance of the DBoW2 method was the worst. Moreover, this method was inferior to other methods based on CNN features. This shows that the traditional BoW method based on manual features was poor in robustness and could only deal with limited scenarios. In addition, the CNNWL method was better than the Conv3 one. This shows that the CNNWL method that combined global and local CNN features had better graph description capabilities than that of the global CNN feature method. Furthermore, the performance of the proposed VSSTC method significantly exceeded that of the CNNWL method. This demonstrates that the proposed method with added spatial constraints had better performance. This also proves that the random walk descriptor based on the semantic topological graph proposed in this paper had an excellent graph description ability. Importantly, our method outperformed the GOCCE one, which shows the advantages of the proposed semantic topology graph in the face of viewpoint changes and dynamic scenes.

The authors of this article conducted three groups of ablation studies to analyze the impact of each component of the proposed method on the overall performance. From Figure 8c and Table 2, we can see that the VSSTC-Label method had better performance than the VSSTC-Pixel approach, which underlines the importance of landmark label information in the topology graph descriptor.

In addition, it could be seen that the performance of VSSTC-Pixel method was inferior to that of the GOCCE method, which shows that the performance of graph descriptors lacking semantic information dropped sharply. Furthermore, the VSSTC method had the best performance, which also reflects that the performance of the proposed complete method was greatly improved by integrating the landmark label and pixel number information.

From Figure 8d and Table 2, we can understand that the performance of the VSSTC-OP method was inferior to that of the VSSTC method, which reveals the superiority of using semantic segmentation to extract landmark regions and employing Hu moments to represent region shape information. As expected, the bounding box extracted by region proposal extracted interference features when facing the presence of a complex background, resulting in performance degradation. The remaining ablation research is given when discussing the mobile robot experiment.

3.2. Mobile Robot Experiment

In order to further verify the robustness of the proposed method to viewpoint changes and dynamic scenes, experiments were carried out in outdoor scenes using the mobile robot of our team.

3.2.1. Experimental Platform

As shown in Figure 9, we used a wheel–leg hybrid hexapod robot with a length of 2 m, a width of 1.7 m, and a height of 1 m as the experimental platform. Moreover, the robot was equipped with an Intel Braswell processor and a centimeter-level integrated navigation system. We used a controller to remotely control the robot and drive 1.5 km in the campus of Xidian University. The data were captured by the front YAMAKO camera (see Figure 9b) for the experiment. In the mobile robot experiment, with a focal length of 10 mm and a working distance of 15 m, a field of view of approximately 9600×7200 mm could be obtained. The detailed parameters of the YAMAKO camera are shown in Table 3. Finally, 108,000 frames of images were collected at a video frame rate of 30 Hz. By setting the distance threshold of the obtained image sequence to 2 m, 720 key frames could be obtained. In addition, the obtained frames were perfectly aligned by the GPS information, which could be used as the ground truth of loop closure detection.

Table 3. Detailed parameters of the used YAMAKO camera.

Product Name	Network Integrated Movement
Model	YM86 × 10M2N
Sensor	1/2″
Focal length	10~860 mm
FOV	42°~0.44°
Resolution	1920 × 1080
Aperture	F2.0~6.8

As seen in Figure 10, the trajectory of the robot was recorded by the GNSS and INS integrated positioning system. The robot drove two laps; the first lap was obtained by driving along the left side of the road, and the second lap was acquired by driving along the right side of the road in the same direction as the first lap. The experimental scene contained changes in viewpoint caused by the lateral displacement and also included a lot of dynamic scenes. In addition, it included a lot of shadows and bright spots caused by light. The experiment selected four types of landmarks: road, building, tree, and sky. Furthermore, the dynamic landmarks of person, bicycle, and car were excluded.

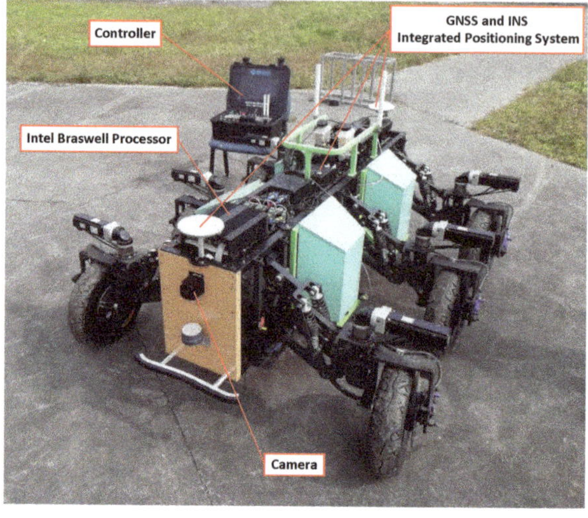

(**a**)

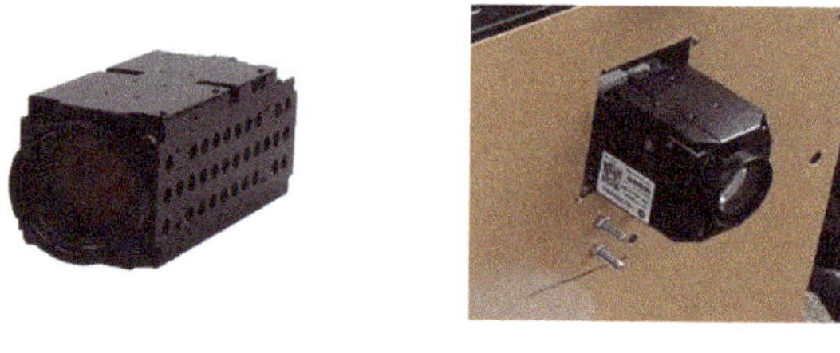

(**b**)

Figure 9. The experimental platform of the mobile robot: (**a**) wheel–leg hybrid hexapod robot and (**b**) the YAMAKO camera used by the mobile robot.

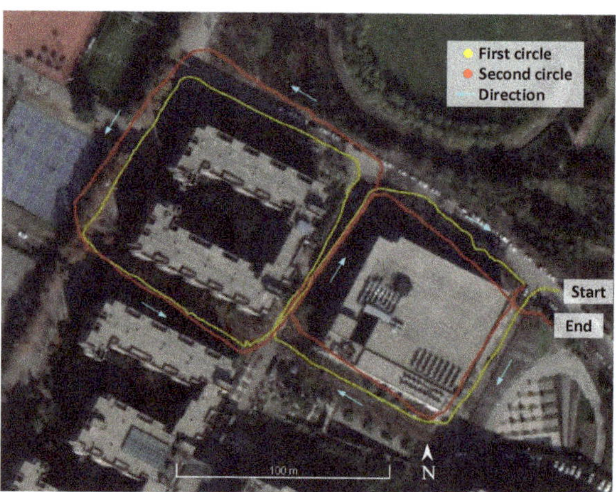

Figure 10. The trajectory of the robot.

3.2.2. Experimental Results and Analysis

Figure 11 shows the P–R curves of the mobile robot, and Table 2 lists the maximum recall rate under the precision of 100% and the area under the P–R curve. According to Section 3.1.2, $t = 5$, $m = 10$, and $n = 3$ were used as the random walk descriptor parameters of the proposed method. In Figure 11 and Table 2, it can be seen that the DBoW2 method performed the worst. A possible reason for this performance is that the experimental scene contained a large number of viewpoint changes and dynamic scenes. Moreover, the CNNWL method had better performance than the Conv3 one, which shows that the CNNWL method was more robust for expressing the image by considering both global and local features in the viewpoint changes and dynamic scenes. Furthermore, the proposed VSSTC method performed better than the CNNWL and GOCCE methods, thus demonstrating that the spatial and semantic information played an important role in improving the loop closure detection performance of changing viewpoint and dynamic scenes.

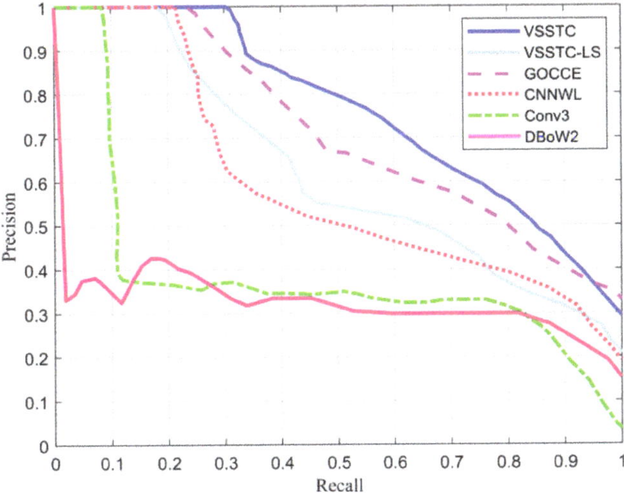

Figure 11. The P–R curves of the mobile robot experiment.

More importantly, from Figure 11 and Table 2, it can be seen from the experimental results that the VSSTC-LS method had a much poorer performance than the VSSTC method that considered spatial information. This shows that the spatial geometric information had a greater impact on loop closure detection performance. In addition, we can see that the performance of the VSSTC-LS method was slightly better than that of the CNNWL method, which reflects that the visual and semantic modules used in the proposed method were superior when there was no geometric information.

Figure 12 shows a loop closure detection result obtained by the proposed method. The blue points are the selected 720 key-frames, and the key-frames connected by the red line indicate the correct loop closure. It can be seen from the figure that the proposed method could still detect a large number of loop closures under the influence of viewpoint changes and dynamic scenes. Figure 13a,b shows the true positive image pairs detected by loop closure detection at 1 and 2 in Figure 12, respectively. Figure 13a,b contains viewpoint changes and pedestrian dynamic scenes, as well as the shadows caused by the changes in illumination. This shows that the proposed method had a better graph description ability in the above-mentioned drastically changing scenes.

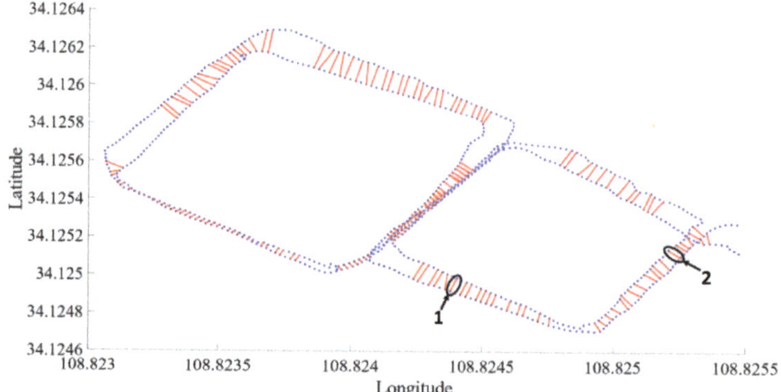

Figure 12. Example of loop closure detection.

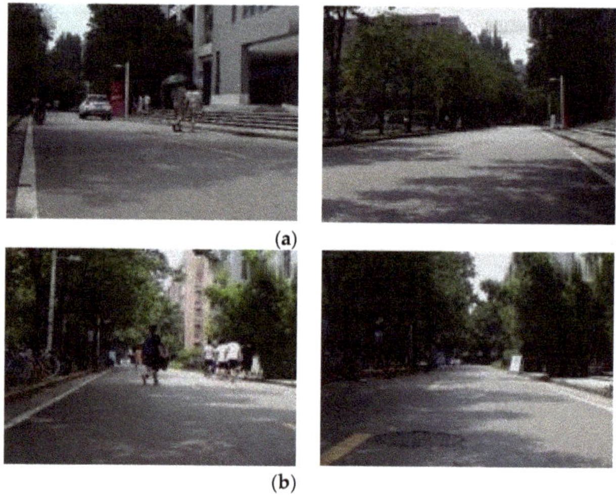

(a)

(b)

Figure 13. Examples of correct matches obtained using our method: (**a**) the true positive image pair at 1 in Figure 12 and (**b**) the true positive image pair at 2 in Figure 12.

4. Conclusions

This paper studied the loop closure detection in visual SLAM and proposed a robust loop closure detection method that integrated visual–spatial–semantic information to deal with viewpoint changes and dynamic scenes. Firstly, semantic topological graphs were employed to represent the spatial geometric relationships of landmarks, and random walk descriptors were applied to represent topological graphs. By adding geometric constraints, the mismatch problem caused by changes in viewpoint was improved. Then, semantic information was utilized to eliminate dynamic landmarks, and distinctive landmarks were selected for loop closure detection, which effectively alleviated the impact of dynamic scenes. Finally, semantic segmentation was used of accurately obtain the landmark region. At the same time, deep learning was adopted to automatically learn the complex internal features of landmarks without the need to manually design features. As a result, it eased the effect of appearance changes. According to the experimental results of the datasets and a mobile robot, the proposed method can effectively cope with changes in viewpoint and dynamic scenes.

However, the proposed method has certain limitations. Firstly, the pros and cons of using semantic segmentation to extract landmark regions depend on the selection of the semantic segmentation model and pre-training datasets. When using this approach, the users need to select a semantic segmentation model according to experimental scenes. At the same time, the model was trained by fine-tuning and transfer learning. Second, this work was offline. It takes a certain amount of time to extract landmark regions and obtain CNN features. In future research, we will try other segmentation models to extract semantic landmarks and use more comprehensive and complete datasets to train the segmentation network so that the model can cope with changing experimental scenarios. Furthermore, this paper used a single image to construct a semantic topology graph. In the future, we will construct a topology graph for sequence images to improve loop closure detection performance. In addition, the proposed strategy for selecting representative landmarks still has room for improvement. To further explore more suitable representative landmark selection strategies, we plan to divide landmarks into the four categories of dynamic, static, unreliable segmentation, and ubiquitous landmarks based on indoor and outdoor scenes while considering the differences between urban and rural scenes. Then, we will assign weights to each type of landmark according to different dataset scenarios to improve our work. Furthermore, in the experimental part, in order to conduct ablation studies, we designed different combinations of the proposed methods, and there is still room for the optimization of these combinations. To further explore the effect of each component of the proposed method on overall performance, we will design more diversified and rigorous ways of measurements to improve the work of this article.

Author Contributions: Conceptualization, Y.W. and Y.Q.; methodology, Y.W., Y.Q. and P.C.; software, Y.W.; resources, X.D. and P.C.; writing—original draft preparation, Y.W.; writing—review and editing, Y.Q. and P.C.; supervision, Y.Q. and X.D. All authors have read and agreed to the published version of the manuscript.

Funding: This research was funded by the National Natural Science Foundation of China under Grant 61871308, the Natural Science Basic Research Plan in Shaanxi Province of China under Grant 2019JM-426.

Acknowledgments: We thank Mark Cummins for providing the City Centre, New College datasets. In addition, we are also very grateful for the Gardens Point dataset provided by QUT. We are also very grateful for the dataset provided by the open source map service of Mapillary.

Conflicts of Interest: The authors declare no conflict of interest.

References

1. Smith, R.C.; Cheeseman, P. On the representation and estimation of spatial uncertainty. *Int. J. Robot. Res.* **1986**, *5*, 56–68. [CrossRef]
2. Palomeras, N.; Carreras, M.; Andrade-Cetto, J. Active SLAM for autonomous underwater exploration. *Remote Sens.* **2019**, *11*, 2827. [CrossRef]
3. Chiang, K.-W.; Tsai, G.-J.; Li, Y.-H.; Li, Y.; El-Sheimy, N. Navigation Engine Design for Automated Driving Using INS/GNSS/3D LiDAR-SLAM and Integrity Assessment. *Remote Sens.* **2020**, *12*, 1564. [CrossRef]
4. Ho, K.L.; Newman, P. Detecting loop closure with scene sequences. *Int. J. Comput. Vis.* **2007**, *74*, 261–286. [CrossRef]
5. Folkesson, J.; Christensen, H. Graphical SLAM-a self-correcting map. In Proceedings of the IEEE International Conference on Robotics and Automation, New Orleans, LA, USA, 26 April–1 May 2004; pp. 383–390.
6. Thrun, S.; Montemerlo, M. The Graph SLAM Algorithm with Applications to Large-Scale Mapping of Urban Structures. *Int. J. Robot. Res.* **2006**, *25*, 403–429. [CrossRef]
7. Grisetti, G.; Kümmerle, R.; Stachniss, C.; Burgard, W. A tutorial on graph-based SLAM. *IEEE Intell. Transp. Syst. Mag.* **2010**, *2*, 31–43. [CrossRef]
8. Triggs, B.; McLauchlan, P.F.; Hartley, R.I.; Fitzgibbon, A.W. Bundle adjustment—A modern synthesis. In Proceedings of the International workshop on vision algorithms, Berlin/Heidelberg, Germany, 21–22 September 1999; pp. 298–372.
9. Williams, B.; Klein, G.; Reid, I. Automatic Relocalization and Loop Closing for Real-Time Monocular SLAM. *IEEE Trans. Pattern Anal. Mach. Intell.* **2011**, *33*, 1699–1712. [CrossRef]
10. Cummins, M.; Newman, P. FAB-MAP: Probabilistic localization and mapping in the space of appearance. *Int. J. Robot. Res.* **2008**, *27*, 647–665. [CrossRef]

11. Gálvez-López, D.; Tardos, J.D. Bags of binary words for fast place recognition in image sequences. *IEEE Trans. Robot.* **2012**, *28*, 1188–1197. [CrossRef]
12. Lowe, D.G. Distinctive Image Features from Scale-Invariant Keypoints. *Int. J. Comput. Vis.* **2004**, *60*, 91–110. [CrossRef]
13. Bay, H.; Ess, A.; Tuytelaars, T.; Gool, L.V. Speeded-Up Robust Features (SURF). *Comput. Vis. Image Underst.* **2008**, *110*, 346–359. [CrossRef]
14. Rublee, E.; Rabaud, V.; Konolige, K.; Bradski, G. ORB: An efficient alternative to SIFT or SURF. In Proceedings of the 2011 International Conference on Computer Vision, Barcelona, Spain, 6–13 November 2011; pp. 2564–2571.
15. Angeli, A.; Filliat, D.; Doncieux, S.; Meyer, J. Fast and Incremental Method for Loop-Closure Detection Using Bags of Visual Words. *IEEE Trans. Robot.* **2008**, *24*, 1027–1037. [CrossRef]
16. Labbe, M.; Michaud, F. Online global loop closure detection for large-scale multi-session graph-based SLAM. In Proceedings of the IEEE/RSJ International Conference on Intelligent Robots and Systems, Chicago, IL, USA, 14–18 September 2014; pp. 2661–2666.
17. Mur-Artal, R.; Tardós, J.D. Orb-slam2: An open-source slam system for monocular, stereo, and rgb-d cameras. *IEEE Trans. Robot.* **2017**, *33*, 1255–1262. [CrossRef]
18. Sünderhauf, N.; Protzel, P. Brief-gist-closing the loop by simple means. In Proceedings of the IEEE/RSJ International Conference on Intelligent Robots and Systems, San Francisco, CA, USA, 25–30 September 2011; pp. 1234–1241.
19. Oliva, A.; Torralba, A. Modeling the Shape of the Scene: A Holistic Representation of the Spatial Envelope. *Int. J. Comput. Vis.* **2001**, *42*, 145–175. [CrossRef]
20. Naseer, T.; Spinello, L.; Burgard, W.; Stachniss, C. Robust visual robot localization across seasons using network flows. In Proceedings of the Twenty-Eighth AAAI Conference on Artificial Intelligence, Québec City, QC, Canada, 27–31 July 2014; pp. 2564–2570.
21. Milford, M.J.; Wyeth, G.F. SeqSLAM: Visual route-based navigation for sunny summer days and stormy winter nights. In Proceedings of the IEEE International Conference on Robotics and Automation, Saint Paul, MN, USA, 14–18 May 2012; pp. 1643–1649.
22. Abdollahyan, M.; Cascianelli, S.; Bellocchio, E.; Costante, G.; Ciarfuglia, T.A.; Bianconi, F.; Smeraldi, F.; Fravolini, M.L. Visual localization in the presence of appearance changes using the partial order kernel. In Proceedings of the European Signal Processing Conference, Rome, Italy, 3–7 September 2018; pp. 697–701.
23. Pepperell, E.; Corke, P.I.; Milford, M.J. All-environment visual place recognition with SMART. In Proceedings of the IEEE international conference on robotics and automation, Hong Kong, China, 31 May–7 June 2014; pp. 1612–1618.
24. Hansen, P.; Browning, B. Visual place recognition using HMM sequence matching. In Proceedings of the IEEE/RSJ International Conference on Intelligent Robots and Systems, Chicago, IL, USA, 14–18 September 2014; pp. 4549–4555.
25. Xia, Y.; Jie, L.; Lin, Q.; Hui, Y.; Dong, J. An Evaluation of Deep Learning in Loop Closure Detection for Visual SLAM. In Proceedings of the 2017 IEEE International Conference on Internet of Things and IEEE Green Computing and Communications and IEEE Cyber, Physical and Social Computing and IEEE Smart Data, Exeter, UK, 21–23 June 2017; pp. 85–91.
26. Hou, Y.; Zhang, H.; Zhou, S. Convolutional neural network-based image representation for visual loop closure detection. In Proceedings of the IEEE International Conference on Information and Automation, Lijiang, China, 8–10 August 2015; pp. 2238–2245.
27. Sünderhauf, N.; Shirazi, S.; Dayoub, F.; Upcroft, B.; Milford, M. On the performance of ConvNet features for place recognition. In Proceedings of the IEEE/RSJ International Conference on Intelligent Robots and Systems, Hamburg, Germany, 28 September–2 October 2015; pp. 4297–4304.
28. Arroyo, R.; Alcantarilla, P.F.; Bergasa, L.M.; Romera, E. Fusion and Binarization of CNN Features for Robust Topological Localization across Seasons. In Proceedings of the IEEE/RSJ International Conference on Intelligent Robots and Systems, Daejeon, South Korea, 9–14 October 2016; pp. 4656–4663.
29. Gao, X.; Zhang, T. Unsupervised learning to detect loops using deep neural networks for visual SLAM system. *Auton. Robot.* **2017**, *41*, 1–18. [CrossRef]
30. Sünderhauf, N.; Shirazi, S.; Jacobson, A.; Dayoub, F.; Pepperell, E.; Upcroft, B.; Milford, M. Place recognition with convnet landmarks: Viewpoint-robust, condition-robust, training-free. *Robot. Sci. Syst.* **2015**, 1–10.

31. Cascianelli, S.; Costante, G.; Bellocchio, E.; Valigi, P.; Fravolini, M.L.; Ciarfuglia, T.A. Robust visual semi-semantic loop closure detection by a covisibility graph and CNN features. *Robot. Auton. Syst.* **2017**, *92*, 53–65. [CrossRef]

32. Finman, R.; Paull, L.; Leonard, J.J. Toward object-based place recognition in dense rgb-d maps. In Proceedings of the ICRA Workshop Visual Place Recognition in Changing Environments, Seattle, WA, USA, 26–30 May 2015.

33. Oh, J.; Jeon, J.; Lee, B. Place recognition for visual loop-closures using similarities of object graphs. *Electron. Lett.* **2014**, *51*, 44–46. [CrossRef]

34. Pepperell, E.; Corke, P.; Milford, M. Routed roads: Probabilistic vision-based place recognition for changing conditions, split streets and varied viewpoints. *Int. J. Robot. Res.* **2016**, *35*, 1057–1179. [CrossRef]

35. Stumm, E.; Mei, C.; Lacroix, S.; Chli, M. Location graphs for visual place recognition. In Proceedings of the IEEE International Conference on Robotics and Automation, Seattle, WA, USA, 26–30 May 2015; pp. 5475–5480.

36. Gawel, A.R.; Don, C.D.; Siegwart, R.; Nieto, J.; Cadena, C. X-View: Graph-Based Semantic Multi-View Localization. *IEEE Robot. Autom. Lett.* **2018**, *3*, 1687–1694. [CrossRef]

37. Stumm, E.; Mei, C.; Lacroix, S.; Nieto, J.; Siegwart, R. Robust Visual Place Recognition with Graph Kernels. In Proceedings of the IEEE Conference on Computer Vision and Pattern Recognition, Las Vegas, NV, USA, 27–30 June 2016; pp. 4535–4544.

38. Han, F.; Wang, H. Learning integrated holism-landmark representations for long-term loop closure detection. In Proceedings of the 32nd AAAI Conference on Artificial Intelligence, New Orleans, LA, USA, 2–7 February 2018; pp. 6501–6508.

39. Chen, Z.; Maffra, F.; Sa, I.; Chli, M. Only look once, mining distinctive landmarks from convnet for visual place recognition. In Proceedings of the IEEE/RSJ International Conference on Intelligent Robots and Systems, Vancouver, BC, Canada, 24–28 September 2017; pp. 9–16.

40. Schönberger, J.L.; Pollefeys, M.; Geiger, A.; Sattler, T. Semantic visual localization. In Proceedings of the IEEE Conference on Computer Vision and Pattern Recognition, Salt Lake City, UT, USA, 18–22 June 2018; pp. 6896–6906.

41. Gao, P.; Zhang, H. Long-Term Loop Closure Detection through Visual-Spatial Information Preserving Multi-Order Graph Matching. In Proceedings of the Thirty-Fourth AAAI Conference on Artificial Intelligence, New York, NY, USA, 7–12 February 2020; pp. 10369–10376.

42. Zitnick, C.L.; Dollár, P. Edge boxes: Locating object proposals from edges. In Proceedings of the European Conference on Computer Vision, Zurich, Switzerland, 6–12 September 2014; pp. 391–405.

43. Chen, L.-C.; Zhu, Y.; Papandreou, G.; Schroff, F.; Adam, H. Encoder-decoder with atrous separable convolution for semantic image segmentation. In Proceedings of the European conference on computer vision, Munich, Germany, 8–14 September 2018; pp. 801–818.

44. Long, J.; Shelhamer, E.; Darrell, T. Fully convolutional networks for semantic segmentation. In Proceedings of the IEEE Conference on Computer Vision and Pattern Recognition, Boston, Mass, USA, 7–13 June 2015; pp. 3431–3440.

45. Ronneberger, O.; Fischer, P.; Brox, T. U-net: Convolutional networks for biomedical image segmentation. In Proceedings of the International Conference on Medical image computing and computer-assisted intervention, Munich, Germany, 5–9 October 2015; pp. 234–241.

46. Badrinarayanan, V.; Kendall, A.; Cipolla, R. Segnet: A deep convolutional encoder-decoder architecture for image segmentation. *IEEE Trans. Pattern Anal. Mach. Intell.* **2017**, *39*, 2481–2495. [CrossRef]

47. Zhou, B.; Zhao, H.; Puig, X.; Xiao, T.; Fidler, S.; Barriuso, A.; Torralba, A. Semantic understanding of scenes through the ade20k dataset. *Int. J. Comput. Vis.* **2019**, *127*, 302–321. [CrossRef]

48. Zhou, B.; Zhao, H.; Puig, X.; Fidler, S.; Barriuso, A.; Torralba, A. Scene parsing through ade20k dataset. Proceedings of IEEE Conference on Computer Vision and Pattern Recognition, Honolulu, HI, USA, 21–26 July 2017; pp. 633–641.

49. Krizhevsky, A.; Sutskever, I.; Hinton, G.E. Imagenet classification with deep convolutional neural networks. In Proceedings of the Advances in neural information processing systems, Lake Tahoe, NV, USA, 3–6 December 2012; pp. 1097–1105.

50. Russakovsky, O.; Deng, J.; Su, H.; Krause, J.; Satheesh, S.; Ma, S.; Huang, Z.; Karpathy, A.; Khosla, A.; Bernstein, M. Imagenet large scale visual recognition challenge. *Int. J. Comput. Vis.* **2015**, *115*, 211–252. [CrossRef]

51. Candes, E.J.; Tao, T. Near-optimal signal recovery from random projections: Universal encoding strategies? *IEEE Trans. Inf. Theory* **2006**, *52*, 5406–5425. [CrossRef]

52. Perozzi, B.; Al-Rfou, R.; Skiena, S. Deepwalk: Online learning of social representations. In Proceedings of the 20th ACM SIGKDD International Conference on Knowledge Discovery and Data Mining, New York, USA, 24–27 August 2014; pp. 701–710.

53. Cascianelli, S.; Costante, G.; Bellocchio, E.; Valigi, P.; Fravolini, M.L.; Ciarfuglia, T.A. A robust semi-semantic approach for visual localization in urban environment. In Proceedings of the IEEE International Smart Cities Conference, Trento, Italy, 12–15 September 2016; pp. 1–6.

54. Hu, M.-K. Visual pattern recognition by moment invariants. *IEEE Trans. Inf. Theory* **1962**, *8*, 179–187.

55. Davis, J.; Goadrich, M. The relationship between Precision-Recall and ROC curves. In Proceedings of the 23rd international conference on Machine learning, Pittsburgh, PA, USA, 25–29 June 2006; pp. 233–240.

56. Abadi, M.; Agarwal, A.; Barham, P.; Brevdo, E.; Chen, Z.; Citro, C.; Corrado, G.S.; Davis, A.; Dean, J.; Devin, M. TensorFlow: Large-scale machine learning on heterogeneous systems. *arXiv* **2015**, arXiv:1603.04467.

57. Jia, Y.; Shelhamer, E.; Donahue, J.; Karayev, S.; Long, J.; Girshick, R.; Guadarrama, S.; Darrell, T. Caffe: Convolutional architecture for fast feature embedding. In Proceedings of the 22nd ACM international conference on Multimedia, Orlando, FL, USA, 3–7 November 2014; pp. 675–678.

Publisher's Note: MDPI stays neutral with regard to jurisdictional claims in published maps and institutional affiliations.

 *remote sensing*

Article

Manhole Cover Detection on Rasterized Mobile Mapping Point Cloud Data Using Transfer Learned Fully Convolutional Neural Networks

Lukas Mattheuwsen *,† and **Maarten Vergauwen** †

Department of Civil Engineering, Geomatics Section, KU Leuven—Faculty of Engineering Technology, 9000 Ghent, Belgium; maarten.vergauwen@kuleuven.be
* Correspondence: lukas.mattheuwsen@kuleuven.be
† These authors contributed equally to this work.

Received: 29 October 2020; Accepted: 17 November 2020; Published: 20 November 2020

Abstract: Large-scale spatial databases contain information of different objects in the public domain and are of great importance for many stakeholders. These data are not only used to inventory the different assets of the public domain but also for project planning, construction design, and to create prediction models for disaster management or transportation. The use of mobile mapping systems instead of traditional surveying techniques for the data acquisition of these datasets is growing. However, while some objects can be (semi)automatically extracted, the mapping of manhole covers is still primarily done manually. In this work, we present a fully automatic manhole cover detection method to extract and accurately determine the position of manhole covers from mobile mapping point cloud data. Our method rasterizes the point cloud data into ground images with three channels: intensity value, minimum height and height variance. These images are processed by a transfer learned fully convolutional neural network to generate the spatial classification map. This map is then fed to a simplified class activation mapping (CAM) location algorithm to predict the center position of each manhole cover. The work assesses the influence of different backbone architectures (AlexNet, VGG-16, Inception-v3 and ResNet-101) and that of the geometric information channels in the ground image when commonly only the intensity channel is used. Our experiments show that the most consistent architecture is VGG-16, achieving a recall, precision and F_2-score of 0.973, 0.973 and 0.973, respectively, in terms of detection performance. In terms of location performance, our approach achieves a horizontal 95% confidence interval of 16.5 cm using the VGG-16 architecture.

Keywords: mobile mapping; manhole cover; point cloud; F-CNN; transfer learning; CAM localization

---

1. Introduction

The mapping and inventory of the public domain is of great importance to many stakeholders. These spatial datasets are used by various government authorities to maintain and update their asset information and by the architecture, engineering and construction (AEC) industry as reference maps for project planning and construction designs. In Belgium, more specifically in the Flemish region, as well as in the Netherlands, well-established spatial databases have already been used for years, and they include the accurate and complete mapping of the public domain [1,2]. These spatial databases, called the GRB and the BGT for Flanders and the Netherlands, respectively, do not only contain general objects such as buildings, bridges and road structures but also more detailed street elements such as light/traffic posts, fire hydrants and manhole covers.

The mapping of manhole covers is of great importance, as manholes are used for many tasks such as rainwater collection, sewage discharge, electricity/gas supply and telecommunication cables.

This information allows utility companies and government authorities to create detailed networks of their underground infrastructure. Additionally, the 3D manhole positions can be used to create drainage system models to evaluate the interaction of the rainwater with the environment and identify high flood risk areas. While this mapping is still done manually using traditional surveying methods, the popularity of mobile mapping systems for the mapping of spatial databases has grown in recent years [3–5]. These systems use a combination of GNSS (global navigation satellite system) and IMU (inertial measurement unit) sensors to accurately determine their position and orientation. Combined with (omnidirectional) cameras and lidar sensors, they are capable of capturing vast amounts of georeferenced data in a short time frame. While the initial cost of a mobile mapping system is high, it is twice as efficient and equally expensive in terms of €/km as traditional surveying methods [3]. However, almost 90% of the total time of the mobile mapping project is spent on data interpretation and mapping. Automating this task reduces the overall costs, including the initial cost of a mobile mapping system, by 22% and results in a time saving of up to 91% compared to the traditional manual surveying techniques [6].

This is why recent research on mobile mapping systems has focused on the automatic detection of different objects such as buildings, road structures or poles using methods such as machine learning and deep learning [4,7]. However, in the case of manhole cover detection, machine learning and deep learning are more difficult to implement. Commonly used image-based object detection methods struggle to detect small objects such as manhole covers; lidar-based methods are even less successful, as a manhole cover has almost no geometric features to stand out from the road surface itself. Therefore, it is still a challenge to develop a manhole detection framework for spatial databases mapping, as this requires high precision and especially high recall performance. Furthermore, deep learning requires a large quantity of training data to achieve high-performance networks.

In this paper, we propose a fully automatic manhole cover detection framework to extract manhole covers from mobile mapping point cloud data. This approach makes use of deep learning networks that only require a small training dataset to achieve good detection results. The point cloud data are first rasterized into a ground image in order to simplify the detection task and use well-established image processing methods. Our ground image consists of three channels based on the lidar data: intensity, minimum height and height variance. While current research only works with intensity channels, our work investigates the use of additional geometric information as additional input channels for the ground image. Our method makes use of pre-trained classification convolutional neural networks (CNNs) which are transfer learned, only requiring a small labeled training dataset. The original network is first modified into a fully convolutional network in order to process larger images in an efficient way. This eliminates the use of a sliding window approach for the manhole cover detection. Additionally, the center of the manhole cover is predicted using the activation maps of the pooling layers of the network. Object detection and localization performance of this approach are evaluated on different CNN backbone architectures (AlexNet [8], VGG-16 [9], Inception-v3 [10] and ResNet-101 [11]). In summary, the main contributions of this paper are:

1. Fully automatic manhole cover detection framework using transfer learned fully convolutional neural networks trained on a small dataset;
2. Influence of additional geometric features as input channels for the CNN is assessed;
3. Different backbone architectures (AlexNet, VGG-16, Inception-v3 and ResNet-101) are investigated for our proposed detection framework.

The remainder of this work is structured as follows. In Section 2, the related work on manhole cover detection using remote sensing data is discussed. This is followed by Section 3, in which we present our methodology. The experiments and results are presented and discussed in Sections 4 and 5. Finally, the conclusions and future work are presented in Section 6.

2. Related Works

There are different methods to map or inventory manhole covers in a spatial database. When remote sensing data such as satellite imagery, UAV imagery or mobile mapping data (image and/or lidar data) are used, the acquisition time can be drastically reduced compared to traditional surveying techniques [3]. Although already more efficient, these methods would benefit more if the mapping of objects such as manhole covers could be automated, as this is still commonly performed manually. This is mainly because mapping for spatial databases requires high recall and precision performance. Research on mapping automation can be split up in three categories: image-based, lidar-based and combined image-/lidar-based methods. Because manhole covers have no distinctive 3D geometric features and object detection using 3D lidar data is more complex, most lidar-based methods convert the point cloud into a 2D intensity ground image [12–15]. By doing so, the point cloud detection problem becomes an image detection problem, making it possible to apply well-established image processing techniques. At first, more basic approaches were investigated using manually designed low-level features [12,16], but more machine and deep learning approaches have emerged in recent years, and their capabilities to learn complex high-level features have been used [13,14,17]. As R-CNN [18], YOLO [19] and SSD [20] are known to struggle with small objects, a more basic classification network and sliding window approach are generally applied. A summary of several manhole cover detection techniques are presented in this section.

In a recent study, manhole cover detection using mobile lidar data was investigated [6]. This method searches for manhole covers by filtering the point cloud with a pre-defined intensity interval, after which a best fitting bounding box is fitted to each cluster. As the dimensions of manhole covers are generally known, bounding boxes that are too small or too big are filtered out. Although this simple method performs well on raw point clouds and achieves usable results in their dataset, this method is not robust for other datasets, as it cannot distinguish the difference between a manhole cover or a dark intensity patch. Additionally, as soon as the manhole cover is partially occluded, the cluster is not square and is regarded as a false positive. Therefore, more complex image processing or deep learning approaches are needed.

In [16], a generic part-based detector model [21] was assessed on single view images from a moving van. These images were projected into an orthographic ground image such that the manhole covers had a circular shape. While the single-view approach resulted in poor recall and precision scores, their multiview approach proved more effective with a recall score of 93%. Such multiview approaches perform object detection on consecutive captured images and utilize their relative position to trace the manhole cover in 3D. This allows their approach to achieve higher recall and predict the 3D position with more accuracy and certainty. A similar single- and multiview approach was assessed on UAV-captured imagery in [22] using Haar-like features as input for a classifier to determine whether the image contained a sewer inlet or not. They compared three classifiers (support vector machine (SVM), logistic regression and neural network) in combination with a sliding window approach to perform the object detection. Their comparison showed that the neural network classifier resulted in the highest precision score compared to the other classifiers.

Yu et al. investigated several approaches to detect manhole covers from lidar data [12–14]. Each of his methods rasterizes the lidar data into an intensity ground image using the improved inverse distance weighted interpolation method proposed in [23]. In [12], Yu assessed a marked point model approach to detect manhole covers and sewer inlets in the ground image. This method, however, assumes that a manhole cover has a round shape and that a sewer inlet has a rectangular shape and attempts to fit these shapes around low-intensity patches of the ground image. While it was effective for clean road surfaces, this method failed on repaired roads where round or dark patches of new asphalt looked like manhole covers or sewer inlets. Their approach was improved in [13] using a machine learning approach. Instead of using low-level manually generated features, they opted to train a deep Boltzmann machine to generate high-level features from a local image patch. Afterward, these features were used in a sliding window approach with a random forest model to classify the patch as "manhole", "sewer inlet" or "background". This new machine learning approach outperformed the method from their previous work. In their most recent work [14], they investigated a deep learning approach. Instead of using a sliding window, the intensity ground image was segmented using a super-pixel-based strategy. Each segment was classified by their own designed convolutional network after which their marked point approach from [12] was used to accurately delineate the edges of the manhole covers. This new deep learning method slightly outperformed their previous machine learning method [13].

Other approaches use modified mobile mapping systems specifically designed to capture the road surface in high detail using lidar/image sensors [15] or a lidar profile scanner [24]. Both methods use a combination of manually designed low-level features such as HOG features, intensity histograms, PCA, etc. as input for their SVM classification method. While [15] used a sliding window approach on both intensity and colored ground images from lidar and imagery, [24] performed a super-pixel segmentation approach on the intensity ground image similar to [14]. Although the approach of [15] resulted in a good recall and precision score, this method uses dense and clear point cloud data from specifically designed mobile mapping systems. Lidar data from commercially available mobile mapping systems are more noisy and less dense, and they capture not only the road surface but the whole of the surroundings, making manhole detection more complex.

In 2019, a fully deep learning approach was assessed on high-resolution satellite imagery using a multilevel convolutional matching network [17]. Although this method achieves better results than traditional object detection methods such as R-CNN, YOLO or SDD, like so many deep learning methods, it needs a large quantity of training data to be successful. As no dedicated manhole cover training data exist, creating this dataset is time-consuming. For example, [15] labeled around 25,000 images to train their SVM classifier, [22] used 6500 labeled images and [13] used 15,000 images. While [17] only needed 1500 images for training, their training dataset did contain around 15,000 manhole cover bounding boxes. With the use of transfer learning on pre-trained networks, it is possible to achieve good results with only a fraction of the training data reducing time needed for manual labeling and the training of the network. This was demonstrated in [25] using ResNet-50 and Resnet-101 together as a backbone for the RetinaNet architecture using only 120 training images. Although it outperformed a Faster-RCNN implementation, these results are questionable, as the testing dataset only contained 36 images with 21 manhole covers, which does not represent a real-world example.

3. Methodology

In this section, a detailed overview is presented of our proposed fully automatic manhole cover detection workflow. It contains two major components: the preprocessing framework and the object detection framework. The preprocessing framework consists of filtering steps that reduce the amount of noisy and unnecessary data for the subsequent processing steps, as well as the ground image conversion step that converts the 3D point cloud into a 2D ground image with intensity, minimum height and height as image channels. These images are then used as input of the object detection framework, which aims to find the manhole cover positions in them. In this framework, the images are processed by a fully convolutional neural network which simulates an internal sliding window producing a spatial classification output in an efficient way. This spatial output indicates where the network expects a manhole cover to be located. Using this result, the center of the manhole cover is predicted using the activation maps from the classification network. The complete workflow is shown in Figure 1, while some examples of intermediate results of the workflow are shown in Figure 2.

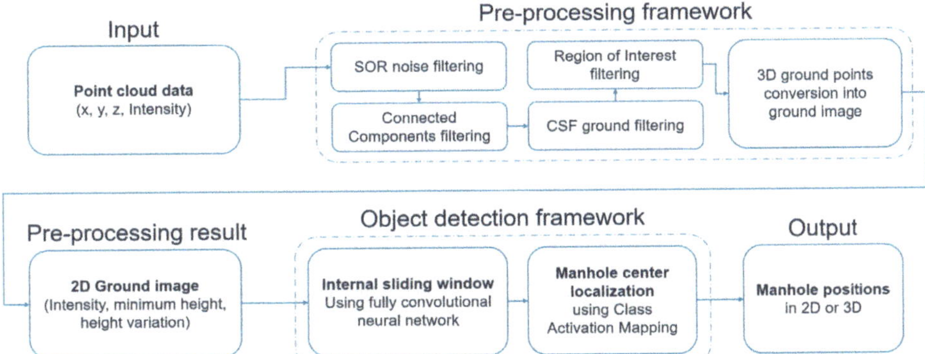

Figure 1. Schematic overview of the proposed method with the preprocessing and the object detection framework. The former filters and converts the mobile mapping point cloud into a 2D ground image. This image is then used by latter to detect the different manhole covers and locate them accurately.

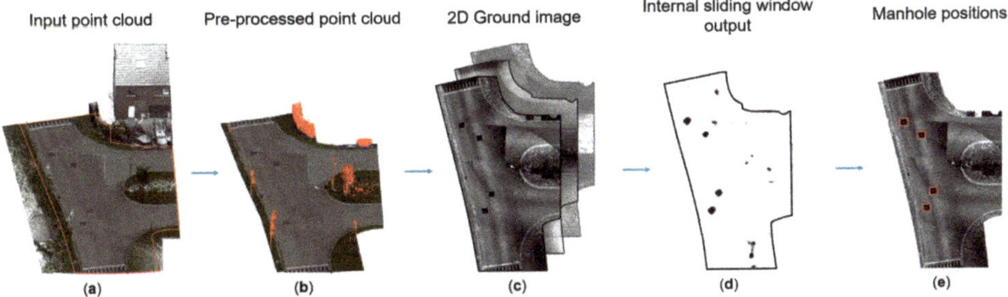

Figure 2. Visualization of the different intermediate results of the proposed method: (**a**) shows the input point cloud with the region of interest indicated by a red outline. (**b**) visualizes the filtered point cloud after preprocessing with the off-ground points colored in red. (**c**) shows the 2D ground image with the intensity, minimum height and height variance channels rasterized from the ground points. (**d**) displays the spatial classification output from the fully convolutional neural network with the colors black and white corresponding to a high or low manhole classification score, respectively. (**e**) visualizes the predicted manhole cover positions in red generated from the spatial classification output.

3.1. Preprocessing Framework

Because of the imperfections of lidar scanners and the influence of reflective surfaces or moving cars/people in the vicinity, mobile mapping point clouds of the public domain contain a considerable amount of noise or ghost points that influence the processing steps. In the preprocessing framework, these points are filtered by statistical outlier removal (SOR) and connected components (CC) filters. Additionally, mobile mapping systems capture all surroundings, including the road surface, road furniture, sidewalk, buildings, vegetation, etc. However, manhole covers only occur on the ground, making a large quantity of data obsolete. Therefore, a cloth simulation filter (CSF) is applied to segment the point cloud into "ground" and "nonground" points. Furthermore, an optional region of interest (RoI) filter is applied to further reduce the data that needs processing. Following the filtering steps, the 3D point cloud is rasterized into a georeferenced 2D ground image with three channels: intensity, height and height variance. These steps are discussed in detail in the following paragraph.

3.1.1. SOR, CC, CSF and RoI Filtering

Mobile mapping point clouds include erroneous measurements caused by the imperfection of the laser scanner and the challenging environments such as reflective surfaces, moving cars and pedestrians. These errors can corrupt the object detection algorithms. This is especially the case for the CSF ground filtering which fails to produce usable results when noisy points are present under the road/ground surface. An example of these points is shown in Figure 3a. These points are commonly removed by applying a statistical, radius or multivariate outlier removal filter. The SOR filter, applied in our method, computes the average Euclidean distance of each point to its k neighbors plus the mean μ and standard deviation σ of the average neighboring distance. A point is classified as outlier/noise when the average neighboring distance is greater than the maximum distance defined by $\mu + \alpha \cdot \sigma$. We found experimentally that the following parameters, $k = 30$ and $\alpha = 2$, removed the subterranean and sky noise points. However, high-density point clusters still remain, such as the ghosting points when measuring through a window (Figure 3b). Filtering these points is done by applying a connected components (CC) filter which segments the point cloud in different clusters separated by a minimum distance d and removes clusters smaller than the minimum cluster size C_{min}. In our implementation, we found that $d = 0.3$ m and $C_{min} = 10,000$ works well to remove the majority of these ghost points without removing other relevant clusters.

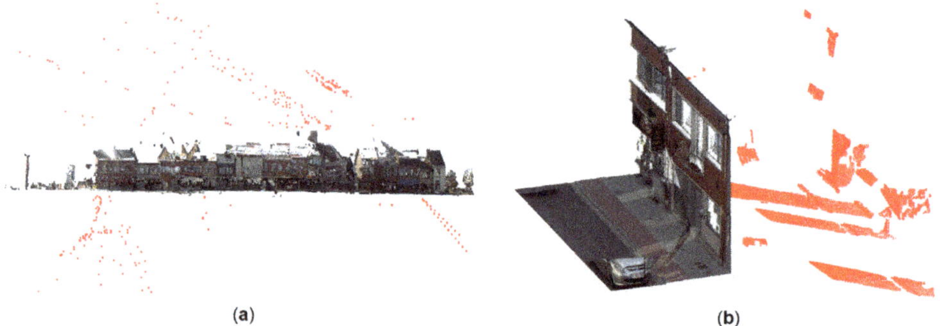

(a)   (b)

Figure 3. Visualization of subterranean/sky noise points (**a**) and ghosting points caused by measuring through glass (**b**). All noise and ghosting points are colored in red.

The ground segmentation is performed by applying a cloth simulation filter [26]. This filter inverts the point cloud along the Z-direction and simulates a cloth being dropped on top of it. Depending on a few parameters such as the rigidity and resolution, the cloth follows the contours of the point cloud representing the digital terrain model (DTM). All points within a specified minimum distance of the DTM are classified as ground points. The main parameters of this filter are the grid resolution, rigidity and the classification threshold. The first determines the resolution of the cloth where a finer cloth will follow the terrain more closely than a coarse one. However, an overly fine resolution causes more nonground to be wrongly classified and results in longer processing times. The rigidity influences the stiffness of the cloth where a soft cloth will follow the terrain better. This parameter can be set to 1, 2 or 3 for steep slopes, terraced slopes or flat terrain, respectively. The classification threshold is the minimum distance to classify a point as ground or nonground. The optimal parameter settings are based on the findings in [26] and a few experimental tests. We chose 1, 3 and 0.3 m for the grid resolution, rigidity and classification threshold, respectively.

In our workflow, an optional region of interest filter can be applied to remove all points from the private domain as many large-scale databases only contain manhole covers in the public domain. The GRB, for example, contains a specific layer that delineates the public domain which is used to filter out points not within this layer. An example of this border is visualized in Figure 2a. This step reduces the quantity of data even more, resulting in faster processing of the following steps. As this is an optional step, the RoI filtering can be skipped when no boundary is available or necessary.

3.1.2. 2D Ground Image Conversion

Manhole covers have almost no height related features and are small objects in a mobile mapping point cloud. This makes it difficult to use point-cloud-based object detection methods. This is why we choose to rasterize the point cloud into a 2D georeferenced ground image. This reduces the amount of data that need processing while also simplifying the task to a image detection problem. As such, it is possible to use well-established image processing methods such as CNNs. Although most mobile lidar point cloud contain RGB data for each point, intensity information is preferred over the RGB information for the channels of the ground image. The difference in reflectivity between manhole covers and the road surface means that a black manhole cover is discernible on a black asphalt surface, which is impossible using RGB channels. However, different capturing positions and frequencies of the omnidirectional camera and laser scanner cause a variable shift between RGB data and intensity data. Although an accurate system calibration can reduce the influence of this error, imperfections of the lidar sensor and omnidirectional camera cause the error to persist. Additionally, passing vehicles or pedestrians result in false coloring of the road surface, which makes an RGB approach unreliable. For that reason, our implementation includes the following point cloud information as channels for the ground image: intensity, minimum height and height variance. The number of channels is limited as pre-trained models are designed for input images with three channels. Altering the number of input channels would require training from scratch and a massive training dataset. Although manhole covers have almost no height-related information, we aim to improve the precision performance of the network by including this information in the input image. As manhole covers occur all over the public domain, this complicates the detection problem. While manhole covers are flat, other areas such as curbs, grass, dirt, etc. have an uneven surface which is captured by the minimum height and height variance information. It is our estimation that the false positive detection rate of the whole workflow will decrease. A comparison between the performance of an intensity image and our IHV (intensity, height, variance) image is discussed in Section 4. The 2D ground image conversion in our implementation works as follows. First, the point cloud is tiled into sections of 50 by 50 m with 5 m overlap. Each tile is rasterized with a ground sampling distance of 2.5 cm which results in a 2000 by 2000 ground image. For each grid cell in the image, the corresponding point cloud position is calculated from which a 2D radius search groups all the points within 2.5 cm of the grid cell. From these points, the intensity values are computed with the method proposed in [23], which is based on an inverse distance weighted interpolation. This method computes a weighted intensity average of all points in the radius search by applying the following rules:

Rule 1: a point with a higher intensity value has a greater weight.
Rule 2: a point farther away from the center point of the grid cell has a smaller weight.

In our implementation, the weight coefficients of these rules α and β are both set to 0.5. Additionally, the minimum height information of the grid cell is the minimum height value from the points in the radius search. The height variance information is the absolute height difference between the lowest and highest point from the corresponding cluster. As a result, the IHV ground image is created with three channels: intensity, minimum height and height variance. An example of such an image with the different channels is shown in Figure 4. Notice how some areas in the intensity channel do not display consistent values. Although the road surface is the same over the entire surface of the intersection, the variance in the intensity channels implies otherwise. This phenomenon commonly occurs at intersections where point cloud segments of different trajectories overlap with each other. As the intensity value of a measured point depends on the angle of incidence, this results in sudden changes in the intensity channels in areas with overlapping point clouds.

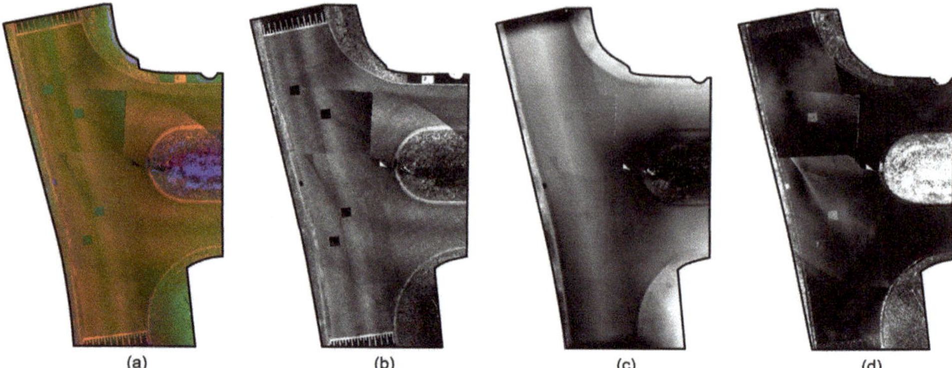

(a)   (b)   (c)   (d)

Figure 4. Example of an IHV (intensity, height, variance) image (**a**) shown as an RGB image and the different channels, intensity (**b**), minimum height (**c**) and height variance (**d**), as gray images. Notice how the intensity and height variance channels indicate sudden changes at the intersection caused by overlap of point cloud segment of different trajectories.

3.2. Object Detection Framework

The second component of our method is the object detection framework, consisting of the internal sliding window part and the manhole center localization. Both make use of the same modified transfer learned classification network to detect and accurately localize manhole covers in the ground image.

3.2.1. Internal Sliding Window

As traditional image object detection methods fail to perform robustly on small objects [17], our method performs a more commonly used sliding window approach that uses simpler classification models and has proved effective in previous research (see Section 2). A pre-trained ImageNet [27] classification network is transfer learned to label an image patch as "manhole" or "background". A common transfer learning workflow is applied with the following steps. The first convolutional layers are frozen: their weights are not adjusted during training. These first layers contain the well-defined basic features trained from millions of images which would be erased or changed when not frozen during training, rendering the benefits of the pre-trained model obsolete. The last "learning" layers are replaced to account for two output classes instead of the original 1000 from the ImageNet dataset [27]. While these layers are typically replaced by similar fully connected layers, we opt to replace them by fully convolutional layers, resulting in a fully convolutional network. This strategy is chosen because networks with fully connected layers are limited to processing images with a fixed size, while networks with fully convolutional layers can process larger images, generating a spatial classification map simulating an internal sliding window. An example of such a spatial classification map is shown in Figure 5. This internal sliding window approach based on Overfeat [28] has proved much more computationally efficient compared to a traditional sliding window.

3.2.2. Manhole Center Localization

The output of the fully convolutional network is a spatial classification map with size $R_o \times C_o$ which depends on the architecture of the network and the size of the input image defined by $R_i \times C_i$. As the fully convolutional network simulates an internal sliding window, each cell from the spatial classification map corresponds to a sliding window position of size $w \times w$ with w being the original input size of the classification network. The value in each cell corresponds to the classification score of the corresponding window located in that position. An example of three grid cells, highlighted in red, green and blue, and their corresponding windows are shown in Figure 5. Using the original image size and the resulting spatial classification map, the step size of the simulated sliding window can be computed using the following equation:

$$Step_{ver} = (R_i - w)/(R_o - 1); \qquad Step_{hor} = (C_i - w)/(C_o - 1) \tag{1}$$

With the horizontal and vertical step size, the sliding window position in the input image of each output cell at position (r, c) in the spatial classification map is computed as follows:

$$r_i = \frac{w}{2} + (r - 1) \cdot Step_{ver}; \qquad c_i = \frac{w}{2} + (c - 1) \cdot Step_{hor} \tag{2}$$

where r_i and c_i are the corresponding row and column coordinates in the input image. Figure 5 visualizes the window positions in the original input image that have a classification score greater than the defined classification threshold T_{class}. As can be seen in this image, there are large clusters of high classification scores (black), indicating a high possibility of a manhole cover, but there are also smaller clusters which are clearly false positives. In our approach, these clusters are detected by applying a clustering algorithm on the spatial classification map. This algorithm considers window positions with a classification score above the threshold T_{class} and considers all adjacent windows to be in the same cluster. For the example in Figure 5, this results in three clusters. As false positive clusters are generally smaller than true positive clusters, clusters smaller than a user-defined cluster threshold $T_{cluster}$ are filtered out. In the subsequent processing steps, it is assumed that each cluster contains a manhole cover.

While common object detection approaches need bounding box training data to train a dedicated location network, our approach uses the same classification network to predict the position of the manhole in the image. This is done by applying a simplified version of class activation mapping, proposed in [29]. This approach uses the activation maps of pooling layers to highlight the region of the image which is most important for classifying the image as "manhole". Additionally, this information can be used to predict the location of the manhole as follows. For each position of a cluster, the activation map of the corresponding window is extracted from the last pooling layer from the classification network. In general, the activation map is a 3D matrix of size $row \times col \times depth$ which is flattened into a 2D matrix of size $row \times col$ by averaging the $depth$ dimension and also min-max normalizing to rescale the results to a value between 0 and 1. Figure 6 shows the normalized activation maps and the corresponding windows in a cluster. Notice how the highest activation values are located around the center of the manhole cover. To predict the center of the manhole, the weighted center of each normalized activation map is computed for each position in a cluster and converted into the image coordinates (I_r, I_c). In the end, the center of the manhole cover (M_r, M_c) is computed with a weighted average using the classification score S and the image coordinates of the activation map center (Img_r, Img_c) of each cluster position, using Equation (3).

$$M_r = \frac{\sum_{i=1}^{s}(S_i \cdot I_r)}{\sum_{i=1}^{s} S_i}; \qquad M_c = \frac{\sum_{i=1}^{s}(S_i \cdot I_c)}{\sum_{i=1}^{s} S_i} \tag{3}$$

where s is the size of the cluster. Doing this for all clusters of the spatial classification map results in the positions of the different manhole covers. The localization algorithm is governed by two main parameters: cluster threshold $T_{cluster}$ and classification threshold T_{class}, which are optimized and analyzed in the results in Section 4.

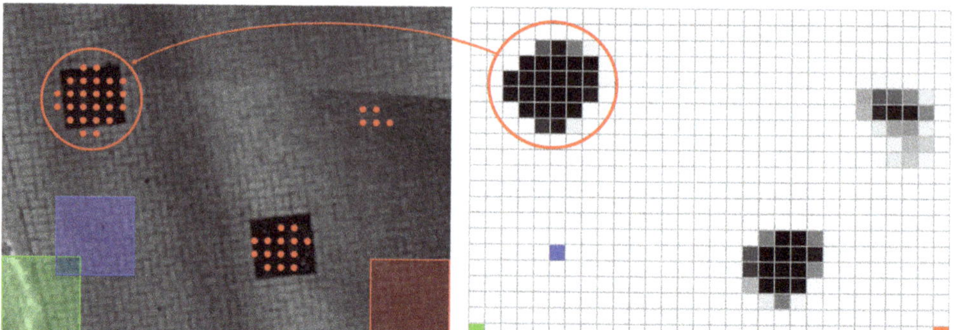

Figure 5. Visualization of the sliding window positions positions of the windows that have a classification score greater than the threshold T_{class} = 0.5 (**left**) and the corresponding clusters in the spatial classification map (**right**). Additionally, the windows of the corresponding highlighted grid cells in red, blue and green from the spatial classification map are shown in the intensity image.

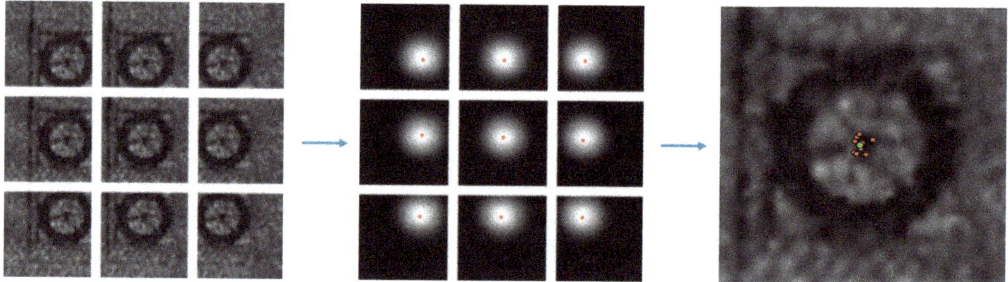

Figure 6. Visualization of the localization workflow, starting with the different sliding window positions (**left**) of which the corresponding activation maps are extracted from the last pooling layer (**middle**). The weighted activation centers are depicted by the red points and are projected onto the original image (**right**). The center of the manhole cover is computed by the weighted average of these activation centers and is indicated by the green point.

4. Results

4.1. Training, Validation and Testing Datasets

All the data used for training and testing is captured by a Lynx mobile mapper SG equipped with the dual lidar system from the Lynx M1 Mobile Mapper. This setup captures points at a rate of 500 kHz with an accuracy of 8 mm up to 200 m to ensure a dense and accurate point cloud. The Lynx system was mounted on the roof of a Mitsubishi SUV driving at a speed of approximately 25–40 km/h during data acquisition. A more detailed summary and analysis of this system is presented in our previous work [30]. As no dedicated training or testing dataset for manhole cover detection is currently publicly available, we captured our own. The three mobile mapping datasets (rural, residential and urban datasets) from [30] were used to create the training and validation dataset to transfer learn the networks. Using the manhole covers positions in the GRB, the ground truth data were automatically extracted in the ground image. Although all images were manually checked and wrong images were removed, this semiautomatic approach was much more efficient compared to manually labeling the images. Additionally, "background" images were extracted, resulting in 443/190 and 455/195 images for the training/validation set for the "manhole" and "background" labels, respectively. Furthermore, a second training set with an additional 545 "background" images was created to investigate if these extra images would decrease false positive detections. In the remainder of the paper, dataset 1 refers to the dataset with equal manhole and background training images while dataset 2 refers to the dataset with the additional background training images. Both dataset 1 and 2 have the same validation dataset. Furthermore, we created two types of ground images for both datasets: one containing only the intensity channels and on with the IHV channels, resulting in four training datasets in total. To evaluate the object detection performance of the workflow, an additional testing dataset was captured with the same mobile mapping system. This dataset contains over 1 km of urban and residential roads and includes 73 manhole covers extracted from the GRB.

4.2. Network Selection and Training Parameters

In this work, we assessed the performance of four different network architectures including AlexNet [8], VGG-16 [9], Inception-v3 [10] and ResNet-101 [11]. From these four networks, AlexNet is considered the simplest with only eight layers compared to the "deeper" architecture of VGG-16 consisting of 16 layers. In contrast, Inception-v3 and ResNet-101 have a more complex architecture, as just adding convolutional layers results in a saturated accuracy at a certain depth that degrades rapidly when going even deeper [11]. Inception-v3 resolves this problem by building a wider network instead of deeper network using side-by-side convolutional computations in the same layer. ResNet, on the other hand, uses skip connections to achieve better results with deep networks with up to 152 layers. Table 1 summarizes the main differences including depth, number of parameters and ImageNet top-5 accuracy. The authors of [31] discovered that better ImageNet accuracy results in better transfer learning performance; thus, Inception is expected to outperform ResNet, VGG and AlexNet in that order.

Each network was transfer learned on their pre-trained ImageNet version using the four different training datasets utilizing the Deep Learning toolbox from MATLAB [32] on a computer with an Intel Xeon W-1233 processor, 32 GB of RAM and a Nvidia GTX 1080. All networks were trained using stochastic gradient descent optimization with a minibatch size of 52 for a total of 30 epochs. The initial learning rate was set between 0.001 and 0.0001, depending on the network/dataset combination, and it decayed after 10 epochs by a factor of 0.3. The momentum was set to 0.9 with a weight decay of 0.0001. Additionally, standard data augmentation such as rotation, scale and translation were performed on the dataset during training. On average, training took 2 min, 17 min, 22 min and 25 min for AlexNet, VGG-16, Inception-v3 and ResNet-101, respectively. Taking into account the depth of each network, it is clear that deeper/more complex networks take longer to train.

Table 1. Summary of the main differences between AlexNet, VGG-16, Inception-v3 and ResNet-101.

Network	Depth	Parameters	ImageNet Top-5 Accuracy	Salient Feature
AlexNet	8	61 M	84.6%	First winning CNN
VGG-16	16	138 M	91.6%	Deeper than AlexNet
Inception-v3	48	24 M	94.4%	Side-by-side convolutions
ResNet-101	101	44.6 M	93.9%	Skip connections

4.3. Classification Performance

The evaluation of the classification performance is done by comparing the accuracy, recall, precision and F-score from each network on the validation dataset. Instead of the commonly used F_1-score, which computes the harmonic mean between recall and precision, we opt to use a weighted F-score with the factor β in which recall is considered β times as important as precision. This is because a high recall score is of more importance than the precision score when mapping manhole covers in spatial databases. The F_β-score can be computed from Equation (4).

$$F_\beta = (1 + \beta^2) \cdot \frac{precision \cdot recall}{(\beta^2 \cdot precision) + recall} \tag{4}$$

In terms of the F_2-score, this means that recall is twice as important as precision. Table 2 shows the validation results for the different networks. When comparing the intensity-trained networks against IHV-trained networks, the former outperform the latter, especially in terms of precision. Our assumption that the extra geometric features would improve the precision score does not hold. When comparing the results of dataset 1 to those corresponding to dataset 2, the intensity-trained networks perform similarly, meaning the extra "background" images have no significant influence. In contrast, the IHV-trained networks show a significant precision improvement of 5.6% and 9.7% when trained on the extra "background" images for the AlexNet and ResNet architectures, respectively. The best performance score for each dataset combination is shown in bold in this table. In general, ResNet-101 performs the best on the validation dataset, with VGG-16 and Inception-v3 close behind, while AlexNet performs the worst for each dataset combination. This indicates that more complex networks outperform the simpler shallow AlexNet architecture.

Table 2. Accuracy, recall, precision and F_2-score for AlexNet, VGG-16, Inception-v3 and ResNet-101 trained on dataset 1 and 2 for intensity and IHV images. The highest scores for each dataset are shown in bold.

Network	Dataset	Image	Accuracy	Recall	Precision	F_2-Score
AlexNet	1	Intensity	0.969	0.963	0.973	0.965
VGG-16	1	Intensity	0.990	**0.995**	0.984	0.993
Inception-v3	1	Intensity	0.979	0.989	0.969	0.985
ResNet-101	1	Intensity	**0.995**	**0.995**	**0.995**	**0.995**
AlexNet	2	Intensity	0.974	0.974	0.974	0.974
VGG-16	2	Intensity	0.992	0.984	**1.000**	0.987
Inception-v3	2	Intensity	0.982	0.979	0.984	0.980
ResNet-101	2	Intensity	**0.997**	**0.995**	**1.000**	**0.996**
AlexNet	1	IHV	0.911	0.963	0.871	0.943
VGG-16	1	IHV	**0.961**	0.974	**0.949**	0.969
Inception-v3	1	IHV	**0.961**	0.979	0.944	**0.972**
ResNet-101	1	IHV	0.914	**1.000**	0.852	0.966
AlexNet	2	IHV	0.948	0.979	0.921	0.967
VGG-16	2	IHV	**0.971**	0.979	**0.964**	0.976
Inception-v3	2	IHV	0.961	0.989	0.935	0.978
ResNet-101	2	IHV	0.961	**0.995**	0.931	**0.982**

In addition to the classification performance, the sliding window performance on ground images is also relevant. This performance is computed for each network on a small test image as is shown in Table 3. Each red point corresponds to a sliding window position with a classification score above 0.5. When analysing the AlexNet results, it is clear that, although the performance on the validation set was relatively high, the network does not perform robustly on larger images. For each dataset combination, there are large clusters of false positive classifications resulting in a low precision score and longer processing times. In contrast, the VGG-16 results look promising with clear clusters around each manhole cover and a few small false positive clusters, which are easily filtered out based on cluster size. However, it should be noted that the majority of these false positive clusters overlap with inspection covers on the side walk. These smaller covers are used for inspecting gas, water or private sewage lines and look similar to manhole covers. Additionally, these are not commonly mapped in large spatial databases and therefore need to be filtered out. Similar inspection cover errors are present in the Inception-v3 results, although these clusters are much bigger. As these have sizes similar to the true positive clusters, filtering based on cluster size is not an option. Additionally, more noisy false positive clusters are present compared to the VGG-16 results. The ResNet-101 results show similar inspection cover errors and large false positive clusters, especially for the IHV-trained networks. Additionally, the networks struggle to classify all manhole covers in the test image while achieving the highest recall scores on the validation dataset in Table 2.

Table 3. Visualization of the sliding window positions with a classification score greater than the classification threshold (T_{class} = 0.5) for the different networks and training datasets. The two manhole covers in the image are indicated by the green bounding box.

Network	Dataset 1 Int.	Dataset 2 Int.	Dataset 1 IHV	Dataset 2 IHV
AlexNet				
VGG-16				
Inception-v3				
ResNet-101				

These results show that a high classification score does not automatically correspond to good sliding window performance on a larger image. In general, the VGG-16 architecture is the only network performing consistently in terms of classification and sliding window across the different dataset combinations. Therefore, only the VGG-16 networks are analyzed on the testing dataset, as these are the most likely to achieve good object detection results.

4.4. Manhole Detection Performance

Our proposed object detection and localization method is evaluated on the additional testing dataset containing 73 manhole covers. A predicted manhole cover position is considered a true positive when it is within a distance of 36 pixels (=90 cm) of the ground truth center point. This distance corresponds to the average size of a manhole cover found in the public domain. For each VGG-16 network, the classification threshold T_{class} and cluster threshold $T_{cluster}$ are optimized for two scenarios (maximum F_2-score and maximum recall score). The optimal localization parameters and corresponding recall, precision and F_2-score are shown in Table 4. From the maximum F_2-score results, it is clear that the intensity-trained networks outperform the IHV-trained networks, as was the case in terms of classification performance. The additional geometric features result in a significantly lower recall score and an unusable precision score of around 0.2. This means that only one in five predicted manhole covers is actually a manhole cover and only 70%–80% of manhole covers are detected. In contrast, both intensity-trained networks achieve a +90% recall score while achieving a 68% and 48% precision score for dataset 1 and 2, respectively. In this case, the additional "background" images resulted in more false positive detections while the opposite was intended. Similar observations are made in the case of maximum recall optimization where the intensity-trained networks outperform the IHV-trained networks. When recall is of utmost importance, the intensity-trained networks are able to detect all manhole covers with a corresponding low precision score of 25%. While the IHV-trained networks also achieve a high recall, the precision score equals to 8%.

Table 4. Summary of the object detection results (recall, precision and F_2-score) with the corresponding classification threshold T_{class} and cluster threshold $T_{cluster}$ for the whole public domain. The results and optimal parameters are displayed for both the maximal F_2-score and maximal recall score.

Dataset	Maximum F_2-Score Optimization					Maximum Recall Optimization				
	Recall	Precision	F_2-Score	T_{class}	$T_{cluster}$	Recall	Precision	F_2-Score	T_{class}	$T_{cluster}$
1 Int.	0.904	**0.680**	**0.848**	0.8	10	**1.000**	0.237	0.608	0.5	2
2 Int.	**0.932**	0.482	0.785	0.9	16	**1.000**	**0.258**	**0.635**	0.8	9
1 IHV	0.699	0.183	0.447	0.2	19	0.973	0.077	0.293	0.7	1
2 IHV	0.781	0.209	0.504	0.3	18	0.973	0.086	0.317	0.7	1

The detection results for the maximum F_2-score configuration are shown in Figure 7 on a small section of the testing dataset. True positive manhole cover predictions are depicted with a green cross, false positives with a red cross. The ground truth manhole covers are indicated with a green or red bounding box depending if they are detected or not. These images illustrate the varying recall scores of the networks where the IHV-trained networks perform slightly worse as is visible in Figure 7d. Additionally, the intensity-trained networks show far fewer false positives compared to the IHV-trained networks. Some of these false positives are caused by the presence of small inspection covers or storm drains. However, the majority of false positives are positioned in areas where it is not clear why they were classified as "manhole". Notice how, in general, these false positives occur on the sidewalks or the side of the road while the networks perform quite well on the road surface. Although the training dataset contains an equal number of "background" images of the road surface or the side of the road, there is a clear performance difference. Therefore, the precision, recall and F_2-score are also computed when only taking into account the detections on the road surface. This is similar to application of the optional region of interest filters during the preprocessing framework as described in Section 3.1. Table 5 summarizes the same performance scores for the different networks only considering the detection on the road surface for the different networks. When compared to Table 4, it is immediately clear how well the networks perform when only the road surface is considered. With the new optimal location parameters, recall achieves 93.2% for the IHV-trained networks with the precision score improving to around 60%–65%. While this is a significant improvement over the previous results, this only matches the performance of the intensity-trained network on dataset 1 on the whole public domain. In contrast, the intensity-trained network outperforms the IHV-trained networks by a large margin. While both intensity-trained networks improved, it is especially the intensity-trained network on the extra "background" images that achieves a high recall and precision score of 97.3%. In the case of the maximum recall optimization when only considering the road surface, recall scores are similar but with a much higher precision score. Similarly, as with the maximum F_2-score optimization, the intensity-trained network on dataset 2 performs the best and achieve an impressive 85% precision score with a 100% recall score. In both the maximum F_2-score and recall optimization, we notice a considerable improvement by training with additional "background" images.

4.5. Manhole Localization Performance

While most research documents detection performance in terms of recall and precision score, the localization accuracy of the center of the manhole cover is generally ignored. This is important as spatial databases usually require a specific level of accuracy for each object. In this section, the location accuracy of our approach is investigated by comparing the predicted center position to the ground truth center position determined from the GRB. These accuracy statistics are computed using the detection results and using the maximum F_2-score optimization from Table 4.

(**a**) VGG16 trained on dataset 1 with intensity images with localization parameters $T_{class} = 0.8$ and $T_{cluster} = 10$.

(**b**) VGG16 trained on dataset 2 with intensity images with localization parameters $T_{class} = 0.9$ and $T_{cluster} = 16$.

(**c**) VGG16 trained on dataset 1 with IHV images with localization parameters $T_{class} = 0.3$ and $T_{cluster} = 17$.

(**d**) VGG16 trained on dataset 2 with IHV images with localization parameters $T_{class} = 0.8$ and $T_{cluster} = 11$.

Figure 7. Visualization of the manhole detection results of the different trained VGG-16 network on a small subsection of the testing dataset. Each square defines the ground truth manhole covers which are colored green when detected and red when undetected. The predicted manhole cover positions are indicated by the crosses. A green cross indicates a true positive detection while a red cross indicates a false positive. The location parameters (classification threshold T_{class} and cluster threshold $T_{cluster}$) are in the caption of each image.

Table 5. Summary of the object detection results (recall, precision and F_2-score) with the corresponding classification threshold T_{class} and cluster threshold $T_{cluster}$ when only considering the detection on the road surface. The results and optimal parameters are displayed for both the maximal F_2-score and maximal recall score.

Dataset	Maximum F_2-Score Optimization					Maximum Recall Optimization				
	Recall	Precision	F_2-Score	T_{class}	$T_{cluster}$	Recall	Precision	F_2-Score	T_{class}	$T_{cluster}$
1 Int.	0.945	0.863	0.927	0.9	5	**1.000**	0.646	0.901	0.5	2
2 Int.	**0.973**	**0.973**	**0.973**	0.9	12	**1.000**	**0.849**	**0.966**	0.8	9
1 IHV	0.932	0.618	0.846	0.7	5	0.973	0.415	0.767	0.7	1
2 IHV	0.932	0.654	0.859	0.9	2	0.973	0.504	0.820	0.7	1

For each network, the mean error, standard deviation, root-mean-square error (RMSE) and 95% confidence interval (=2 × RMSE) are computed. The results are shown in Table 6, both in pixels and centimeters, taking the GSD of 2.5 cm into account. From these results, it is clear that, like the object detection performance, the intensity-trained networks outperform the IHV-trained networks with an average RMSE of 8.7 cm and 15.8 cm, respectively. Additionally, there is a noticeable performance difference between the intensity-trained networks on dataset 1 or 2. Although both datasets contain the same "manhole" training images, the additional "background" training images in dataset 2 degrade the 95% confidence interval from 16.5 cm to 18.1 cm. Because of the difference in sample size, this is not the case for the IHV-trained networks. When only taking the manhole covers detected by both IHV-trained networks into account, a similar performance difference is observed. These differences in accuracy become clearer when plotted on a scaled picture of a manhole cover as in Figure 8. As reference, a manhole cover is, in general, 90 cm or 36 pixels wide. Figure 8 allows to observe the performance difference between the intensity-/IHV-trained networks and the dataset-1-/2-trained networks by comparing the 95% confidence interval displayed by the red circle. There is a slight systematic deviation on the location prediction toward the upper-right corner. This unpredictable error is the downside of using a CAM-based localization approach. While the activation maps indicate the region of interest most important to classify the image and can be used for coarse localization of an object, this region does not necessarily correspond with the center position of the manhole cover, causing a systematic error on the location performance. Nevertheless, our approach is able to predict the manhole center with a 95% confidence interval of 16.5 cm using the intensity-trained VGG-16 network on dataset 1. In combination with the best manhole detection performance on the public domain, this network ensures the best results to map manhole covers for a spatial database.

Table 6. Summary of the error results (mean, standard deviation, root-mean-square error and 95% confidence interval) for the different trained VGG-16 networks. Both the errors, in pixels and centimeters, are displayed, taking into account the GSD of 2.5 cm.

Dataset	Sample Size	Mean Error	Std. Dev	RMSE	95% Conf. Int.
1 Int.	66	2.8 pix/7.1 cm	1.7 pix/4.3 cm	3.3 pix/8.3 cm	6.6 pix/16.5 cm
2 Int.	68	3.1 pix/7.7 cm	1.9 pix/4.8 cm	3.6 pix/9.1 cm	7.3 pix/18.1 cm
1 IHV	51	5.1 pix/12.7 cm	3.9 pix/9.6 cm	6.4 pix/15.9 cm	12.7 pix/31.8 cm
2 IHV	57	5.4 pix/13.6 cm	3.1 pix/7.7 cm	6.2 pix/15.6 cm	12.5 pix/31.2 cm

In order to determine if the results are accurate enough for the GRB, a more detailed analysis must be performed. The GRB specification state that the errors in the X and Y directions between the control/ground truth positions and predicted position must follow a certain distribution [1]. This distribution is defined by an object-specific standard deviation σ_{GRB} computed as follows:

$$\sigma_{GRB} = \sqrt{\sigma_m^2 + \sigma_{cm}^2 + 2\sigma_i^2} \tag{5}$$

where σ_m is the requested measurement accuracy in X and Y of 0.03 m, σ_{cm} the control/ground truth measurement accuracy in X and Y of 0.03 m and σ_i the object-specific accuracy which is 0.007 m for manhole covers. All together, this results in a manhole-specific standard deviation σ_{GRB} of 0.044 m. Using this standard deviation, the different GRB specified accuracy intervals of the distribution are defined as shown in Table 7. As an example, the GRB requires at least 60% of X and Y errors between ground truth and measured manhole center are smaller than $\sigma_{GRB} \times 1$ which deviates slightly from a normal distribution. Additionally, no less than 95% of X and Y errors must be smaller than $\sigma_{GRB} \times 3$. For a dataset of manhole covers to be accepted to update the GRB, all six accuracy interval requirements must be met. The distribution percentages for each VGG-16 network are presented in Table 7, with each result shown in bold when conforming to the GRB requirement. From these results, it is clear that only the intensity-trained networks comply with a few accuracy intervals. Although these results are not accurate enough for the GRB, an improved localization framework or additional fine-tuning of the prediction in postprocessing would improve these results and ensures the accuracy needed for the GRB. A few suggestions are presented in Section 5.

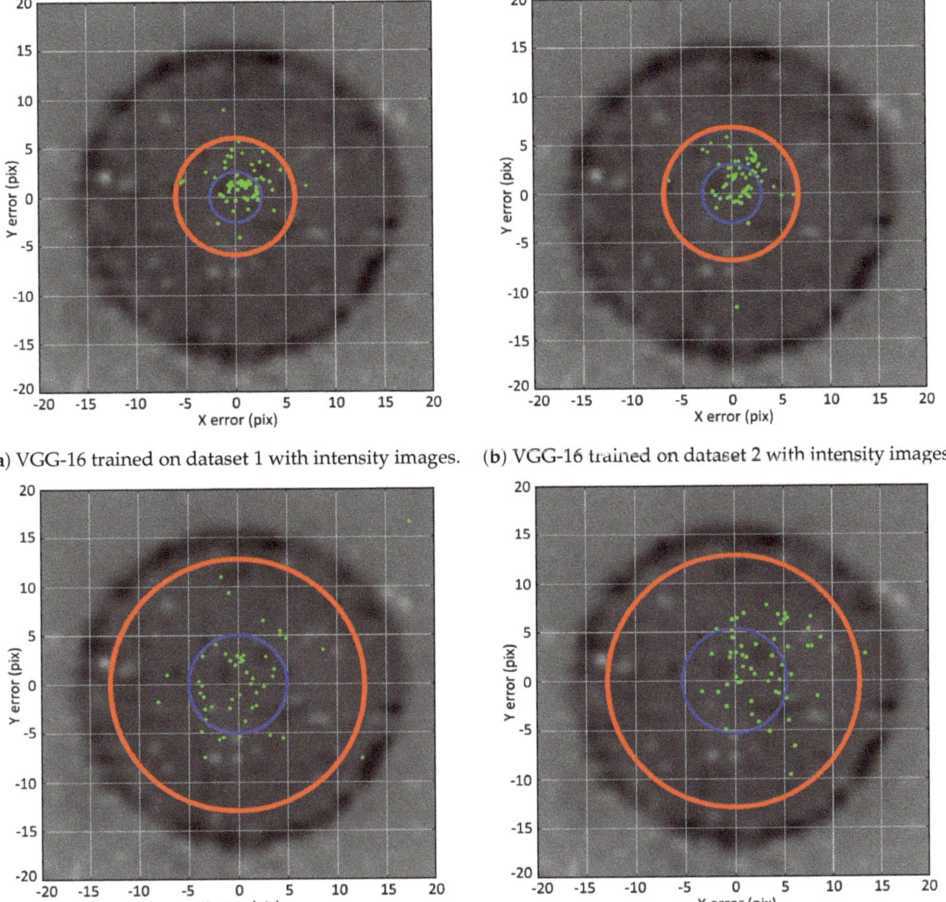

(**a**) VGG-16 trained on dataset 1 with intensity images.　　(**b**) VGG-16 trained on dataset 2 with intensity images.

(**c**) VGG-16 trained on dataset 1 with IHV images.　　(**d**) VGG-16 trained on dataset 2 with IHV images.

Figure 8. Visualization of the location predictions (green points) on a manhole cover for the different trained VGG-16 networks (**a**–**d**). The mean error and 95% confidence interval circles are shown in blue and red, respectively. The manhole cover in the background is up to scale and gives a sign of reference on how accurate each network performs.

Table 7. The distribution results per accuracy interval for the different VGG-16 networks. Results conforming to the GRB requirements are shown in bold.

Dataset	Sample Size	$\sigma_{GRB} \times 1$ = 4.4 cm	$\sigma_{GRB} \times 1.2$ = 5.2 cm	$\sigma_{GRB} \times 1.5$ = 6.5 cm	$\sigma_{GRB} \times 2$ = 8.7 cm	$\sigma_{GRB} \times 3$ = 13.1 cm	$\sigma_{GRB} \times 4$ = 17.4 cm
1 Int.	66	**63.6 %**	**70.5 %**	77.3 %	87.1 %	**96.2 %**	98.5 %
2 Int.	68	53.8 %	62.9 %	73.5 %	87.1 %	**96.2 %**	99.2 %
1 IHV	51	35.0 %	40.0 %	51.0 %	63.0 %	83.0 %	90.0 %
2 IHV	57	35.1 %	37.7 %	44.7 %	53.5 %	73.7 %	91.2 %
GRB requirements:		**60 %**	**70 %**	**80 %**	**90 %**	**95 %**	**100 %**

5. Discussion

Although multiple studies have looked into mapping manhole covers from mobile mapping point cloud data, no dedicated testing dataset exists to easily compare different methods. Fortunately, the research conducted by Yu et al. [12–14] has several similarities with our approach, such as the detection, which is performed on intensity ground images rasterized from mobile point cloud data. These different methods are discussed in detail in Section 2 and can be described as the marked-points-based method [12], the deep Boltzmann machine/random forest and sliding window method [13] and the super-pixel and CNN method [14]. In the following, only the detection results from Table 5 are considered as all methods of Yu et al. only detect manhole covers on the road surface. Note that the methods of Yu et al. are trained and evaluated on a much larger dataset compared to ours. Because of this, our results are not as reliable compared to previous research. However, our test dataset does contain a diverse environment, varying from urban to residential areas, representing a real-world example, and it is sufficiently large to evaluate our proposed method. Table 8 lists the manhole detection results for each approach, including our maximum F_2-score and recall optimization of the intensity-trained VGG-16 network on dataset 2. Of these methods, the more traditional model-based detection approach achieves the lowest recall, F_1- and F_2-score. The DBM/RF machine learning approach significantly improves these results by using high-level feature generation. The deep learning approaches improve these results even more. Our proposed method with the VGG-16 intensity-trained network on dataset 2 and F_2-score optimization achieves the best performance results compared to the other methods. This is quite impressive, considering that our method only needs a fraction of the positive "manhole" training images because of transfer learning. Although our approach using intensity-trained networks achieves high detection scores on the road surface, the same cannot be said for the detection results on the whole public domain or the IHV-trained networks. A few remarks and possible solutions to improve these results are discussed below.

Table 8. Summary of the detection results of different manhole cover detection approaches on intensity ground images.

Method	# Training Images *	# Manhole Covers	Recall	Precision	F_1-Score	F_2-Score
Yu et al. [12]	NA	491	0.896	0.903	0.900	0.898
Yu et al. [13]	7820/7820	491	0.953	0.955	0.954	0.954
Yu et al. [14]	2200/2200	491	0.965	0.961	0.963	0.965
Proposed	443/1000	73	0.973	**0.973**	**0.973**	**0.973**
Proposed	443/1000	73	**1.000**	0.849	0.918	0.966

* (manhole training images)/(background training images).

Our assumption that the additional geometric channels in the ground image would help detection does not hold up. We still believe geometric features can be used to enhance the precision performance, although not with transfer learning using a small training dataset. Instead, training a network from scratch, which would require much more data, could result in better detection results as the network would learn specific features from the IHV ground images. An alternative solution is to use the intensity-trained networks to detect the manhole covers with a high recall score and postprocess these results using common geometric features to filter out false positives. Additionally, there was a significant detection performance difference between the road surface and the rest of the public domain. This is mainly because the appearance of the public domain in the ground image has much more variation. Our results also indicated that training on additional "background" images slightly improves the results. Although this dataset contains a slight class imbalance, 1:2 manhole/background ratio, adding more "background" images to the training dataset is simply not going to further improve the results. This causes a severe class imbalance, 1:5 or 1:10 manhole/background ratio, and drastically degrades the manhole cover detection performance. Fortunately, different methods exist to address class imbalance on a dataset level or classifier level [33].

In addition to the object detection performance, the manhole center location also needs improvement to comply with the accuracy requirements of large-scale spatial databases such as the GRB. A dedicated localization network can be trained to determine the center of a manhole based on the ground image. This approach requires additional manual tagging of the manhole centers in our dataset as it only contains labeled images. As a network will always be less accurate than the dataset it was trained on, the manhole positions in the GRB are not accurate enough to automatically create this dataset. On the other hand, a postprocessing step to fine-tune the center position can be performed on the detection results of our approach. Some examples of such methods are the marked point [12,14] and GraphCut segmentation [16] approach to accurately delineate each manhole cover. If not successful, a semiautomatic approach can be employed where the user fine-tunes the predicted position manually. This way, the user can also filter out any false positive results and check all detection results, which will most likely be necessary anyway in a production environment, no matter how accurate the detection methods become.

6. Conclusions

In this paper, a fully automatic manhole cover detection method is presented to extract manhole covers from mobile mapping lidar data consisting of two components. First, the preprocessing framework removes the noisy and ghosting points, segments the point cloud into "ground" or "nonground" and rasterizes the "ground" points into a ground image with channels intensity, minimum height and height variance (IHV). Second, the object detection framework uses these ground images as input for a transfer learned fully convolutional network which simulates an internal sliding window and outputs a spatial classification map. This map indicates where the network expects a manhole cover to be located and is used to accurately determine the center position of each manhole cover using the activation maps of the last pooling layer. In this work, different backbone architectures (AlexNet, VGG-16, Inception-v3 and ResNet-101) are assessed after transfer learning on relatively small datasets with dataset 1 only containing the intensity channel while dataset 2 contains IHV images. Furthermore, the influence of additional "background" training images is investigated.

Our method is tested in a variety of experiments. First, the classification and sliding window performance of each network is compared which reveals that a high classification score does not automatically results in a good sliding window performance. The VGG-16 architecture performs the most consistent in both tasks. Next, the object detection performance of the VGG-16 networks is assessed on a dedicated testing dataset containing 73 manhole covers. Although our intention was to detect manhole covers all over the public domain, we noticed a significant detection performance difference between the false positive detection rate on the road surface and the rest of the public domain. When only taking into account the detection on the road surface, the best detection results are achieved with the intensity trained network on dataset 1, achieving a recall, precision and F_2-score of 0.973, 0.973 and 0.973, respectively. Overall, the experiments show that the networks trained on the IHV channels with geometric information degrade the detection performance instead of improving it. Furthermore, training with additional "background" images improves the precision score slightly. Last, the localization performance is compared, using the ground truth manhole center positions. Again, the intensity-trained networks outperform the IHV-trained networks with a RMSE of around 8.7 cm and 15.8 cm, respectively. Additionally, training on more "background" images resulted in a poorer localization accuracy. Our approach achieves a horizontal 95% confidence interval of 16.5 cm for the intensity-trained VGG-16 network on dataset 1, which almost complies with the GRB accuracy requirements.

Our future work will focus on improving the localization performance accuracy by implementing a dedicated-localization-network- or model-based approach. While currently our approach only uses the mobile point cloud data, future work will also focus on manhole cover detection on omnidirectional mobile mapping images. This image-based approach has the advantage that a manhole cover can be detected in multiple images, resulting in better detection results. Additionally, a combined image- and lidar-based approach will be investigated to further enhance the results. This approach will improve the recall performance, as a manhole cover can be detected in both the omnidirectional image and the lidar data. Additionally, detecting a manhole cover in both the image and lidar data increases the reliability of the result, creating a more robust detection framework.

Author Contributions: L.M. conceptualized the research and M.V. supervised the work. All authors have read and agreed to the published version of the manuscript.

Funding: This project has received funding from the FWO research foundation (FWO PhD SB fellowship 1S87120N) and the Geomatics section of the Department of Civil Engineering of KU Leuven (Belgium).

Acknowledgments: We would like to thank the surveying company Teccon BVBA for the acquisition and use of the mobile mapping datasets.

Conflicts of Interest: The authors declare no conflict of interest.

References

1. Informatie Vlaanderen. Basiskaart Vlaanderen (GRB). Available online: https://overheid.vlaanderen.be/informatie-vlaanderen/producten-diensten/basiskaart-vlaanderen-grb (accessed on 31 January 2020).
2. Dutch Digital Government. Basisregistratie Grootschalige Topografie (BGT). Available online: https://www.digitaleoverheid.nl/overzicht-van-alle-onderwerpen/basisregistraties-en-afsprakenstelsels/inhoud-basisregistraties/bgt (accessed on 31 January 2020).
3. Jalayer, M.; Zhou, H.; Gong, J.; Hu, S.; Grinter, M. A Comprehensive Assessment of Highway Inventory Data Collection Methods for Implementing Highway Safety Manual. *J. Transp. Res. Forum* **2014**, *53*, 73–92, doi:10.1016/j.jconrel.2007.08.001.
4. Guan, H.; Li, J.; Cao, S.; Yu, Y. Use of mobile LiDAR in road information inventory: A review. *Int. J. Image Data Fusion* **2016**, *7*, 219–242, doi:10.1080/19479832.2016.1188860.
5. Sairam, N.; Nagarajan, S.; Ornitz, S. Development of Mobile Mapping System for 3D Road Asset Inventory. *Sensors* **2016**, *16*, 367, doi:10.3390/s16030367.
6. Alshaiba, O.; Núñez-Andrés, M.A.; Lantada, N. Automatic manhole extraction from MMS data to update basemaps. *Autom. Constr.* **2020**, *113*, 103110, doi:10.1016/j.autcon.2020.103110.
7. Ma, L.; Li, Y.; Li, J.; Wang, C.; Wang, R.; Chapman, M.A. Mobile laser scanned point-clouds for road object detection and extraction: A review. *Remote Sens.* **2018**, *10*, 1531, doi:10.3390/rs10101531.
8. Krizhevsky, A.; Sutskever, I.; Hinton, G.E. ImageNet Classification with Deep Convolutional Neural Networks. *Commun. ACM* **2017**, *60*, 84–90, doi:10.1201/9781420010749.
9. Simonyan, K.; Zisserman, A. Very deep convolutional networks for large-scale image recognition. In Proceedings of the International Conference on Learning Representations, San Diego, CA, USA, 7–9 May 2015; pp. 1–14.
10. Szegedy, C.; Vanhoucke, V.; Ioffe, S.; Shlens, J.; Wojna, Z. Rethinking the Inception Architecture for Computer Vision. In Proceedings of the IEEE Conference on Computer Vision and Pattern Recognition, Las Vegas, NV, USA, 27–30 June 2016; pp. 2818–2826,
11. He, K.; Zhang, X.; Ren, S.; Sun, J. Deep residual learning for image recognition. In Proceedings of the IEEE Conference on Computer Vision and Pattern Recognition, Las Vegas, NV, USA, 27–30 June 2016; pp. 770–778, doi:10.1109/CVPR.2016.90.
12. Yu, Y.; Li, J.; Guan, H.; Wang, C.; Yu, J. Automated detection of road manhole and sewer well covers from mobile LiDAR point clouds. *IEEE Geosci. Remote Sens. Lett.* **2014**, *11*, 1549–1553, doi:10.1109/LGRS.2014.2301195.
13. Yu, Y.; Guan, H.; Ji, Z. Automated Detection of Urban Road Manhole Covers Using Mobile Laser Scanning Data. *IEEE Trans. Intell. Transp. Syst.* **2015**, *16*, 3258–3269, doi:10.1109/TITS.2015.2399492.
14. Yu, Y.; Guan, H.; Li, D.; Jin, C.; Wang, C.; Li, J. Road Manhole Cover Delineation Using Mobile Laser Scanning Point Cloud Data. *IEEE Geosci. Remote Sens. Lett.* **2020**, *17*, 152–156.
15. Wei, Z.; Yang, M.; Wang, L.; Ma, H.; Chen, X.; Zhong, R. Customized Mobile LiDAR System for Manhole Cover Detection and Identification. *Sensors* **2019**, *19*, 2422.

16. Timofte, R.; Van Gool, L. Multi-view Manhole Detection, Recognition, and 3D Localisation. In Proceedings of the IEEE International Conference on Computer Vision Workshop, Barcelona, Spain, 6–13 November 2011; pp. 188–195.

17. Liu, W.; Cheng, D.; Yin, P.; Yang, M.; Li, E.; Xie, M.; Zhang, L. Small Manhole Cover Detection in Remote Sensing Imagery with Deep Convolutional Neural Networks. *ISPRS Int. J. Geo-Inf.* **2019**, *8*, 49, doi:10.3390/ijgi8010049.

18. Girshick, R.; Donahue, J.; Darrell, T.; Malik, J. Rich feature hierarchies for accurate object detection and semantic segmentation. In Proceedings of the IEEE Conference on Computer Vision and Pattern Recognition, Columbus, OH, USA, 23–28 June 2014; pp. 580–587,

19. Redmon, J.; Divvala, S.; Girshick, R.; Farhadi, A. You only look once: Unified, real-time object detection. In Proceedings of the IEEE Conference on Computer Vision and Pattern Recognition, Las Vegas, NV, USA, 27–30 June 2016; pp. 779–788,

20. Liu, W.; Anguelov, D.; Erhan, D.; Szegedy, C.; Reed, S.; Fu, C.Y.; Berg, A.C. SSD: Single shot multibox detector. In Proceedings of the European Conference on Computer Vision, Amsterdam, The Netherlands, 8–16 October 2016; Volume 9905,

21. Felzenszwalb, P.F.; Girshick, R.B.; McAllester, D.; Ramanan, D. Object Detection with Discriminatively Trained Part Based Models. *IEEE Trans. Pattern Anal. Mach. Intell.* **2010**, *32*, 1627–1645.

22. Moy de Virty, M.; Schnidler, K.; Rieckermann, J.; Leitão, J.P. Sewer Inlet Localization in UAV Image Clouds: Improving Performance with Multiview Detection. *Remote Sens.* **2018**, *10*, 706, doi:10.3390/rs10050706.

23. Guan, H.; Li, J.; Yu, Y.; Wang, C.; Chapman, M.; Yang, B. Using mobile laser scanning data for automated extraction of road markings. *ISPRS J. Photogramm. Remote Sens.* **2014**, *87*, 93–107, doi:10.1016/j.isprsjprs.2013.11.005.

24. Sultani, W.; Mokhtari, S.; Yun, H.b. Automatic Pavement Object Detection Using Superpixel Segmentation Combined With Conditional Random Field. *IEEE Trans. Intell. Transp. Syst.* **2018**, *19*, 2076–2085, doi:10.1109/TITS.2017.2728680.

25. Santos, A.; Junior, J.M.; Silva, J.D.A.; Pereira, R.; Matos, D.; Menezes, G.; Higa, L.; Eltner, A.; Ramos, A.P.; Osco, L.; Gonçalves, W. Storm-Drain and Manhole Detection Using the RetinaNet Method. *Sensors* **2020**, *20*, 4450, doi:10.3390/s20164450.

26. Zhang, W.; Qi, J.; Wan, P.; Wang, H.; Xie, D.; Wang, X.; Yan, G. An Easy-to-Use Airborne LiDAR Data Filtering Method Based on Cloth Simulation. *Remote Sens.* **2016**, *8*, 501, doi:10.3390/rs8060501.

27. Russakovsky, O.; Deng, J.; Su, H.; Krause, J.; Satheesh, S.; Ma, S.; Huang, Z.; Karpathy, A.; Khosla, A.; Bernstein, M.; Berg, A.C.; Fei-Fei, L. ImageNet Large Scale Visual Recognition Challenge. *Int. J. Comput. Vis.* **2015**, *115*, 211–252,

28. Sermanet, P.; Eigen, D.; Zhang, X.; Mathieu, M.; Fergus, R.; LeCun, Y. Overfeat: Integrated recognition, localization and detection using convolutional networks. In Proceedings of the International Conference on Learning Representations, Banff, AB, Canada, 14–16 April 2014.

29. Zhou, B.; Khosla, A.; Lapedriza, A.; Oliva, A.; Torralba, A. Learning Deep Features for Discriminative Localization. In Proceedings of the IEEE Conference on Computer Vision and Pattern Recognition, Las Vegas, NV, USA, 27–30 June 2016; pp. 2921–2929,

30. Mattheuwsen, L.; Bassier, M.; Vergauwen, M. Theoretical accuracy prediction and validation of low-end and high-end mobile mapping system in urban, residential and rural areas. *Int. Arch. Photogramm. Remote Sens. Spat. Inf. Sci.* **2016**, *42*, 121–128.

31. Kornblith, S.; Shlens, J.; Le, Q.V. Do Better ImageNet Models Transfer Better? In Proceedings of the IEEE Conference on Computer Vision and Pattern Recognition, Long Beach, CA, USA, 15–20 June 2019.

32. Matlab, version 9.7.0.1216025 (R2019b); The MathWorks Inc.: Natick, Massachusetts, 2019.

33. Buda, M.; Maki, A.; Mazurowski, M.A. A systematic study of the class imbalance problem in convolutional neural networks. *Neural Netw.* **2018**, *106*, 249–259.

Publisher's Note: MDPI stays neutral with regard to jurisdictional claims in published maps and institutional affiliations.

Article

Novel Approach to Automatic Traffic Sign Inventory Based on Mobile Mapping System Data and Deep Learning

Jesús Balado [1,2,*] , **Elena González** [3] , **Pedro Arias** [1] and **David Castro** [4]

[1]   Applied Geotechnologies Group, Department Natural Resources and Environmental Engineering, School of
    Mining and Energy Engineering, University of Vigo, Campus Lagoas-Marcosende, 36310 Vigo, Spain;
    parias@uvigo.es
[2]   Faculty of Architecture and the Built Environment, Delft University of Technology,
    2628 BL Delft, The Netherlands
[3]   School of Industrial Engineering, University of Vigo, Campus Lagoas-Marcosende, 36310 Vigo, Spain;
    elena@uvigo.es
[4]   Department Innovation, INSITU Engineering, CITEXVI, Street Fonte das Abelleiras, 36310 Vigo, Spain;
    david.boga@ingenieriainsitu.com
*   Correspondence: jbalado@uvigo.es

Received: 10 January 2020; Accepted: 29 January 2020; Published: 1 February 2020

Abstract: Traffic signs are a key element in driver safety. Governments invest a great amount of resources in maintaining the traffic signs in good condition, for which a correct inventory is necessary. This work presents a novel method for mapping traffic signs based on data acquired with MMS (Mobile Mapping System): images and point clouds. On the one hand, images are faster to process and artificial intelligence techniques, specifically Convolutional Neural Networks, are more optimized than in point clouds. On the other hand, point clouds allow a more exact positioning than the exclusive use of images. The false positive rate per image is only 0.004. First, traffic signs are detected in the images obtained by the 360° camera of the MMS through RetinaNet and they are classified by their corresponding InceptionV3 network. The signs are then positioned in the georeferenced point cloud by means of a projection according to the pinhole model from the images. Finally, duplicate geolocalized signs detected in multiple images are filtered. The method has been tested in two real case studies with 214 images, where 89.7% of the signals have been correctly detected, of which 92.5% have been correctly classified and 97.5% have been located with an error of less than 0.5 m. This sequence, which combines images to detection–classification, and point clouds to geo-referencing, in this order, optimizes processing time and allows this method to be included in a company's production process. The method is conducted automatically and takes advantage of the strengths of each data type.

Keywords: LiDAR; RetinaNet; inception; Mobile Laser Scanning; point clouds; data fusion

---

1. Introduction

Communication and mobility of people and goods are key elements of modern societies and developing countries. Economic growth has a huge dependence on and a big relationship with transport networks. Infrastructures such as ports (maritime and river), airports, railways, highways, and roads are among the most relevant transport systems to guarantee the quality of life of people. This relevance is well known by the EU. Proof of that are substantial national and EU funds spent on transport infrastructures every year [1]. These policies are developed based on annual budgets dedicated to new project construction and maintenance of existing infrastructures. In recent times,

EU infrastructure policies are changing, focusing more on keeping existing infrastructure in good working condition and less on new construction [2]. This goal has been gathered, among other ways, through different national and EU research work programs (e.g., smart, green, and integrated transport in the case of H2020) [3]. This has promoted numerous activities to improve the service given to society in fields such as monitoring, resilience, reduction of fatal accidents, traffic disruption, maintenance costs, and improvement of network capacity.

Highways and roads are the most used infrastructures for mobility in short distances. As a consequence, their conservation and maintenance show high relevance in terms of safety and secure mobility and reducing associated costs [4,5]. New concepts, called digital infrastructure and Intelligent Transport System (ITS), are being developed in parallel with new concepts for mobility: resilient and fully automated infrastructures; electric, connected and autonomous cars [6].

Concepts of digital infrastructure and ITS are connected. Digital road and permanent monitoring are the bases of any ITS applied to highways and roads to ensure safe mobility and good service conditions. There are different techniques and technologies to achieve digital road monitoring. Depending on the effectiveness and applicability, the most used are based on satellite images, aerial images, and Mobile Mapping System (MMS) solutions.

The low resolution of satellite images makes it impossible to extract certain information from linear infrastructures [7]. Roads, highways, or railways are detectable in the satellite images, but it is not feasible to know the state of the pavement, rails, or their signalling. As a consequence, the scale of work is too small to get effective results with the ITS.

Aerial solutions have grown hugely in recent times based on civil drones and remotely piloted civil systems [8]. This is an emerging market of huge interest. However, drones still have many legal limitations related to the safety and protection of people. These drawbacks limit their use in many fields, among them transport infrastructures.

The MMS solutions, based on Light Detection and Ranging (LiDAR) technology, images (video and panoramic), and GNSS (Global Navigation Satellite System) technologies (for data geolocation) [9], are mature solutions that saw limited growth for mainly two reasons: the high price of the technology and high cost of processing the captured data (in terms of labour cost). Notwithstanding, the market is showing novel and very active emerging low-cost and multiplatform solutions for autonomous vehicles. On the other side, big data and artificial intelligence techniques allow efficient data processing [10].

This work is focused on developing a technical solution to generate infrastructure digital models and road infrastructure inventory based on the MMS. This work is applied to a specific component of roads, i.e., traffic signs (TSs), which are very relevant in transport infrastructures for the safety and security of people. The objective is the fast TS detection, recognition, and classification with accurate localization. But the application field of the proposed method is not limited to TSs. The proposed method shows high relevance for autonomous mobility solutions and urban planning. Based on them, the solution provides key information about on existing traffic signs, including accurate geolocation parameters.

This paper is organized as follows: Section 2 collects related work about traffic sign detection, recognition, and mapping in images and point clouds. Section 3 explains the designed method. Section 4 presents and discusses the results obtained from applying the method to case studies, and Section 5 concludes the work.

2. Related Work

The interest in the off-line automation of traffic sign inventory has increased in recent years. Previously, proposed approaches tackled the particular properties of traffic signs (i.e., retro-reflectivity for night-time visibility, colour, shape, size, height, orientation, planarity, and verticality), usually following safety standards. These properties require traffic signs to be treated as different objects from traffic lights [11–13], poles [14], lanes [15–17], trees [18], and other objects present in roads. A review of approaches depending on the object can be found in [19].

Traffic sign (TS) current technology provides mainly two sources of data: 3-D georeferenced point clouds acquired through Mobile Laser Scanning (MLS) techniques; and digital images from a still camera or as a frame extracted from a video. 3-D point-cloud data contains precise information related to 3-D location and geometrical properties of the TS, as well as intensity. However, resolution of most MMS techniques under normal use is not accurate enough to recognise all TS classes. Images are used to overcome that weakness as they contain visual properties, despite the lack of spatial information. Since the objective in automated traffic sign inventory is to accurately determine placement in global coordinates and the specific type of each traffic sign on the road, point cloud and image become complementary [20–22].

For TS inventory to be automated it is required to follow four main steps: traffic sign detection (TSD), segmentation (TSS), recognition or classification (TSR), and TS 3-D location. TSD aims to identify regions of interest (ROI) and boundaries of traffics signs. In TSS, a segment corresponding to the object is separated from the set of input data. TSR consists of determining the meaning of the traffic sign. Meanwhile, TS 3-D location deals with estimating 3-D position and orientation, or pose, of the TS. A variety of approaches for these steps have been proposed in literature directly or indirectly related to TS inventory.

One group of these approaches defines techniques focused on detecting and segmenting the set of points with spatial information of the TS from 3-D laser scanner point clouds. These techniques are based on the a priori knowledge of 3-D location, geometrical and/or retro-reflective properties. All approaches are conditioned by the huge amount of information contained in point clouds (see, for instance, [23–30]). With the aim of accurate TSR, aforementioned approaches combine point clouds with images to extract features. As a previous step to TSR, segmented points can be projected onto the corresponding 2-D image in the traffic-sign-mapping (TSM) step. A review of methods for TSR in point-cloud and image approaches can be found in [31].

These types of techniques based on TSD in 3-D point cloud and TSR in image are accurate and reliable for TS inventory. However, they entail high time and computational costs, mainly for the TSD and TSS steps. As an alternative, images can be used not only for TSR but also for TSD without making use of the 3-D point cloud. Some authors have used TSD in image for coarse segmentation of the 3-D point cloud [32,33].

TSD, TSS, and TSR in image, which become TSDR, have been extensively studied for TS inventory as well as for other applications such as advanced driver assistance systems (ADAS). The vast variety of techniques proposed by the computer–vision community have been reviewed and compared, detailing advantages and drawbacks, in [34–38]. Recently, Wali et al. [39] provided a comprehensive survey on vision-based TSDR systems.

According to them, in TSDR image-based techniques detection consists in finding the TS bounding box, while recognition involves classification by giving an image a label. Common TSD methods are: colour-based, on different colour spaces, i.e., RGB, CIELab, and HIS [40]; shape-based, such as Hough Transform (HT) and Distance Transform (DT); texture-based, such as Local Binary Patterns (LBP) [41]; and hybrid. By these methods a feature vector is extracted from image with lower computational cost than from 3-D point cloud. Then, the class label of the feature vector is obtained using a classifier such as Support Vector Machine (SVM) or with Deep Learning-based (DL) methods [42–44]. Among the latter, Convolutional Neural Networks (CNN) have been widely adopted, given their high performance in both TSD and TSR in images [45–48] and in point clouds [49].

Regarding TS inventory, TSDR in image requires the TS 3-D location to be completed. TS 3-D location, after TSDR, has been considered by several authors in image-based 3-D reconstruction approaches without making use of a 3-D point cloud. These techniques require prior accurate camera calibration and removement perspective distortion. In [50], 3-D localization is based on epipolar geometry of multiple images, while Hazelhoff et al. [51] calculated the position of the object from panoramic images referenced to northern direction and horizon. Balali et al. [52] built a point cloud by photogrammetry techniques using a three parameter pinhole model. Wang et al. [53] used stereo vision and triangulation techniques. In [54], 3-D reconstruction is conducted by geometric epipolar, taking into account geometric shape of TS.

While TSDR in image is proved as high-performance, reconstruction models for TS 3-D location from image are overcome in precision by 3-D point-cloud-based location. However, little research has paid attention to techniques for TS inventory that jointly takes advantage of TSDR in image and TS 3-D location from the 3-D MLS point cloud. In [32], a method to combine DL with retro-reflective properties is developed for TS extraction and location from a point cloud. In [55], TS candidates are detected on images based on colour and shape features and filtered with point-cloud information. Most authors use point clouds for TSD and images only for TSR [23–26,28,29].

In contrast to other approaches, in this work a data flow is implemented to minimize processing times by taking advantage of each type of data. First, images are used for TSD and TSR. Image processing is faster than point-cloud processing and allows the application of DL techniques, which right now are state of the art. In addition, the design of a modular workflow allows each network to be replaced in the future as its success rates increase. To maximize a correct TS identification, different networks for TSD and TSR are used, unlike other works that use the same network to detect and classify, see also [44,48]. After image processing, point clouds are used to filter out multiple TS detections and false positives. Point clouds allow more precise geolocation than the use of epipolar geometry of multiple images. Point clouds are not used for detection and classification since:

- DL point-cloud processing techniques are computationally more expensive than their equivalent in image processing.
- The addition of point-cloud data to images increases processing times.
- TSs are not always in good condition to be detected by their reflectivity, as other authors have proposed [23,56].
- The low point density does not provide useful information for TSR.

3. Method

The method consists of four main processes. First, TSs are detected in images. Second, detected TSs are recognized. Third, TSs are 3-D geolocated by the projection of detected signs to the point cloud. Fourth, multiple TS detections of the same sign in different images are filtered. The input data of the method are MMS data: images from a 360° camera, point clouds, GPS-IMU positioning data, and camera calibration data. Figure 1 shows the workflow of the method.

3.1. Traffic Sign Detection

TSD is based on object detection in images. No point-cloud data are used at this stage to speed up the detection process. The input images are acquired with a 360° RGB camera mounted on the MMS during acquisition. The panoramic image is converted and rectified into six images oriented according to cube sides. Images in trajectory direction I_T provide TS information in front of the MMS. Images in the opposite direction I_{To} provide TS information in back of the MMS, either in the same lane or in different lanes. Lateral images are perpendicular to trajectory direction $I_{\perp T}$ and provide information about signs located on MMS sides. Lateral images $I_{\perp T}$ are particularly relevant for detecting no-parking signs or no-entry signs. The images forming the top and bottom of the cube are not relevant for TSD. Bottom images are occupied by the camera support. TSs that could be detected on top images are already detected by front images I_T.

The object detector implemented in this method is RetinaNet [57]. This detector has been chosen because it is state of the art in standard accuracy metrics, memory consumption, and running times. RetinaNet is a one-stage detector that has good behaviour with unbalanced classes and in images with a high density of objects at several scales, key factors for traffic sign detection. RetinaNet uses ResNet [58] as a basic feature extractor, and in this work is used the ResNet 50.

The RetinaNet detector is applied to each cube-side image I of the set acquired with the MMS during the acquisition. As a result, an array is obtained for each TS detected $S(l,I_x,I_y,w,h)$ where l indicates the label, I_x and I_y indicate top left corner position of the bounding box, w indicates TS width and h indicates TS height. In order to obtain maximum classification accuracy, the number of classes has been reduced to coincide with shapes of traffic sings. The classes for detection with RetinaNet are five: yield, stop, triangular, circular, and square (Figure 1). In the recognition phase (Section 3.2) these classes will be classified.

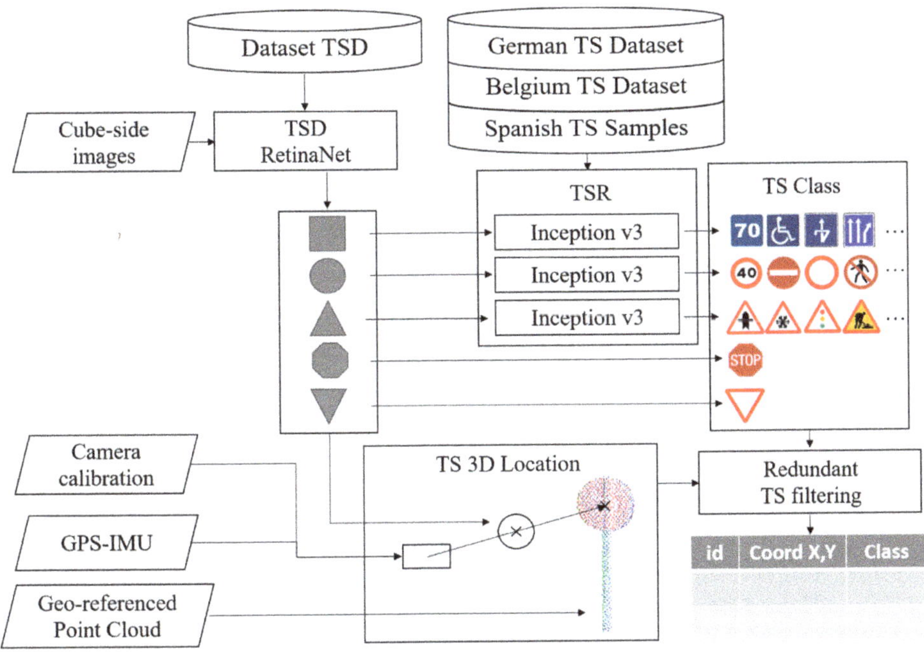

Figure 1. Workflow of the method.

3.2. Traffic Sign Recognition

In this phase, TSs previously detected by their shape are classified with their final label. In some TSs, their shape coincides with their final class, as in the case of stop signs (octagonal) and yield sign (inverted triangle). TSs of obligation (circular), recommendation–information (square), and danger (triangular) encompass multiple classes that must be classified for a correct inventory. For each of these three sign shapes, an InceptionV3 network [59] has been trained and implemented. The InceptionV3 network needs input samples of fixed size 299 * 299 * 3 pixels, the bounding boxes images of detected signs S are resized to adapt them to the network input.

3.3. Traffic Sign 3-D Location

The projection of TSs detected in images onto the georeferenced point cloud is done using the pinhole model [60]. While in other works the four vertices of the detection polygon have been projected, in this work, only the central TS point is projected S_c. This saves processing time and minimizes calibration error. Another alternative would be detecting the pole directly or after TS detection. Pole detection would mean more precise positioning, but it has the following limitations: (1) poles may not have enough points to be easily detected, (2) some TSs share a pole, and (3) some TSs are located on traffic lights, light posts, or buildings, so specific detection methods are needed for each case. In view of the above, the authors have chosen to consider the error of positioning the TS to the pole as negligible, and obtain a simpler and faster method based on TS location and positioning. The TS location in point cloud P_S is done by projecting a line $\vec{CS_c}$ defined by the camera focal point C and the central sign-point S_c (pinhole model in Equation (1)).

$$s \cdot S_c = K\,[R|t]P_s \tag{1}$$

where s is the scalability factor; S_c is the centre of traffic signal S detected in an image I, $S_c = [u, v, 1]^T$ with $u = I_x + w$ and $v = I_y - h/2$ w; K is the intrinsic camera parameters matrix provided by the manufacturer; $[R|t]$ is the extrinsic camera parameters matrix; P_s is the centre of the point-cloud traffic signal $P_s = [X_w, Y_w, Z_w, 1]^T$.

The rotation and translation matrix $[R|t]$ positions the camera in the same coordinate system as the point cloud P_s, which is already georeferenced. $[R|t]$ is formed by two rotation–translation matrices $[R|t] = [R_1|t_1][R_2|t_2]$ relates the positioning of the pixels with the image by calibration prior to implementation of the method. Once the matrix $[R_1|t_1]$ for one image is obtained, it is valid for all images acquired with the same equipment. The calibration is done by manually selecting the four pairs of pixels in images and points in the point cloud per image. $[R_2|t_2]$ positions the camera in the optical centre C of each image I.

The TS points in the point cloud P_s form a plane T_s. The TS is located in the intersection between the projection of the line $\vec{CS_c}$ following the pinhole model and plane T_s (Figure 2), $P_s = \vec{CS_c} \cap T_s$.

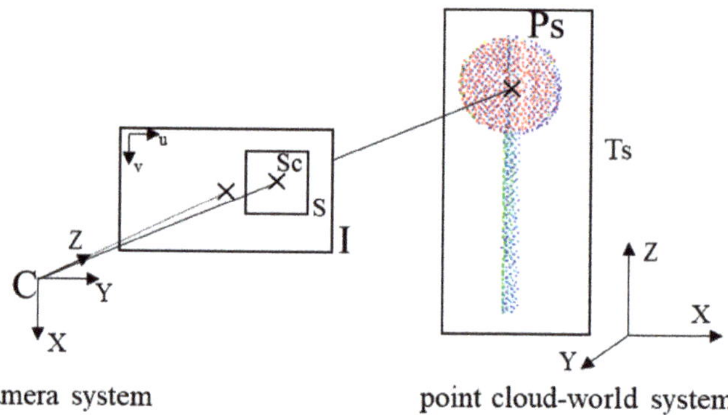

Figure 2. Pinhole model used to project traffic signs (TSs) detected in the image onto the point cloud.

In order to reduce processing time, a region of interest (ROI) is delimited in the point cloud to calculate possible T_s planes (Figure 3). First, points located at a distance more than d from the MMS location at the time of taking the image are discarded. Distant TSs from the MMS are considered to have very low point density for correct location. Distant TSs also are detected in successive images captured near the MMS. Second, points located at a larger distance than r from line $\vec{CS_c}$ are discarded. Third, points not located in the image orientation are discarded. TSs detected in images cannot be in a point cloud in a different orientation. For remaining points, planes are detected in order to discard point not in planes. Since TSs are planar elements, planar estimation avoids false locations due to noise points crossing the projection line $\vec{CS_c}$.

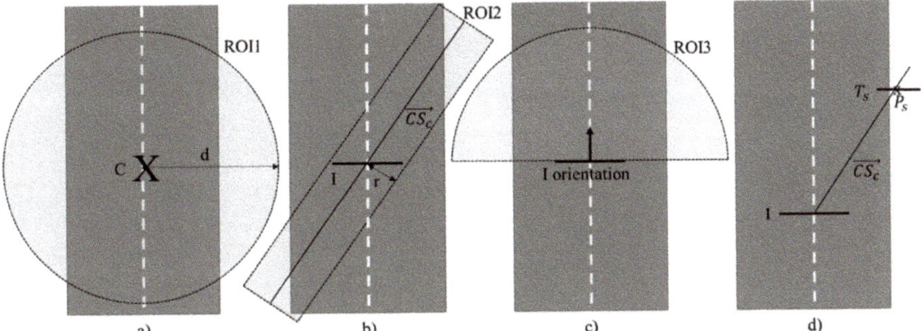

Figure 3. Top view of ROI delimitation in a point cloud road environment: (**a**) delimitation by distance *d* from camera location; (**b**) delimitation by distance from projection TS line; (**c**) delimitation by image orientation; (**d**) location of P_s in the first S of points forming a plane and crossed by $\vec{CS_c}$.

3.4. Redundant Traffic Sign Filtering

Since the same TS can be detected in multiple images, multiple detections of the same TS must be simplified. The filtering is done with information of the classified TS, because one post can contain TSs of different classes. TSs of the same class grouped in a smaller radius than *f* are eliminated, leaving only the first detected (Figure 4).

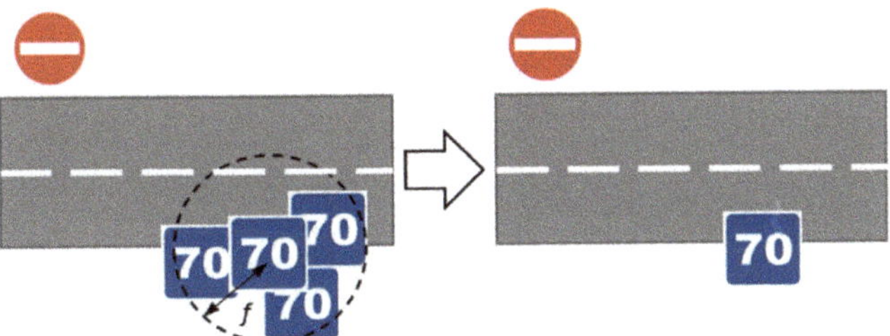

Figure 4. Filtering of the same TS class in a radius *f*.

4. Experiments

4.1. Equipment and Parametes

The MMS equipment used for this work consisted of a Lynx Mobile Mapper, with a Ladybug5 360° camera and a GPS-IMU Applanix POS LV 520. The cube-images had a resolution of 2448 × 2024 pixels and they were captured with a frequency of 5 m in MMS trajectory. The point cloud was a continuous acquisition over time. The values of parameters *d* and *r* to delimit the ROI were set at 15 m and 2 m, respectively. The value of parameter *f* was set to 1 m in order to simplify duplicate signals.

For the RetinaNet training, 9500 images were used with 12,036 TSs, obtained by the 360° camera and labelled. The training of the InceptionV3 networks was carried out with data sets of Belgium [50], Germany [61], and images of Spanish traffic signs. The whole process (training and testing in real case studies) was executed on a CPU computer Intel i7 6700, 32 GB RAM, GPU Nvidia 1080ti. The code was combined TensorFlow–Python for TSD and TSR and C++ for 3-D location and filtering.

The RetinaNet training consumed 70 h with hyper-parameter optimization method *adam*, learning rate 1e-5, L2 Regularization 0.001, Max Epochs 50 and Batch Size 1. The hyper-parameters for the three Inceptionv3 training were optimization method *sgdm*, learning rate 1e-4, Momentum 0.9, Max Epochs 126 and Batch Size 64. The training of the triangular signs required 50 min, 12,995 samples for training and 407 for validation. The training of the circular signs required 80 min, 25,000 samples for training and 743 for validation. The training of the squared signs required 7 min, 1094 samples for training and 243 for validation. The training process in terms of loss per epoch is shown in Figure 5.

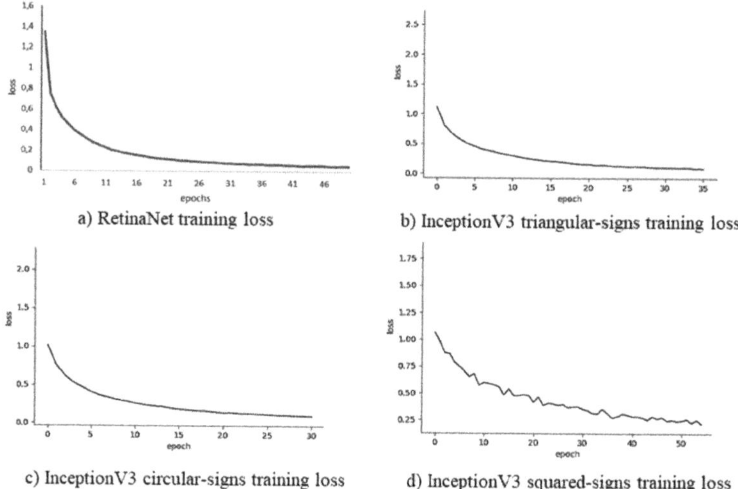

a) RetinaNet training loss

b) InceptionV3 triangular-signs training loss

c) InceptionV3 circular-signs training loss

d) InceptionV3 squared-signs training loss

Figure 5. Evolution of the loss during training processes: (**a**) RetinaNet detector training loss, (**b**) InceptionV3 triangular-signs classifier training loss, (**c**) InceptionV3 circular-signs classifier training loss, (**d**) InceptionV3 squared-signs classifier training loss.

4.2. Case Studies

The methodology was tested in two real case studies: two secondary roads located in Galicia (Spain) denominated EP9701 and EP9703. Road EP9701 case study was 9.2 km long, the point cloud contained 350 million points and was acquired with 7392 images. Road EP9703 case study was 5.5 km long, the point cloud contained 180 million points and was acquired with 4520 images. Both roads were located in rural areas where houses, fields, and wooded areas were interspersed. The roads had frequent crossings and curves. The sign-posting of both roads was abundant and in good condition, with few samples that were damaged or partially occluded. The case studies were processed in 30 and 20 min, respectively.

The acquisition was performed at the central hours of the day (to minimize shadows) and on a sunny day without fog, so as not to affect visibility. The MLS maintained a constant driving speed of approximately 50 km/h, although this speed was reduced by following rules at intersections or traffic lights. Point density increased as the driving speed decreased. It was estimated that the points in acquisition direction were 1 cm closer for every 10 km/h that the speed was reduced.

4.3. Results

TS accounting was done manually by reviewing acquired images, detected signals, classified signs and their locations in the point cloud. Table 1 shows the image count for each case study.

TSs were correctly detected at 89.7%, while 10.3 % were not detected. The use of the 360° camera and the cube-images made it possible to locate TSs in the opposite and lateral directions to the MMS movement. Some of the undetected TSs were partially occluded or were eliminated in the redundant TS filtering process (Section 3.4), since they were traffic signs of the same class separated within a distance f. Figure 6 shows examples of detected TSs.

Table 1. Results.

	EP9701		EP9703		TOTAL	
Total detections	113		137		250	
TS total	98		116		214	
TS detected	86	87.8%	106	91.4%	192	89.7%
TS undetected	12	12.2%	10	8.6%	22	10.3%
False detections	22	19.5%	27	19.7%	49	19.6%
TS duplicated	5	5.1%	4	3.4%	9	4.2%
TS correctly classified	84	92.3%	102	92.7%	186	92.5%
TS uncorrectly classified	7	7.7%	8	7.3%	15	7.5%

Figure 6. TSs detected: case study 1 with frontal cube-image (**a**), and with lateral cube-image (**b**), case study 2 with frontal cube-image (**c**) and with lateral cube-image (**d**). Detected TSs remarked in red boxes and filtered TSs remarked in green boxes.

A high percentage of false detections was counted (19.6%). Of these, traffic mirrors represented 81.8% and 37% of false detections in case studies 1 and 2 respectively. Mirrors have a circular shape surrounded by a red ring, so they were detected as false circular signals. Although the use of the point cloud has been considered to eliminate these false positives, since mirrors should not have points due to their high reflectivity, in the case studies the mirrors contained points due to their dirt or deterioration. Nor have any characteristics been found that differentiate mirror points from TS points. The remaining false detections corresponded to different objects on roadsides. Figure 7 shows some examples of false detections.

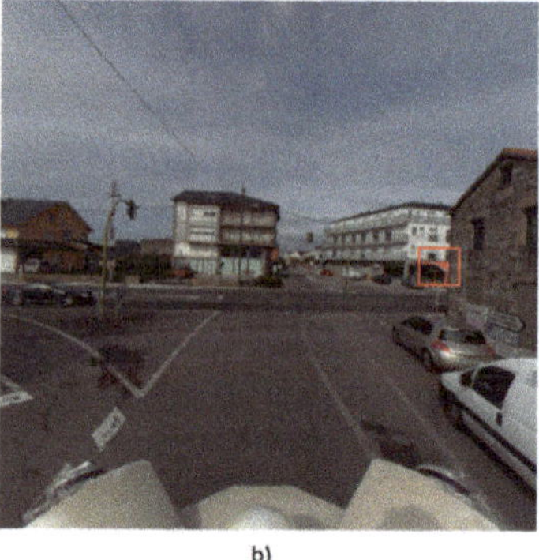

Figure 7. False detections (red boxes) caused by a road mirror in case study 1 (**a**) and by an awning in case study 2 (**b**).

Duplicate TSs were not filtered due to incorrect positioning by the TS 3-D location process (Section 3.3). In the input cube-side images, 382 TSs were detected in case study 1 and 441 signs in case study 2. After the 3-D localization and redundant filtering processes, the set of detections was reduced to 113 and 137 TSs, respectively. Duplicated TSs were 4.2% of the total.

The positioning of a TS was based on the georeferenced point cloud, where the authors assumed that the location of the point cloud corresponded precisely to reality, as in [23]. Authors also considered that the TSs positioned in the correct TS point cloud were correctly located (0 m error). A total 97.5% of the detected TSs corresponded to points belonging to TSs (Figure 8). Only five TSs were positioned with an error of between 0.5 m and 8 m to the real location of the sign. These TSs not correctly positioned in the corresponding TS point clouds were manually measured from their incorrect detected location to the real TS location in the point cloud.

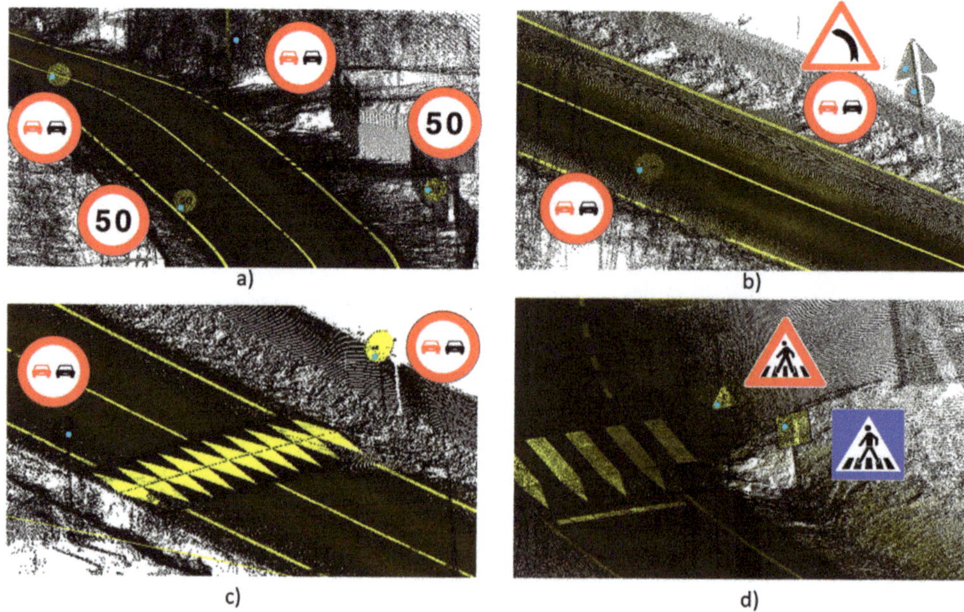

Figure 8. Traffic sign location in point cloud (blue point) and labelled (**a–d**).

With regard to signal recognition, 92.5% of the detected TSs were correctly classified, both in good condition and partially erased. Since the methodology was tested in real case studies, it was not possible to test all the existing classes of training data. The main classes in the case studies were TSs of dangerous curves, speed bumps, no overtaking, and speed limits. To a lesser extent, there were also traffic signs of yield, stop, roundabouts, no entry, roadworks, and pedestrian crossings. No significant confusion was detected among classes. Errors in confusion were isolated and were corresponded to the results of training.

4.4. Discussion

In general, most TSs were detected and positioned correctly, although the algorithm showed a tendency to over-detection. This behaviour was chosen to facilitate monitoring by a human operator. In a correction process, it was considered easier to eliminate false detections than to check all input data for undetected signals. In terms of false positives per image, the false detection rate was low, 0.004 FP/image, compared to [54], where 0.32 and 0.07 were reached in the cases with images of better resolution. Regarding undetected TSs (false negatives), the neural network did not detect 10.3% of all TSs, which was similar to other artificial intelligence works: 10% in [51] and 11% in [28], but far from the best of the state of the art: 6% in [62], based on laser scanner intensity; 5% in [32], based on combining two neural networks; and 4% in [29], based on bag-of-visual-phrases.

The authors are aware that the detection success rate was not as high as in other applications using RetinaNet [63]. This was due to the relative small size of the data set for TSD and the great variability of elements that existed in the road environment. Generating a data set for detection is a costly work and was not the final objective of this work, which was focused on presenting a methodology composed of a series of processes to inventory signals, and not on optimizing the success rates of Deep Learning networks such as RetinaNet and InceptionV3.

The methodology did not reach detection rates as high as reference works in TSD and TSR, such as [50,52], although it is worth mentioning that the latter classifies TS grouped by type. By contrast, the proposed methodology is adaptable for mapping different objects, as it does not focus on exclusive TS features. Particularly, by not using reflectivity, it was possible to detect TSs whose reflectivity had diminished due to the passage of time and incorrect maintenance. With the use of Deep Learning techniques, although they do not explain exactly why false detections occur, it is possible to intuit the underlying problem. Deep Learning techniques allow continuous improvement and updates to the training database with new samples that, in this case, may be the wrong detections once corrected. In this way, the algorithm will be able to avoid them.

The combination of images to TSD and TSR with a point cloud to TS 3-D locations allowed a precise positioning of 97.5% of detected TSs in points belonging to TS point clouds, which was not reached by other works based exclusively on epipolar geometry of multiple images, such as [50], which only achieved a positioning with 26 cm of average error, [53] with 1-3 m of average error using dual cameras, and [64] with 3.6 m of average error using Google Street View images.

While point clouds provide valuable information for locating objects, they also require much more processing time than images. The methodology designed in [23] for TSD and TSR in point clouds was implemented in the two case studies. Processing times using point clouds has reached 45 and 30 min, respectively. The time increment is 50% more than performing TSD and TSR on images and 3-D location in point clouds, as proposed in this work. No relation was found between inventory quality and driving speed changes during acquisition. The work maintained a driving acquisition speed similar to other point-cloud mapping works.

5. Conclusions

In this work, a methodology for the automatic inventory of road traffic signs was presented. The methodology consists of four main processes: traffic sign detection (TSD), recognition (TSR), 3-D location and filtering. For the TSD and TSR phases, cube-images acquired with a 360° camera were used and processed by Deep Learning techniques. Five shapes of traffic signs were detected in the cube-side images (stop, yield, triangular, circular and square) applying RetinaNet. Since the stop and yield forms each corresponded to only one TS, in order to recognize the other forms in their respective classes, an InceptionV3 network was trained for each classification. For the 3-D location and filtering phases, the georeferenced point cloud of the environment was used. TSs detected in the images were projected onto the cloud using the pinhole model for correct 3-D geolocation. Finally, the duplicate signals detected in different images were filtered based on the coincidence between classes and distance between them. The methodology was tested in two real case studies with a total of 214 TSs, 89.7% of the TSs were correctly detected, of which 92.5% were correctly classified. The false positive rate per image was only 0.004 and main false detections were due to road mirrors. 97.3% of the detected signals were correctly 3-D geolocated with less than 0.5 m of error.

The effectiveness in the combination of data images and point clouds was demonstrated in this work. Images allow the use of artificial intelligence techniques for detection and classification, which improve their success rates day by day with new networks and designs. In addition, image processing is much faster and more efficient than point cloud processing. The use of a 360° camera does not require the passage of the MMS in two road directions. Furthermore, point clouds allow a more precise geolocation of signals than only using images.

The entire process of TS inventorying from processing images (first) and point cloud (continued) ensures speed and effectiveness in processing time, 50% faster than other proposals that first treat point clouds and then images with much higher computational costs which, although they provide satisfactory results in terms of success rates, make their inclusion in production processes unfeasible due to cost of time and computer equipment. Due to these advantages, the presented methodology is suitable to be included in the production process of any company. Also, it is conducted automatically without human intervention.

Future work will focus on extending the methodology to more objects important for road safety and for the inventory of objects, as the methodology does not depend on any exclusive feature of TSs. In addition, it is proposed to feed back the network to improve the success rate of detections with corrected images that present the main types of error. It is also considered to test the methodology in other case studies such as highways and urban roads, to analyse the influence of driving speed during acquisition on 3-D point cloud location.

Author Contributions: Conceptualization, P.A. and D.C.; method and software, D.C.; validation, D.C. and J.B.; investigation, E.G.; resources, P.A.; writing, J.B., E.G. and P.A.; visualization, J.B.; supervision, E.G. and P.A. All authors have read and agreed to the published version of the manuscript.

Funding: This research was funded by Xunta de Galicia given through human resources grant (ED481B-2019-061, ED481D 2019/020) and competitive reference groups (ED431C 2016-038), the Ministerio de Ciencia, Innovación y Universidades -Gobierno de España (RTI2018-095893-B-C21). This project has also received funding from the European Union's Horizon 2020 research and innovation programme under grant agreement No 769255. This document reflects only the views of the authors. Neither the Innovation and Networks Executive Agency (INEA) or the European Commission is in any way responsible for any use that may be made of the information it contains. The statements made herein are solely the responsibility of the authors.

Acknowledgments: Authors would like to thank to those responsible for financing this research.

Conflicts of Interest: The authors declare no conflict of interest.

References

1. European Union EU Funding for TEN-T. Available online: https://ec.europa.eu/transport/themes/infrastructure/ten-t-guidelines/project-funding_en (accessed on 1 June 2019).

2. European Commission EC Road and Rail Infrastructures. Available online: https://ec.europa.eu/growth/content/ec-discussion-paper-published-maintenance%5C%5C-road-and-rail-infrastructures_en (accessed on 28 June 2019).

3.   European Union Horizon 2020 Smart, Green and Integrated Transport. Available online: https://ec.europa.eu/programmes/horizon2020/en/h2020-section/smart-green-and-integrated-transport (accessed on 3 December 2019).

4.   Balado, J.; Díaz-Vilariño, L.; Arias, P.; Lorenzo, H. Point clouds for direct pedestrian pathfinding in urban environments. *ISPRS J. Photogramm. Remote Sens.* **2019**, *148*, 184–196. [CrossRef]

5.   Soilán, M.; Riveiro, B.; Sánchez-Rodríguez, A.; Arias, P. Safety assessment on pedestrian crossing environments using MLS data. *Accid. Anal. Prev.* **2018**, *111*, 328–337. [CrossRef] [PubMed]

6.   European Commission Innovating for the Transport of the Tuture. Available online: https://ec.europa.eu/transport/themes/its_pt (accessed on 3 December 2019).

7.   Quackenbush, L.; Im, I.; Zuo, Y. Road extraction: A review of LiDAR-focused studies. *Remote Sens. Nat. Resour.* **2013**, *9*, 155–169.

8.   Leonardi, G.; Barrile, V.; Palamara, R.; Suraci, F.; Candela, G. Road Degradation Survey Through Images by Drone. In *International Symposium on New Metropolitan Perspectives*; Springer: Cham, Switzerland, 2019; pp. 222–228. ISBN 978-3-319-92101-3.

9.   Soilán, M.; Sánchez-Rodríguez, A.; del Río-Barral, P.; Perez-Collazo, C.; Arias, P.; Riveiro, B. Review of laser scanning technologies and their applications for road and railway infrastructure monitoring. *Infrastructures* **2019**, *4*, 58. [CrossRef]

10. Wirges, S.; Fischer, T.; Stiller, C.; Frias, J.B. Object Detection and Classification in Occupancy Grid Maps Using Deep Convolutional Networks. In Proceedings of the IEEE Conference on Intelligent Transportation Systems, ITSC, Maui, HI, USA, 4–7 November 2018; Volume 2018-Novem.

11. Fairfield, N.; Urmson, C. Traffic light Mapping and Detection. In Proceedings of the 2011 IEEE International Conference on Robotics and Automation, Shanghai, China, 9–13 May 2011; pp. 5421–5426.

12. Levinson, J.; Askeland, J.; Dolson, J.; Thrun, S. Traffic Light Mapping, Localization, and State Detection for Autonomous Vehicles. In Proceedings of the 2011 IEEE International Conference on Robotics and Automation, Shanghai, China, 9–13 May 2011; pp. 5784–5791.

13. Diaz-Cabrera, M.; Cerri, P.; Medici, P. Robust real-time traffic light detection and distance estimation using a single camera. *Expert Syst. Appl.* **2015**, *42*, 3911–3923. [CrossRef]

14. Li, Y.; Wang, W.; Tang, S.; Li, D.; Wang, Y.; Yuan, Z.; Guo, R.; Li, X.; Xiu, W. Localization and Extraction of Road Poles in Urban Areas from Mobile Laser Scanning Data. *Remote Sens.* **2019**, *11*, 401. [CrossRef]

15. Chen, X.; Kohlmeyer, B.; Stroila, M.; Alwar, N.; Wang, R.; Bach, J. Next Generation Map Making: Geo-referenced Ground-level LIDAR Point Clouds for Automatic Retro-reflective Road Feature Extraction. In Proceedings of the GIS, Seattle, WA, USA, 4–6 November 2009; pp. 488–491.

16. Wan, R.; Huang, Y.; Xie, R.; Ma, P. Combined Lane Mapping Using a Mobile Mapping System. *Remote Sens.* **2019**, *11*, 305. [CrossRef]

17. Gu, X.; Zang, A.; Huang, X.; Tokuta, A.; Chen, X. Fusion of Color Images and LiDAR Data for Lane Classification. In Proceedings of the 23rd SIGSPATIAL International Conference on Advances in Geographic Information Systems, Washington, DC, USA, 3–6 November 2015.

18. Kang, Z.; Yang, J.; Zhong, R.; Wu, Y.; Shi, Z.; Lindenbergh, R. Voxel-based extraction and classification of 3-D pole-like objects from mobile LiDAR point cloud data. *IEEE J. Sel. Top. Appl. Earth Obs. Remote Sens.* **2018**, *11*, 4287–4298. [CrossRef]

19. Che, E.; Jung, J.; Olsen, M.J. Object recognition, segmentation, and classification of mobile laser scanning point clouds: A state of the art review. *Sensors* **2019**, *19*, 810. [CrossRef]

20. Meierhold, N.; Spehr, M.; Schilling, A.; Gumhold, S.; Maas, H.-G. Automatic feature matching between digital images and 2D representations of a 3D laser scanner point cloud. *Int. Arch. Photogramm. Remote Sens. Spat. Inf. Sci.* **2010**, *38*, 446–451.

21. Gómez, M.J.; García, F.; Martín, D.; de la Escalera, A.; Armingol, J.M. Intelligent surveillance of indoor environments based on computer vision and 3D point cloud fusion. *Expert Syst. Appl.* **2015**, *42*, 8156–8171. [CrossRef]

22. García, F.; García, J.; Ponz, A.; de la Escalera, A.; Armingol, J.M. Context aided pedestrian detection for danger estimation based on laser scanner and computer vision. *Expert Syst. Appl.* **2014**, *41*, 6646–6661. [CrossRef]

23. Soilán, M.; Riveiro, B.; Martínez-Sánchez, J.; Arias, P. Traffic sign detection in MLS acquired point clouds for geometric and image-based semantic inventory. *ISPRS J. Photogramm. Remote Sens.* **2016**, *114*, 92–101. [CrossRef]

24. Arcos-García, Á.; Soilán, M.; Álvarez-García, J.A.; Riveiro, B. Exploiting synergies of mobile mapping sensors and deep learning for traffic sign recognition systems. *Expert Syst. Appl.* **2017**, *89*, 286–295. [CrossRef]

25. Guan, H.; Yan, W.; Yu, Y.; Zhong, L.; Li, D. Robust Traffic-Sign Detection and Classification Using Mobile LiDAR Data With Digital Images. *IEEE J. Sel. Top. Appl. Earth Obs. Remote Sens.* **2018**, *11*, 1715–1724. [CrossRef]

26. Wen, C.; Li, J.; Luo, H.; Yu, Y.; Cai, Z.; Wang, H.; Wang, C. Spatial-related traffic sign inspection for inventory purposes using mobile laser scanning data. *IEEE Trans. Intell. Transp. Syst.* **2016**, *17*, 27–37. [CrossRef]

27. Wu, S.; Wen, C.; Luo, H.; Chen, Y.; Wang, C.; Li, J. Using Mobile LiDAR Point Clouds for Traffic Sign Detection and Sign Visibility Estimation. In Proceedings of the 2015 IEEE International Geoscience and Remote Sensing Symposium (IGARSS), Milan, Italy, 6–31 July 2015; pp. 565–568.

28. Bruno, D.R.; Sales, D.O.; Amaro, J.; Osório, F.S. Analysis and Fusion of 2D and 3D Images Applied for Detection and Recognition of Traffic Signs Using a New Method of Features Extraction in Conjunction with Deep Learning. In Proceedings of the 2018 International Joint Conference on Neural Networks (IJCNN), Rio de Janeiro, Brazil, 8–13 July 2018; pp. 1–8.

29. Yu, Y.; Li, J.; Wen, C.; Guan, H.; Luo, H.; Wang, C. Bag-of-visual-phrases and hierarchical deep models for traffic sign detection and recognition in mobile laser scanning data. *ISPRS J. Photogramm. Remote Sens.* **2016**, *113*, 106–123. [CrossRef]

30. Balado, J.; Sousa, R.; Díaz-Vilariño, L.; Arias, P. Transfer Learning in urban object classification: Online images to recognize point clouds. *Autom. Constr.* **2020**, *111*, 103058. [CrossRef]

31. Ma, L.; Li, Y.; Li, J.; Wang, C.; Wang, R.; Chapman, A.M. Mobile laser scanned point-clouds for road object detection and extraction: A review. *Remote Sens.* **2018**, *10*, 1531. [CrossRef]

32. You, C.; Wen, C.; Wang, C.; Li, J.; Habib, A. Joint 2-D-3-D Traffic sign landmark data set for geo-localization using mobile laser scanning data. *IEEE Trans. Intell. Transp. Syst.* **2019**, *20*, 2550–2565. [CrossRef]

33. Zhang, R.; Li, G.; Li, M.; Wang, L. Fusion of images and point clouds for the semantic segmentation of large-scale 3D scenes based on deep learning. *ISPRS J. Photogramm. Remote Sens.* **2018**, *143*, 85–96. [CrossRef]

34. Mogelmose, A.; Trivedi, M.M.; Moeslund, T.B. Vision-based traffic sign detection and analysis for intelligent driver assistance systems: Perspectives and survey. *IEEE Trans. Intell. Transp. Syst.* **2012**, *13*, 1484–1497. [CrossRef]

35. Gudigar, A.; Chokkadi, S.; U, R. A review on automatic detection and recognition of traffic sign. *Multimed. Tools Appl.* **2016**, *75*, 333–364. [CrossRef]

36. Wali, S. Comparative survey on traffic sign detection and recognition: A review. *PRZEGLĄD ELEKTROTECHNICZNY* **2015**, *1*, 40–44. [CrossRef]

37. Dewan, P.; Vig, R.; Shukla, N. An overview of traffic signs recognition methods. *Int. J. Comput. Appl.* **2017**, *168*, 7–11. [CrossRef]

38. Saadna, Y.; Behloul, A. An overview of traffic sign detection and classification methods. *Int. J. Multimed. Inf. Retr.* **2017**, *6*, 193–210. [CrossRef]

39. Wali, S.B.; Abdullah, M.A.; Hannan, M.A.; Hussain, A.; Samad, S.A.; Ker, P.J.; Mansor, M. Bin Vision-based traffic sign detection and recognition systems: Current rrends and challenges. *Sensors (Basel)* **2019**, *19*, 2093. [CrossRef]

40. Bello-Cerezo, R.; Bianconi, F.; Fernández, A.; González, E.; Maria, F. Di Experimental comparison of color spaces for material classification. *J. Electron. Imaging* **2016**, *25*, 1–10. [CrossRef]

41. González, E.; Bianconi, F.; Fernández, A. An investigation on the use of local multi-resolution patterns for image classification. *Inf. Sci. (Ny)* **2016**, *361–362*, 1–13.

42. Kaplan Berkaya, S.; Gunduz, H.; Ozsen, O.; Akinlar, C.; Gunal, S. On circular traffic sign detection and recognition. *Expert Syst. Appl.* **2016**, *48*, 67–75. [CrossRef]

43. Ellahyani, A.; El Ansari, M.; Lahmyed, R.; Treméau, A. Traffic sign recognition method for intelligent vehicles. *J. Opt. Soc. Am. A* **2018**, *35*, 1907–1914. [CrossRef] [PubMed]

44. Arcos-García, Á.; Alvarez-Garcia, J.; Soria Morillo, L. Evaluation of Deep Neural Networks for traffic sign detection systems. *Neurocomputing* **2018**, *316*, 332–344. [CrossRef]

45. Qian, R.; Zhang, B.; Yue, Y.; Wang, Z.; Coenen, F. Robust Chinese Traffic Sign Detection and Recognition with Deep Convolutional Neural Network. In Proceedings of the 2015 11th International Conference on Natural Computation (ICNC), Zhangjiajie, China, 15–17 August 2015; pp. 791–796.
46. Zhu, Z.; Liang, D.; Zhang, S.; Huang, X.; Li, B.; Hu, S. Traffic-Sign Detection and Classification in the Wild. In Proceedings of the 2016 IEEE Conference on Computer Vision and Pattern Recognition (CVPR), Las Vegas, NV, USA, 27–30 June 2016; pp. 2110–2118.
47. Wu, Y.; Liu, Y.; Li, J.; Liu, H.; Hu, X. Traffic Sign Detection Based on Convolutional Neural Networks. In Proceedings of the The 2013 International Joint Conference on Neural Networks (IJCNN), Dallas, TX, USA, 4–9 August 2013; pp. 1–7.
48. Hussain, S.; Abualkibash, M.; Tout, S. A Survey of Traffic Sign Recognition Systems Based on Convolutional Neural Networks. In Proceedings of the 2018 IEEE International Conference on Electro/Information Technology (EIT), Rochester, MI, USA, 3–5 May 2018.
49. Balado, J.; Martínez-Sánchez, J.; Arias, P.; Novo, A. Road environment semantic segmentation with deep learning from MLS point cloud data. *Sensors* **2019**, *19*, 3466. [CrossRef] [PubMed]
50. Timofte, R.; Zimmermann, K.; Van Gool, L. Multi-view traffic sign detection, recognition, and 3D localisation. *Mach. Vis. Appl.* **2014**, *25*, 633–647. [CrossRef]
51. Hazelhoff, L.; Creusen, I.; With, P.H.N. de Robust detection, classification and positioning of traffic signs from street-level panoramic images for inventory purposes. In Proceedings of the 2012 IEEE Workshop on the Applications of Computer Vision (WACV), Breckenridge, CO, USA, 9–11 January 2012; pp. 313–320.
52. Balali, V.; Jahangiri, A.; Machiani, S.G. Multi-class US traffic signs 3D recognition and localization via image-based point cloud model using color candidate extraction and texture-based recognition. *Adv. Eng. Inform.* **2017**, *32*, 263–274. [CrossRef]
53. Wang, K.C.P.; Hou, Z.; Gong, W. Automated road sign inventory system based on stereo vision and tr; pp. 570–573acking. *Comput. Civ. Infrastruct. Eng.* **2010**, *25*, 468–477. [CrossRef]
54. Soheilian, B.; Paparoditis, N.; Vallet, B. Detection and 3D reconstruction of traffic signs from multiple view color images. *ISPRS J. Photogramm. Remote Sens.* **2013**, *77*, 1–20. [CrossRef]
55. Li, Y.; Shinohara, T.; Satoh, T.; Tachibana, K. Road signs detection and recognition utilizing images and 3D point cloud acquired by mobile mapping system. *ISPRS Int. Arch. Photogramm. Remote Sens. Spat. Inf. Sci.* **2016**, *XLI-B1*, 669–673. [CrossRef]
56. You, C.; Wen, C.; Luo, H.; Wang, C.; Li, J. Rapid Traffic Sign Damage Inspection in Natural Scenes Using Mobile Laser Scanning Data. In Proceedings of the 2017 IEEE International Geoscience and Remote Sensing Symposium (IGARSS), Fort Worth, TX, USA, 23–28 July 2017.
57. Lin, T.-Y.; Goyal, P.; Girshick, R.; He, K.; Dollar, P. Focal Loss for Dense Object Detection. In Proceedings of the The IEEE International Conference on Computer Vision (ICCV), Venice, Italy, 22–29 October 2017.
58. He, K.; Zhang, X.; Ren, S.; Sun, J. Deep Residual Learning for Image Recognition. In Proceedings of the IEEE Conference on Computer Vision and Pattern Recognition, Las Vegas, NV, USA, 27–30 June 2016; pp. 770–778.
59. Szegedy, C.; Vanhoucke, V.; Ioffe, S.; Shlens, J.; Wojna, Z.B. Rethinking the Inception Architecture for Computer Vision. In Proceedings of the The IEEE Conference on Computer Vision and Pattern Recognition (CVPR), Las Vegas, NV, USA, 27–30 June 2016; pp. 2818–2826.
60. Miyamoto, K. Fish eye lens. *J. Opt. Soc. Am.* **1964**, *54*, 1060–1061. [CrossRef]
61. Stallkamp, J.; Schlipsing, M.; Salmen, J.; Igel, C. The German Traffic Sign Recognition Benchmark: A Multi-Class Classification Competition. In Proceedings of the The 2011 International Joint Conference on Neural Networks, San Jose, CA, USA, 31 July–5 August 2011; pp. 1453–1460.
62. Soilán, M.; Riveiro, B.; Martínez-Sánchez, J.; Arias, P. Automatic road sign inventory using mobile mapping systems. *Int. Arch. Photogramm. Remote Sens. Spat. Inf. Sci. ISPRS Arch.* **2016**, *41*, 717–723. [CrossRef]
63. Wang, Y.; Wang, C.; Zhang, H.; Dong, Y.; Wei, S. Automatic ship detection based on RetinaNet using multi-resolution Gaofen-3 imagery. *Remote Sens.* **2019**, *11*, 531. [CrossRef]
64. Campbell, A.; Both, A.; Sun, Q. (Chayn) Detecting and mapping traffic signs from Google Street View images using deep learning and GIS. *Comput. Environ. Urban Syst.* **2019**, *77*, 101350. [CrossRef]

MDPI

St. Alban-Anlage 66

4052 Basel

Switzerland

Tel. +41 61 683 77 34

Fax +41 61 302 89 18

www.mdpi.com

Remote Sensing Editorial Office

E-mail: remotesensing@mdpi.com

www.mdpi.com/journal/remotesensing

9 783036 534909